SMALLER SATELLITES: BIGGER BUSINESS?

SPACE STUDIES

VOLUME 6

Editors
Prof. MICHAEL RYCROFT and Dr. Norma Crosby

International Space University
Excellence in space education for a changing world

The International Space University (ISU) is dedicated to the development of outer space for peaceful purposes through international and interdisciplinary education and research. ISU works in association with a number of Affiliates (universities, research institutes, consortia ...) around the world and in partnership with space agencies and industry.

For young professionals and postgraduate students, ISU offers an annual ten-week Summer Session in different countries and a one-year master of Space Studies (MSS) program based at its Central Campus in Strasbourg, France. ISU also offers short courses and workshops to professionals working in space-related industry, government and academic organizations.

Independent of specific national and commercial interests, ISU is an ideal forum for discussion of issues relating to space and its applications. The network of alumni, faculty, guest lecturers, Affiliate representatives and professional contacts which characterizes the ISU Community makes it possible to bring together leading international specialists in an academic environment conductive to the exchange of views and to the creation of innovative ideas. ISU aims to promote productive dialogue between space-users and providers. In addition to the Annual Symposium, ISU supports smaller forum activities, such as workshops and roundtables, for constructive discussions which may help to chart the way forward to the rational international utilization of space.

SMALLER SATELLITES: BIGGER BUSINESS?

Concepts, Applications and Markets for Micro/Nanosatellites in a New Information World

Edited by

M. RYCROFT

and

N. CROSBY

for the
International Space University
Strasbourg, France

KLUWER ACADEMIC PUBLISHERS

DORDRECHT / BOSTON / LONDON

A C.I.P. Catalogue record for this book is available from the Library of Congress.

ISBN 1-4020-0199-1

Published by Kluwer Academic Publishers,
P.O. Box 17, 3300 AA Dordrecht, The Netherlands.

Sold and distributed in North, Central and South America
by Kluwer Academic Publishers,
101 Philip Drive, Norwell, MA 02061, U.S.A.

In all other countries, sold and distributed
by Kluwer Academic Publishers,
P.O. Box 322, 3300 AH Dordrecht, The Netherlands.

Printed on acid-free paper

Printed in the Netherlands

ISU is most grateful for the sponsorship provided by

Lockheed Martin Space Operations
(General Sponsor)

Atos Origin Nederland B.V.

Canadian Space Agency

Centre national d'études spatiales

European Space Agency

International Launch Services

New Skies Satellites N.V.

City of Strasbourg, Department of Bas-Rhin and the

Region of Alsace

Table of Contents

Acknowledgements

ISU wishes to thank the following people:

Program Committee Chair:

Y. Fujimori, ISU Faculty, NASDA Invited Scholar

Advisory Committee Members:

M. Albrecht, President, ILS (International Launch Services), USA

J. Blamont, Conseiller du Directeur Général, CNES, France

W. Kröll, Chairman of the Board, DLR, Germany

M. Othman, Director, United Nations, Office of Outer Space Affairs, Austria

H. Rath, ZARM, Center of Applied Space Technology and Microgravity, University of Bremen, Germany

A. Rodotà, Director General, ESA

J. Stocky, Chief Technologist, New Millennium Program, Micro-Nano Sciencecraft Thrust, NASA Jet Propulsion Laboratory, USA

A. Valenzuela, President, Media Lario S.r.l. and IAA Committee Chair on Small Satellite Missions, Italy

P. Worden, USAF, USA

Program Committee Members:

D. Bearden, Systems Director, Civil & Commercial Division, The Aerospace Corporation, USA

M. Bousquet, Director, Aerospace Electronics & Comm. Prog., ENSAE/SUP'Aero, France

J. Broquet, Deputy Director, Technology, Strategy and Planning, ASTRIUM, France

A. Browne, Chief Financial Officer, New Skies Satellites N.V., The Netherlands

M. Courtois, Technical Director, Alcatel Space Industries, Cannes, France

P. Eckart, Division of Astronautics, Technical University of Munich, Germany

R. Fleeter, President, AeroAstro Corporation, USA

T. Grant, Principal Consultant, Atos Origin Nederland B.V., The Netherlands

J. van der Ha, Consultant, Cost-Effective Space Mission Design & Operations

H. Jie-Cai, Dean of Astronautics School, Harbin Institute of Technology, China

D. Hernandez, Delegué à la Recherche Technique et Innovation, CNES, France

T. Iida, Director General, Communications Research Laboratory, Japan

V. Jha, Director General, Space Technologies, Canadian Space Agency, Canada

T. Lawrence, Chief, Space Technology, USAF European Office of Aerospace Research and Development, UK

R. Lindberg, Vice President and Deputy General Manager, Advanced Programs Group, Orbital Sciences Corporation/Dulles HQ, USA

P. Lionnet, Chargé d'études, Eurospace, France

K. Maeda, Satellite Mission Planning Department, NASDA HQ, Japan

G. Malyshev, Head of "Astra" – Lab, Moscow Aviation Institute, Russia

S. Matousek, Group Supervisor, Mars Program Office Advanced Studies, Microspacecraft and Advanced Development, NASA Jet Propulsion Laboratory, USA

F. Ongaro, Directorate of Strategy and Technical Assessment, ESA HQ, France

M. Sweeting, CEO, Surrey Satellite Technology Ltd., University of Surrrey, United Kingdom

E. Trinh, Director, Microgravity Research Division, NASA HQ, USA

Local Organizing Team

I. Pryke, Head, Washington Office, ESA

M. Stephens, Vice President, Business Development, USA

P. Finarelli, Vice President, North American Operations, ISU

L. Laidet, Delegate, CNES, France

L. Chestnutt, Symposium Coordinator

ISU Resident Faculty and Staff

Foreword

Y. Fujimori, Symposium Programme Committee Chair, and Faculty Member, International Space University

e-mail: fujimori@isu.isunet.edu

M.Rycroft, Faculty Member, International Space University

e-mail: rycroft@isu.isunet.edu

N. Crosby, International Space University

e-mail: norma@bock-crosby.fsbusines.co.uk

For the sixth annual ISU Symposium the theme was "Smaller Satellites: Bigger Business? Concepts, Applications and Markets for Micro/Nanosatellites in a New Information World". Thus, the Symposium addressed the crucial question: are small satellites the saviour of space programmes around the world today? It did this from the unique perspective of the International Space University – the interdisciplinary, international and intercultural perspective.

This Symposium brought together a variety of people working on small satellites – engineers, scientists, planners, providers, operators, policy makers and business executives, together with representatives from regulatory bodies, from national and international organizations, and from the finance sector, and also entrepreneurs. Discussion and debate were encouraged, based on the papers presented and those published here.

The views presented were so diverse that strictly speaking it is impossible to sum up. However, there was a general feeling that small satellites are the means to an end, where the end is a service of value to humankind. The key to a successful service is useful, new information, with competition playing a significant role in the market place. So let us now initiate a new space race, for the prize of using space in innovative ways. And the race track is delineated by small satellites, Robust, Adaptable and Cost Effective, a RACE indeed !

Almost 200 people attended this Symposium, including members of the sixth Master of Space Studies class of ISU, young professionals and postgraduate students who are developing an aspect of the Symposium's theme in their Team Project – PISCES, Proposal for International Spacecraft Cooperation for Education and Science included here. Their final report will be completed at the end of July 2001, and published separately.

Keynote Address

Micro/Nanosatellites - The New World

M. N. Sweeting, Surrey Space Centre, University of Surrey, Guildford, Surrey GU2 7XH, UK

e-mail: m.sweeting@ee.surrey.ac.uk

Abstract

Changing world politics and military emphasis has brought considerable pressure on space agency financial budgets and a shift to increasing commercialisation of space activities. Budgetary pressure, coupled with the rapid advancement of commercial and consumer micro-electronics, has catalysed the use of smaller and more computationally capable satellites as a "faster, cheaper, better" means of realising space missions - complementary to conventional large satellite systems. Affordable small satellites, however, require a very different approach compared with established space engineering techniques. Since 1979, the University of Surrey (UoS., UK) has pioneered cost-effective satellite engineering techniques for small satellites, and has developed a series of highly sophisticated, yet inexpensive, microsatellites built for around US $ 3 million each. Sixteen such microsatellites have been launched by Surrey into low Earth orbit carrying a wide range of satellite communications, space science, remote sensing and in-orbit technology demonstration payloads - both civil and military. Seven microsatellite missions have also provided the focus and means for highly successful satellite "know-how" transfer and training programmes between Surrey and emerging space nations such as Korea, Portugal, Pakistan, South Africa, Chile, Singapore, Malaysia and Thailand and now P.R. China - leading to the launch of their first national satellites (e.g., KITSAT-1, PoSAT-1, BADR-1, Thai-Phutt and FASat-Bravo) - providing rapid response, low risk and affordable access to space.

The Surrey Space Centre (SSC) offers a unique environment combining both academic research and postgraduate teaching with commercial development and manufacturing of satellites through its spin-off company, Surrey Satellite Technology Ltd (SSTL). The Centre is almost entirely funded from commercial contracts. In April 1999, Surrey launched a new 315 kg minisatellite (UoSAT-12) designed and built at SSTL carrying high resolution Earth observation, advanced digital communications and technology demonstration payloads within the same cost-effective satellite engineering philosophy. Surrey's latest research and development project, a 6.5 kg nanosatellite, was launched in June 2000. SNAP-1 demonstrates the use of commercial micro-miniature technologies and, for the first time, orbital formation flying. Surrey is currently preparing a six-satellite communications constellation (ESAT) and a five satellite daily disaster monitoring network which will use Surrey's latest enhanced 130 kg microsatellites. Surrey has also commenced its first low-cost interplanetary mission, to the Moon, scheduled for launch in 2002.

1. Introduction

Following the first space launch in 1957, satellites rapidly grew increasingly large and enormously expensive. Initially, the "space race" was an effective catalyst for the development of advanced technology as the super-powers strove to out-do each other and gain the advantageous "high ground" of space - irrespective of budget. However, as costs escalated and timescales lengthened,

M. Rycroft and N. Crosby (eds.), Smaller Satellites: Bigger Business?, 1–19.

with satellites generally taking many years to mature from concept to useful orbital operation, this process limited access to space to only a relatively few nations or international agencies.

Changing world politics and military emphasis in the last decade has brought about a quiet revolution in space. Pressure on space agency financial budgets has increasingly meant that fewer (and bigger) satellites have been commissioned and that new ideas, technologies and scientific experiments have found it difficult to gain timely access to space.

The staggering developments in micro-electronics, stimulated increasingly by the consumer market rather than military requirements, and the dramatic pace of consumer product development, caused space technology often to lag considerably behind that now taken for granted on Earth. The combination of reducing budgets for space and increasing capability of low-power micro-electronics have enabled a new breed of highly capable "smaller, faster, cheaper" satellites to realise many space missions - complementing the conventional large satellite systems still necessary for large-scale space science and communications services to small terminals. Indeed, in the field of commercial satellite communications, numerous constellations of small satellites have been proposed and are being built to provide a range of global services to hand-held terminals either for real-time voice or non-real-time data.

However, whether a particular satellite is "large" or "small" depends somewhat upon viewpoint. For instance the "small" satellites for the Iridium mobile communications system weigh in at over 600 kg each – whereas those for the HealthNet e-mail network are a mere 50 kg! In view of this potential for confusion, the classification given in Table 1 has become widely adopted.

Class	Mass (kg)	Cost (US $ million)
Large satellite	> 1000	> 140
Small satellite	500 - 1000	50 – 140
Mini-satellite	100 - 500	10 – 30
Micro-satellite	10 - 100	3 – 6
Nano-satellite	1 - 10	0.3 – 1.5
Pico-satellite	< 1	< 0.3

Table 1. Classification of satellites according to mass and typical cost

Although there have been many examples of large, small and even mini-satellites, it is only relatively recently that capable *microsatellites* have shown that it is possible to execute both civil and military missions very effectively, rapidly, and at low cost and risk, for the following applications:

- Specialised communications services and research
- Earth observation (EO) and remote sensing
- Small-scale space science
- Technology demonstration/verification
- Education and training.

2. Surrey Microsatellites

The University of Surrey (UoS) has pioneered microsatellite technologies since beginning its *UoSAT* programme in 1978. From very modest experimental beginnings, its space-related research, post-graduate teaching and international commercial activities are now housed within a purpose-built *Surrey Space Centre.* There are over 150 academic and professional staff and postgraduate research students. The objectives of the Centre's programmes are:

- to research cost-effective small satellite techniques
- to demonstrate the capabilities of micro/minisatellites
- to catalyse the commercial use of micro/minisatellites
- to promote space education and training.

Over the last decade, Surrey has established an international reputation as pioneers of innovative small satellites in a uniquely combined academic research teaching and commercial environment.

Surrey's first experimental microsatellites (UoSAT-1 and 2) were launched free-of-charge as "piggy-back" payloads through a collaborative arrangement with NASA on DELTA rockets in 1981 and 1984, respectively. Since then, a further eleven low cost yet sophisticated microsatellites have been placed in low Earth orbit using Ariane, Tsyklon and Zenit launchers for a variety of international customers and carrying a wide range of payloads.

Surrey's new 315 kg minisatellite UoSAT-12 was launched from the Baikonur Cosmodrome during April 1999 and four new microsatellites for Malaysia, China and the USAF are currently being prepared for launch in 2000. Summary information on the Surrey small satellite programme is given in Table 2.

Microsatellite	Launch	Orbit	Customer	Payloads
UoSAT-1	1984-D	560 km	UoS	Research
UoSAT-2	1984-D	700 km	UoS	S&F, EO, radiation
UoSAT-3	1990-A	900 km	UoS	S&F
UoSAT-4	1990-A	900 km	UoS/ESA	Technology
UoSAT-5	1991-A	900 km	SatelLife	S&F,EO, radiation
S80/T	1992-A	1330 km	CNES	LEO Comms
KitSat-1	1992-A	1330 km	Korea	S&F,EO, radiation
KitSat-2	1993-A	900 km	Korea	S&F,EO, radiation
PoSAT-1	1993-A	900 km	Portugal	S&F,EO, radiation
HealthSat-2	1993-A	900 km	SatelLife	S&F
Cerise	1995-A	735 km	CNES	Military
FASat-Alfa	1995-T	873 km	Chile	S&F,EO
FASat-Bravo	1998-Z	835 km	Chile	S&F,EO
Thai-Phutt	1998-Z	835 km	Thailand	S&F,EO
Clementine	1999-A	735 km	CNES	Military
UoSAT-12	1999-S	650 km	SSTL & Singapore	EO, Comms
SNAP-1	2000-C	650 km	SSTL	Technology
Tsinghua-1	2000-C	750 km	PR China	EO, Comms
TiungSAT-1	2000-S	1020 km	Malaysia	EO, Comms
PicoSAT	2001-At	650 km	USAF	Military
BiltenSAT	2002-?	650 km	Turkey	EO, Comms
AISAT	2002-?	650 km	Algeria	EO, Comms

Table 2. University of Surrey Microsatellite Missions
KitSat-2 was built in Korea using a SSTL platform and a KAIST payload. D=Delta, A= Ariane, T=Tsyklon, Z=Zenit, S=SS18/Dnepr, LM=Long March, C=Cosmos, At= Athena indicate the launch vehicles.

UoSAT-1 and 2 used a rather conventional structure - a framework "skeleton" onto which were mounted module boxes containing the various electronic subsystems and payloads with a complex 3-dimensional inter-connecting wiring harness. Following UoSAT-1 and UoSAT-2, the need to accommodate a variety of payload customers within a standard (ASAP) launcher envelope (400x400x600 mm and 50 kg), coupled with increased demands on packing density, electromagnetic compatibility, economy of manufacture and ease of integration, catalysed the development at Surrey during 1986 of a novel modular design of multi-mission microsatellite platform. This modular microsatellite platform has been used successfully on *seventeen* missions, each with different payload requirements, with the spacecraft proceeding from order-to-orbit in typically 10-12 months!

The SSTL modular microsatellite has no "skeleton" but is rather a series of identical-in-outline, machined module boxes stacked one on top of the other to form a body onto which solar panels and instruments may be mounted. Each module box houses the various microsatellite subsystems, e.g., batteries, power conditioning, on-board data handling, communications and attitude control. Payloads are housed either in similar modules or on top of the platform alongside antennas and attitude sensors as appropriate.

Electronically, the microsatellite uses modern, sophisticated, but not necessarily space-proven, electronic circuits to provide a high degree of capability. These are underpinned by space-proven subsystems - resulting in a layered architecture that achieves high performance with operational redundancy by using alternative technologies rather than by simple duplication.

Communications and Earth observation payloads require an Earth-pointing platform. Thus the microsatellite is maintained to within 0.3° of nadir by employing a combination of gravity-gradient stabilisation (using a 6-metre boom) and closed-loop active damping using electromagnets operated by the on-board computer. Attitude determination is provided by Sun and geomagnetic field sensors, and star field cameras, whilst orbital position is determined autonomously to within ±50 metres by an on-board GPS receiver. Electrical power is generated by four body-mounted GaAs solar array panels, each generating ~35 W, and is stored in a 7Ah NiCd rechargeable battery.

Communications are supported by VHF uplinks and UHF downlinks, using fully error-protected AX.25 packet link protocols operating at 9.6 to 76.8 kbps. These are capable of transferring several hundred kbytes of data to brief-case sized communications terminals.

It is the On-Board Data Handling (OBDH) system that is the key to the sophisticated capability of the microsatellite. At the heart of the OBDH system is a 80C386 On-Board Computer (OBC), which runs a 500 kbyte real-time multi-tasking operating system with a 128 Mbyte solid-state CMOS RAM DISK. In addition, there is a secondary 80C186 OBC with 16 Mbyte SRAM, two 20 MHz T805 Transputers with 4 Mbytes SRAM, and a dozen other microcontrollers. A primary feature of the OBDH philosophy is that all the software on-board the microsatellite is *loaded after launch*, and can be upgraded and reloaded by the

Control Groundstation at will thereafter. Normally, the satellite is operated via the primary OBC-386 and the real-time multi-tasking operating system. All telecommand instructions are formulated into a "diary" at the groundstation and then transferred to the satellite OBC for execution either immediately or, more usually, at some future time. Telemetry from on-board platform systems and payloads is similarly gathered by the OBC-386 and either transmitted immediately and/or stored in the RAM DISK whilst the satellite is out of range of the Control Station. The OBC's also operate the attitude control systems according to control algorithms that take input from the various attitude sensors and then act accordingly. Thus it is this OBDH environment that allows such a tiny microsatellite to operate in a highly complex, flexible and sophisticated manner, enabling fully automatic and autonomous control of the satellites systems and payloads.

The latest SSTL microsatellite platforms have enhanced sub-systems supporting the following major new features for greatly increased performance and payload capacity. The first version of the enhanced microsatellite platform was used on the FASat-A mission for Chile launched into LEO by a Ukrainian Tsyklon launcher in 1995. It has since been further developed and used on the new FASat-Bravo, Thai-Puht, TiungSat-1, PICOSat and now Tsinghua-1 missions providing:

- distributed telemetry/ telecommand for easy expansion
- 5 MIPS (386) OBC + 256 MB RAMDISK
- 1 Mbps on-board LAN
- autonomous GPS navigation (50m)
- attitude determination (0.005^{o})
- 0.25^{o} rms nadir pointing pitch/roll axes, 2° yaw
- 128 kbps BPSK downlinks.

3. Applications of Micro/Minisatellites

The early UoSAT missions demonstrated the potential capabilities of microsatellites, and generated considerable interest in applications such as digital store-and-forward (S&F) file transfer and in-orbit technology verification. However, wider applications such as space science and Earth observation were slow to develop. Emerging space nations, however, were quick to recognise the benefits of entering the space-faring community with an affordable, low-risk "first step" via an extremely inexpensive yet realistic microsatellite programme.

The need to handle this growing interest, to catalyse wider industrial and commercial applications, and to generate regular income to sustain a research activity in satellite engineering at the University of Surrey without dependence on government funding, stimulated the formation in 1985 of a University company - *Surrey Satellite Technology Ltd (SSTL)*. SSTL provides a formal mechanism to handle the transfer of small satellite technologies from the University's academic research laboratories into industry in a professional manner via commercial contracts.

Since UoSAT-5, all the Surrey microsatellites have been designed and built for individual commercial customers. The income generated by SSTL is then re-invested to support the academic activities of the Surrey Space Centre, which has now become the European "centre of excellence" in small satellite engineering research, teaching and applications.

The range of applications of microsatellites can be demonstrated by reviewing recent examples of payloads carried by the UoSAT/SSTL microsatellite platform.

4. LEO Communications

Various constellations of small satellites in LEO have been proposed to provide world-wide communications using only hand-held portable terminals; these broadly fall into two main categories:

- real-time voice/data services (e.g., Iridium, Globalstar)
- non-real-time data transfer (e.g., Orbcom, HealthNet).

The close proximity of satellites in LEO to the user and the consequent small transmission loss and delay time appear attractive when compared to traditional communications satellites in a distant geostationary orbit. Such satellites also hold out the promise of less expensive ground terminals and regional frequency reuse. The communications characteristics associated with a LEO constellation pose, however, quite different and demanding problems, such as varying communications paths and links, high Doppler shifts, and hand-over from satellite to satellite.

Currently, there is only one small LEO satellite service in full operation (HealthNet) which employs a "constellation" of just two microsatellites, HealthSat-1 & 2, both built by SSTL for the network operator, SatelLife (USA). HealthNet, and the services waiting to be implemented (e.g., TemiSat, OrbComm, VITASat, GEMStar) use the narrow-band frequencies (at VHF/UHF) recently allocated to the "little-LEO" services to provide digital data store-and-forward "e-mail" capabilities for use with small, low-power user ground terminals. These can be located in remote regions where existing the telecommunications infra-structure is inadequate or non-existent.

These VHF/UHF frequency allocations exhibit such deleterious effects as multi-path propagation and, in particular, man-made co-channel interference which can very significantly reduce the performance that can be achieved in practice by the satellite communications system compared with that expected from a simple theoretical model. A thorough understanding of the real LEO communications environment at VHF/UHF is therefore necessary in order to select optimum modulation and coding schemes.

Several SSTL satellites/microsatellites carry a Digital Signal Processing Experiment (DSPE). This has been designed to provide a sophisticated in-orbit test bed for research into optimising communications links with satellites in LEO. The DSPE comprises a TMS320C25 and a TMS320C30 with data interfaces to the spacecraft communications sub-systems - enabling it to replace the hardware onboard modems with an in-orbit, re-programmable modem. The DSPE is being used in a research programme to evaluate adaptive communications links, continuously optimising modulation/demodulation techniques, data rates and coding schemes.

The interference characteristics of the VHF LEO frequency allocations have been measured using an experimental communications payloads on the S80/T and HealthSat-2 microsatellite missions. In conjunction with a mobile groundstation, S80/T has measured the VHF spectrum noise and interfering signals, in order to evaluate the use of VHF frequencies for a full-scale LEO communications service (S80). S80/T was completed by SSTL for the Centre National d'Etudes Spatiales (France), from proposal to launch, within 12 months!

5. Space Science

Microsatellites can offer a very quick turn-around and inexpensive means of exploring well-focused, small-scale science objectives (e.g., monitoring the space radiation environment, updating the International Geomagnetic Reference Field) or providing an early proof-of-concept prior to the development of large-scale instrumentation in a fully complementary manner to expensive, long-gestation, large-scale space science missions. SSTL missions have demonstrated that it is possible to progress from concept through to launch and orbital operation within 12 months and within a budget of less than US $ 5 million. This not only yields early scientific data but also provides opportunities for young scientists and engineers to gain "real-life" experience of satellite and payload engineering (an invaluable experience for later, large-scale missions). Students can initiate a programme of research, propose and build an instrument, and retrieve orbital data for analysis and present a thesis within a normal period of post-graduate study.

UoSAT-3 and 5, KITSAT-1 and 2, and PoSAT-1 provide examples of the use of a microsatellite platform for collaborative space science research between the University of Surrey, the UK Defence Research Agency, UK AEA, the UK Science and Engineering Research Council, KAIST and Portugal.

A "Cosmic Ray Effects and Dosimetry (CREDO)" payload monitors the near-Earth radiation environment. It provides an important opportunity to validate ground-based numerical models with flight data yielding simultaneous measurements of the radiation environment and its effects upon on-board systems (especially single event upsets (SEUs) in VLSI devices).

6. Earth Observation

Conventional Earth observation and remote sensing satellite missions are extremely costly - typically in excess of US $ 250 million each. There are relatively few such missions and the resulting data, whilst impressive, are correspondingly expensive. The development of high-density two-dimensional semi-conductor Charge-Coupled Device (CCD) optical detectors, coupled with low-power consumption yet computationally powerful microprocessors, presents a new opportunity for remote sensing using inexpensive small satellites.

UoSAT-1 and 2 both carried the first experimental 2D-CCD Earth imaging cameras, and led to the development of the CCD Earth Imaging System on-board UoSAT-5. That was intended to demonstrate the potential of inexpensive, rapid-response microsatellite missions to support remote sensing applications. Clearly, the limited mass, volume, attitude stability and optics that can be achieved with a tiny microsatellite means that a different approach must be taken to produce worthwhile Earth observations. For these reasons, SSTL employs electronic cameras with 2-dimensional CCD arrays to gather imagery from its microsatellites. Because cameras capture whole images in a single snap-shot, they preserve scene geometry; they are therefore immune to the residual attitude drift experienced on microsatellites.

The Earth Imaging System (EIS) on-board the microsatellites comprises a 1024x1024 pixel area CCD digitised to 256 grey levels. The digitised data are stored within a CMOS RAM which is accessed by two Transputers. This allows the image data to be processed to enhance quality and compressed to reduce storage and transmission requirements. The EIS data are transferred via the microsatellite's local area network (LAN) to the 80C186/386 on-board computers and stored as files within the 32-128 Mbyte RAM DISK for later transmission to ground - some 60 images can be stored within the RAM DISK at any one time. The EIS is commanded to collect an image of a particular area of the Earth's surface by the on-board computer which operates a multi-tasking, real-time operating system responsible for the automatic operation of the microsatellite mission.

Ground controllers select a sequence of images of areas of interest anywhere on the Earth's surface. Checking the time and position of the microsatellite using an on-board GPS receiver and orbital model, they instruct the on-board computer to collect the images according to a "diary" that is loaded periodically in advance to the microsatellite.

The PoSAT-1 microsatellite carries two independent cameras providing a wide-field view (ground resolution 2 km) for meteorological imaging and a narrow-field view (ground resolution 200 metres) for environmental monitoring; the 650 nm (±40 nm) optical filters provide good separation of arid/vegetation and land/sea boundaries and disaster warning information.

The latest generation of small Earth observation satellites using the SSTL platform, such as Thai-Puht, support EIS cameras yielding better than 90-metre ground sampling distance. Their 3 spectral bands provide multispectral image data in LANDSAT-compatible bands.

The Tsinghua-1 microsatellite launched in June 2000 provides 35 m 4-band multi-spectral imaging with the capability of ±15° (±200 km) off-nadir imaging coverage upon demand.

7. In-Orbit Technology Verification

Microsatellites also offer an attractive and low-cost means of demonstrating, verifying and evaluating new technologies or services rapidly in a realistic orbital environment, and within acceptable risks, prior to a commitment to a full-scale, expensive mission.

UoSAT-based microsatellites have supported a wide range of such in-orbit technology demonstrations, covering:

- new solar cell technologies
- modern VLSI devices in space radiation
- demonstration of advanced communications

- pilot demonstrations of new communications methods.

Satellites depend upon the performance of solar cell arrays for the production of primary power to support on-board housekeeping systems and payloads. Knowledge of the long-term behaviour of different types of cells in the radiation environment experienced in orbit is, therefore, essential. The continuing development of solar cell technology, based upon a variety of materials and different process techniques, yields a range of candidate cells potentially suitable for satellite missions.

Unfortunately, ground-based, short-term radiation susceptance testing does not necessarily yield accurate data on the eventual in-orbit performance of the different cells. Hence, there is a real need for evaluation in an extended realistic orbital environment. UoSAT-5 carries a Solar Cell Technology Experiment (SCTE) designed to evaluate the performance of a range of 27 samples of GaAs, Si and InP solar cells, and from a variety of manufacturers. When the Sun passes directly overhead the panel mounted on the body of the microsatellite, the monitoring electronics are triggered automatically and measure typically 100 current/voltage points for each cell sample. These data are then sent in a burst to the microsatellite's on-board computer, together with associated temperature and radiation dose data, for storage prior to later transmission to ground. SCTE measurements are taken repeatedly immediately after launch, when the radiation degradation is most rapid, and then at increasing intervals thereafter.

8. Military Applications

The demands of a military-style satellite procurement and the cost-effective approach to microsatellite engineering might, at first sight, appear incompatible! However, whilst retaining the essential characteristics of low cost and rapid response, a military version of the SSTL microsatellite platform with deployable solar panels has been developed to support various military payloads. The main differences between the commercial and military versions of the platform is in the specification of components and, particularly, in the amount of paperwork that traces hardware and procedures. An optimum trade-off between the constraints of a military programme and economy has been sought. This has resulted in an increase by factor of 1.5 for cost and timescale when compared with the "commercial" procurement process.

The first use of the SSTL military microsatellite platform was on the CERISE mission designed and built for the French MoD and launched a 700km low Earth orbit by Ariane in July 1995. After a year of perfect operations, Cerise made history as the first operational satellite to be struck by a piece of space debris (rocket fragment) which severed its stabilisation boom. However, due to the flexibility of the microsatellite systems, SSTL engineers were able to re-stabilise Cerise by uploading new attitude control algorithms and return it to operations. A second microsatellite for the French (Clementine) was put into LEO in 1999, and a microsatellite (PICOsat) for the USAF FCT programme is awaiting launch.

The Geostationary Transfer Orbit (GTO) provides a good opportunity to study the effects of a severe radiation environment on satellite components - especially solar cells and VLSI components. Surrey has provided platform sub-systems and payloads to the UK Defence Research Agency (DRA - Farnborough) for their two Space Technology Research Vehicle (STRV-1) microsatellites which were launched into GTO in early 1994. STRV-1a and b carry a range of in-orbit technology demonstration payloads, particularly to study the effects of the space radiation environment on military satellite components.

9. Know-How Transfer, and Training Using Microsatellites

Although microsatellites are physically very small, they are nevertheless complex and exhibit virtually all the characteristics of a large satellite - but in a microcosm. This makes them particularly suitable as a focus for the education and training of scientists and engineers by providing a means for direct, hands-on experience of all stages and aspects (both technical and managerial) of a real satellite mission - from design, construction, test and launch through to orbital operation.

The very low cost, rapid timescale and manageable proportions makes this approach very attractive to emerging space nations who wish to develop and establish a national expertise in space technology through an affordable small satellite programme. Each know-how transfer and training (KHTT) programme is carefully structured according to the specific requirements or circumstances of the country or organisation concerned, but the first phase typically comprises:

- Academic education (MSc, PhD degrees)
- Hands-on training (seconded to SSTL)
- Groundstation (installed in country)
- Microsatellites (first at SSTL, second in country)
- Know-how transfer (satellite design licence).

Nine highly successful international know-how transfer programmes have been completed by Surrey and SSTL, as shown in Table 3. Two new programmes, with Algeria and Turkey, commenced recently.

Country	Dates	Satellites
Pakistan	1985-89	BADR-1
South Africa	1989-91	UoSAT-3/4/5
South Korea	1990-94	KITSat-1/2
Portugal	1993-94	PoSAT-1
Chile	1995-97	FASat-Alfa/Bravo
Thailand	1995-98	TMSAT-1
Singapore	1995-99	Merlion payload
Malaysia	1996-98	TiungSAT-1
China	1998-99	Tsinghua-1
Turkey	*2000-01*	*BiltenSAT-1*
Algeria	*2000-01*	*AITSAT*

Table 3. Surrey's know-how transfer programmes

In total, 70 engineers have been trained through these in-depth KHTT programmes at Surrey — a further 320 students from countries world wide have graduated from the MSc course in Satellite Communications Engineering unrelated to these KHTT programmes.

Once developing space nations have mastered microsatellite technology, the minisatellite provides a logical next step in the development of an increasingly capable national space infrastructure.

10. "Surrey Space Club"

As each of these international know-how transfer and training programmes has been based on an SSTL microsatellite, the participating organisations share a common experience with Surrey - a common spacecraft design heritage, common spacecraft-groundstations and common communications protocols. After returning to the home country, each organisation has commenced its own national space activities and research projects. In order to be able to share new ideas, discuss common interests, and work together to achieve common space goals, the *"Surrey Space Club"* has been formed. It is a regular forum for this "commonwealth" of developing space nations to meet and learn from each other, and thus to:

- share satellite recourses - there are 8 Surrey-designed microsatellites currently in operation in orbit from partners
- gain greater access time to spacecraft in orbit through co-ordinated and shared access to the network of 9 Surrey-based groundstations around the world
- exchange new ideas on small satellites research and development
- build small satellite constellations together (e.g., the Disaster Network of EO Microsatellites)
- co-ordinate low cost launch opportunities for members
- help new members to maintain the momentum of a long term national space programme
- contribute to planetary exploration by international co-operation (e.g., Surrey's Lunar Mission 2002).

It is believed that all the members will derive great benefit through closer collaboration via the "Surrey Space Club" whilst at the same time, importantly, maintaining their own clear independence.

The first forum of the "Surrey Space Club" was held during the visit of Her Majesty The Queen to Surrey in December 1998 when she opened the new Surrey Space Centre building and presented SSTL with the Queen's Award for Technological Achievement.

All past (and future!) Surrey technology transfer partners are welcome to join this unique opportunity for international co-operation.

11. Microsatellite Groundstations

Compact and low-cost mission control groundstations have been developed by SSTL to operate the microsatellites once in orbit. These groundstations are based on PCs and are highly automated - interacting autonomously with the microsatellite in orbit - to reduce manpower requirements and to increase reliability. SSTL has installed nine such groundstations for KHTT partners and other customers worldwide.

The SSTL Mission Control Centre at Surrey operates eleven microsatellites in LEO with just a single operator.

12. SSTL Minisatellites

In response to growing payload demands for power, volume and mass - but still within a small-scale financial budget - UoSAT and SSTL are developing an enhanced, modular, multi-mission minisatellite platform capable of supporting missions up to 400 kg.

The SSTL minisatellite platform has been designed according to similar cost-effective principles that have proved so successful on the UoSAT/SSTL microsatellites. The basic platform costs around US $ 6 million. It is compatible with a range of affordable launch options - on Ariane, CIS, Chinese and US (Pegasus) launchers to meet a variety of mission objectives and capable of operating in different orbits. Its primary features are:

- 400 kg total mass
- 150 kg payload capacity
- 1.2m diameter, 1m height
- 3-axis, 0.1 degree, attitude control
- GPS autonomous orbit and attitude determination
- 1 Mbps L/S-band communications links
- on-board propulsion for orbit manoeuvres
- cold gas thrusters for attitude control
- 300 W orbit-average power, 1 kW peak power.

13. UoSAT-12 Minisatellite

UoSAT-12 demonstrates the capability of this minisatellite platform. Three-axis control is provided by a combination of magnetorquers, momentum wheels and cold gas (N_2) thrusters. An experimental electric N_2 "resisto-jet" thruster will provide orbit trimming and maintenance demonstrations for future network constellations. UoSAT-12 was launched successfully by a converted SS18 ICBM (Dnepr) launcher into LEO in April 1999. The UoSAT-12 minisatellite carries 35-metre resolution multispectral and 8-m panchromatic CCD Earth observation cameras. Sophisticated frequency-agile VHF/UHF and L/S-band DSP regenerative transponders provide both real-time and store-and-forward communications to small terminals.

The SSTL minisatellite platform can be used to support a range of other missions, such as a light-weight synthetic aperture radar (SAR).

14. New SSTL Enhanced Microsatellite

A new SSTL 100 kg "enhanced microsatellite", based on the proven flight heritage of the 50 kg microsatellites, supports missions requiring larger payloads at low cost. The 600 x 600 x 500 mm bus structure accommodates larger solar panels, and provides flexible internal and external payload accommodation. Nevertheless, this bus fits comfortably on low-cost secondary payload carriers such as the new Ariane-V ASAP.

Four body-mounted GaAs solar panels provide 38 W average power. A full 3-axis attitude control system allows payloads to be pointed at any terrestrial or celestial target with an accuracy better than 0.2°. A network of on-board computers and embedded controllers automates all telemetry and telecommand, communications, payload control and data management functions. Communications links in the VHF and UHF bands use standard packet protocols to provide connection to the mission control centre and to any remote communications or data-collection terminal. A 1 Mbit/s S-band downlink provides high-speed communications for the remote sensing payload, using international standard CCSDS ground/space link techniques.

The SSTL enhanced-microsatellite can carry 5 CCD remote sensing cameras, providing 3-band multispectral imaging with 50 metres ground resolution and panchromatic imaging with 20 metres ground resolution. A panchromatic or colour-mosaic wide area meteorological imager with 1 km ground resolution may also be carried into LEO.

A clear payload volume measuring approximately 300 x 300 x 200 mm is available on the Earth-pointing facet of the satellite, and additional volume is available within the space frame. Payloads up to 30 kg can be accommodated in this enhanced small satellite.

15. Nanosatellites

A tiny, 6.5 kg "nanosatellite" SNAP-1, designed and built at Surrey in under one year, was successfully launched in June 2000. SNAP-1 is a highly-integrated and sophisticated spacecraft carrying advanced micro-miniature GPS navigation, camera technology, onboard computing, propulsion and attitude control technologies, all developed in the UK by Surrey.

SNAP's primary payload is a machine vision system (MVS) to inspect other spacecraft in orbit. The MVS consists of three ultra-miniature wide-angle CMOS video cameras and one narrow-angle CMOS video camera, together with sophisticated image processing electronics. The MVS also provides medium resolution images of the Earth from SNAP's 650km altitude, near-polar orbit. The near infra-red filter in the narrow angle camera provides good discrimination between land, sea and clouds. It allows the camera to produce clear images through the otherwise hazy atmosphere at 500m ground resolution.

Future applications for the nanosatellite are for the remote inspection of satellites or the International Space Station, and for monitoring deployment systems in orbit, and for small space science instruments requiring measurements with spatial diversity.

16. Low Cost Launches

A sustained, commercial, low-cost small satellite programme must also be matched by correspondingly inexpensive and regular access to orbit through formal launch service contracts, because it makes little sense to construct sophisticated yet inexpensive microsatellites if the launch costs remain prohibitively high. Early microsatellites were launched virtually free, on a "favour" basis, by the US and the USSR, but these launch opportunities were infrequent and unpredictable. The breakthrough came in 1988 when Arianespace developed the Ariane Structure for Auxiliary Payloads (ASAP) specifically to provide, for the first time, regular and affordable launch opportunities for 50 kg microsatellites into both LEO and GTO on a commercial basis. To date, some 18 microsatellites have been launched via ASAP, including SSTL Microsatellites (KitSat-1 and S80/T).

Ariane alone cannot now provide the number of launch opportunities into LEO needed to meet the burgeoning growth of small satellites. Thus, alternative, inexpensive launch options from the CIS (on Tsyklon, Zenit or Cosmos) are now increasingly being used for micro/minisatellites.

Within the last few years, the large stockpiles of ICBMs in the CIS have become available for use as small launchers through the demilitarilisation programme (e.g., SS-18/Dnepr, SS-19/Rockot, and SS-25/START). SSTL co-operated with ISC Kosmotras (Moscow) to convert the SS18 ICBM into the first Dnepr small satellite launcher for the successful launch of the UoSAT-12 minisatellite from a silo at Baikonur in April 1999.

17. Project Management

Solving the technical challenges associated with the design, construction, test and operation of a microsatellite is less than half of the story - in parallel with the technical considerations of the mission, effective project management is crucial to the realisation of a successful low-cost, sophisticated small satellite project.

Affordable small satellites require a very different approach to management (as well as to technology) if cost, performance and delivery targets are to be met. Several attempts at taking a traditional aerospace organisation to produce such satellites have failed because of the rigidity of management structure and the existing "mind-set". Small teams (25 persons), working in close proximity with good communications, with well-informed and responsive management, are essential.

These characteristics are best found in small companies or research teams rather than in large aerospace organizations. Such organisations may find it difficult to adopt or modify procedures necessary to produce affordable small satellites using staff and structures used to conventional aerospace projects. The main ingredients for a successful small satellite project can be summarised as:

- highly innovative technical staff
- small, motivated teams
- personal responsibility for work rigour and quality
- good team communications, and close physical proximity
- well-defined mission objectives and constraints
- knowledgeable use of volume, modern components
- layered, failure-resilient system architecture
- thorough burn-in
- technically-competent project management
- short timescale (prevents escalation of objectives!).

18. Summary and Conclusions

The University of Surrey embarked upon the design of its first experimental microsatellite in 1978 and UoSAT-1 was launched by NASA in 1981. Since then, a further seventeen low cost yet highly sophisticated microsatellites have been built and launched into low Earth orbit. The UoSAT/SSTL missions have demonstrated that microsatellites can play a useful role in supporting specialised communications, Earth observation, small-scale space science, and in-orbit technology verification missions.

Whilst obviously limited in payload mass, volume and power, but with very real and attractive advantages in terms of cost and response time, microsatellites offer a complementary role to traditional large satellites. They can be a "gap-filler" for affordable quick-response, or for exploratory missions with either civil or military objectives. Developing space nations have used rapid and inexpensive microsatellite projects to act as the focus for effective technology transfer and for an affordable first step into orbit.

Surrey has established itself firmly as the international centre of excellence in academic research, teaching and commercial applications of small satellites.

Acknowledgements

The author would like to recognise the contributions of colleagues at and support from Surrey Satellite Technology Ltd, UK Defence Research Agency, Science and Engineering Research Council, Arianespace (France), Alcatel Espace (France), KAIST (Korea), PoSAT Concorcium (Portugal), Matra-Marconi Space (France), SatelLife (USA) and the University of Surrey. FASat-Alfa and FASat-Bravo is owned by the Chilean Air Force (FACH) and has been developed, constructed and will be operated under a Technical Collaboration Program between FACH and Surrey Satellite Technology Ltd. (SSTL), UK. Thai-Puht is a collaborative programme between Mahanakorn University of Technology (Thailand), the Thai Microsatellite Company (TMSC) and SSTL. UoSAT-12 is funded by SSTL and the L/S-band communications payload on UoSAT-12 is supported by Nanyang Technological University, Singapore. TiungSAT-1 is supported by ASTB, Malaysia. PICOSat is supported by the USAF Foreign Comparative Test program. Tsinghua-1 is supported by Tsinghua University, Beijing, PR China, through a joint venture company with SSTL.

Keynote Address

Nanotechnology - Tools for the Satellite World

S. W. Janson, Mail Stop M2/241, The Aerospace Corporation, P.O. Box 92957, Los Angeles, California, 90009-2957, USA

e-mail: siegfried.w.janson@aero.org

Abstract

Microelectronics uses batch-fabrication techniques to produce planar, interconnected sub-micron structures. Complimentary metal oxide semiconductor (CMOS) nanoelectronics will be mass-produced by the year 2004 when lateral structures smaller than 100-nm will become common. Microelectromechanical systems (MEMS) are fabricated using similar processes and will benefit from decreasing minimum feature size over time. Nanoelectronics and nanoelectromechanical systems will evolve over the next decade to provide ever-higher levels of functional density per unit area. Beyond the next decade the future is less certain, but other technologies such as nanometer-scale single electron transistors provide an idea of what may be possible in 20-25 years. The continuing increase in functional density will enable decreased spacecraft system size, and ultra-small spacecraft.

Micro/nano/picosatellites (100-to-0.1 kg mass) have flown, and femto/atto/zeptosatellites (100-to-0.1 gram mass) are possible. Small spacecraft enable distributed spacecraft architectures using modest launch systems, but they will require mm-to-cm scale attitude determination and control systems, along with propulsion systems, to enable constellation maintenance, control, and deorbit capability. Examples of prototype microengineered attitude determination and propulsion systems are given, along with comments on how nanofabrication could improve their performance.

1. CMOS Mico/Nanoelectronics

Nanotechnology is the systematic use of sub-micron structures over more than one dimension to achieve a practical function. By current convention, nanotechnology covers the size range of individual atoms to about 100 nm. Bulk micro/nanotechnology uses conventional semiconductor fabrication techniques to produce micron and sub-micron scale structures. The ability to make transistors smaller and smaller with time, with associated increases in operating speed and density, are now taken for granted. In 1965, Gordon Moore noticed that memory chips were doubling in capacity roughly every 18-to-24 months. That trend, now known as Moore's Law, still holds and also applies to microprocessor performance doubling every 24-to-30 months. The Intel 4004 microprocessor, introduced in 1971 with 3,200 transistors, was fabricated using 10-μm technology, while the 42 million transistor Intel Pentium IV, introduced in 2000, was fabricated using 0.18-μm technology [Reference 1]. This is a doubling of the number of transistors on a chip every 25 months over the last three decades.

M. Rycroft and N. Crosby (eds.), Smaller Satellites: Bigger Business?, 21–30.

Complementary metal oxide semiconductor (CMOS) technology is used in most of the microelectronic circuits sold today. They are fabricated a layer at a time; individual metal, dielectric, or polycrystalline silicon (polysilicon) layers are deposited across an entire 5-to-30 cm-diameter wafer and excess material is removed. Patterning of layers typically includes deposition of a photoresist layer, exposure to light through a two-dimensional mask, chemical removal of the exposed (or unexposed) regions of photoresist, and chemical or plasma removal of newly-exposed material. The density of active devices is determined by minimum feature size, or linewidth, which is directly proportional to the wavelength of light used in the photoresist exposure step. Wavelengths of 365 nm (mercury I-line) and 248 nm (KrF laser) are now common, and 193 nm (ArF laser) is coming on-line. The latter wavelength is used to produce 0.13-μm minimum feature size. Optical lithography will likely produce 110-nm minimum feature size, while X-ray or extreme ultra violet (EUV) lithography should produce 70-nm minimum feature sizes with high productivity [Reference 2]. The 0.7-to-1.3 nm EUV light is typically synchrotron radiation produced by an electron storage ring [Reference 3].

Fig. 1 shows an idealized schematic drawing of a metal oxide semiconductor (MOS) transistor. A polysilicon electrode is electrically isolated from a single crystal silicon substrate by an extremely thin gate oxide, and two impurity-doped regions straddle the gate/oxide stack. These regions have much higher electrical conductivity than the silicon substrate. The drain is typically at a higher positive potential than the source, and the potential applied to the gate determines the magnitude of current flow from source to drain. The length of the polysilicon gate (horizontal dimension in Fig. 1) is usually equal to the minimum feature size or linewidth, and the width (vertical dimension in the "Top View" of Fig. 1) is adjusted for current-carrying capability.

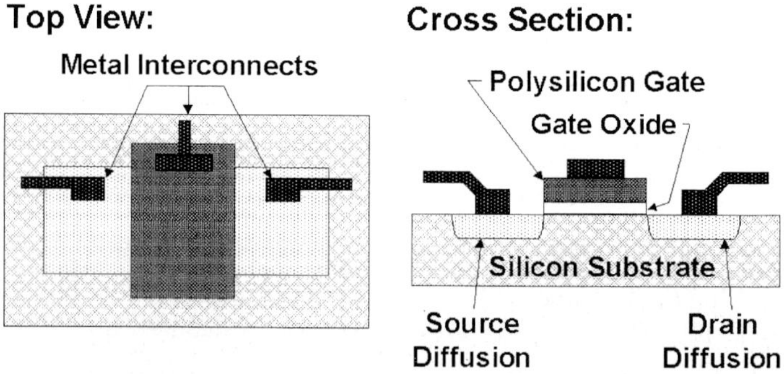

Figure 1. Schematic drawing of an idealized metal-oxide semi-conductor (MOS) transistor; multiple dielectric and interconnect layers have been omitted for clarity

Table 1 gives projected (in 1995) minimum linewidth, die size, and number of active elements per die for newly-introduced dynamic random access memory (DRAM) chips and microprocessors. As of early 2001, 0.18-μm CMOS processes are readily available and 0.13-μm CMOS processes will appear by year's end [Reference 4]. Intel reports that the gate oxide thickness will be only 1.5 nm for its 0.13-μm process [Reference 5]. According to Table 1, we are already a year ahead of the projections made in 1995. More details on expected linewidths, die size, etc., for the next 14 years can be found in the 1999, and 2000 update, of the International Technology Roadmap for Semiconductors (ITRS) [Reference 6]. From the year 2000 update, we can expect 90-nm minimum feature size by 2004 and 35-nm minimum feature sizes by 2014. These roadmaps show significant technology challenges starting about 2005, and assume that these challenges will be met as time goes on.

Year	Linewidth (microns)	DRAM size (sq. cm)	Bits/DRAM (billions)	CPU size (sq. cm)	Transistors/die (millions)
1995	0.35	1.9	0.064	2.5	4
1998	0.25	2.8	0.256	3.0	7
2001	0.18	4.2	1	3.6	13
2004	0.13	6.4	4	4.3	25
2007	0.10	9.6	16	5.2	50
2010	0.07	14	64	6.2	90

Table 1. Estimated (in 1995) CMOS minimum linewidth, die size, and number of active elements per die for new DRAMs and microprocessors (CPUs) [Reference 7]

One of the major limitations to CMOS scaling, assuming that fabrication obstacles such as infinitely-fine photolithography can be overcome, is a practical minimum equivalent gate oxide thickness of ~1 nm (a few atoms thick!) imposed by quantum-mechanical tunneling and fabrication uniformity. This limitation should come into play about 2005 for mass-produced nanoelectronics [Reference 8]. Reference 9 gives an excellent review of CMOS scaling limitations in more depth. On a positive note, Intel researchers have already demonstrated 0.07-μm CMOS transistors (mass production predicted for ~2009) with sub-1.0 nm gate oxide [Reference 10]. As the scaling limits draw near during the next 5-to-10 years, a number of technology changes should occur. One option is to use silicon-on-insulator (SOI) technology to reduce current leakage into the substrate and allow closer packing of transistors. Another option is to depart from two-dimensional MOS transistor layout by using a three-dimensional transistor geometry. We can expect Moore's law to continue for at least another decade, and enjoy the benefits of mass-produced CMOS nanoelectronics. Beyond the next decade, CMOS may give way to other technologies.

Ultimately, gate electrodes start to become "quantum dots" at the ~30 nm size. For a 30-nm sphere embedded in a 7-nm dielectric, the threshold voltage shifts by about 0.04 V per electron while at 10-nm diameter, the threshold voltage shifts by 0.26 V per electron [Reference 9]. Only a few electrons are needed to turn the transistor on at room temperature in the latter case. The ultimate limit is single-electron switching which occurs at diameters of about 5 nm. Quantum dots can be used to create single electron transistors (SETs) using just dielectrics and conductors [Reference 11]. References 12 and 13 provide more detailed information on single electron devices. If Moore's Law holds over the next 30 years, with a doubling of transistor density every 2.3 years, which is mere speculation at this time, SETS will be mass-produced by the year 2026. CMOS may be approaching its limits within the next decade, but nanoelectronics has plenty of room and time to grow.

2. MEMS/NEMS

Microelectromechanical Systems (MEMS) are micron-to-millimeter scale sensors and actuators that can be produced using modified semiconductor fabrication techniques. Integration of sensors, actuators, and control electronics at the die or semiconductor substrate level results in higher reliability and reduced cost, a reduction in the number of piece-parts, and reduction in manual assembly steps. In addition, reduced mass and power typically result from the small size of sensors and actuators, with increased signal-to-noise levels produced by shortened power, control, and signal lines. References 14 and 15 provide good overviews of MEMS fabrication processes, devices and applications.

Fig. 2 shows an example of a MEMS switch. The 60-μm diameter hexagonal central polysilicon plate is mechanically, but not electrically, anchored to the substrate by three folded polysilicon beams which act as springs. It can move up and down (perpendicular to the surface) by a few microns. A planar polysilicon electrode, anchored to the substrate, is underneath the hexagonal plate with a nominal 1.5-micron separation. When a potential difference of 50 Volts is applied between the plate and the electrode, the plate is attracted to the electrode, closing electrical contacts located around the periphery of the plate. This device can also act as an accelerometer by monitoring the capacitance between the hexagonal plate and electrode in real time. Adding a thin metal layer to the top of the central polysilicon plate produces a piston mirror useful for adaptive optics; it has a stroke larger than the wavelengths of visible light.

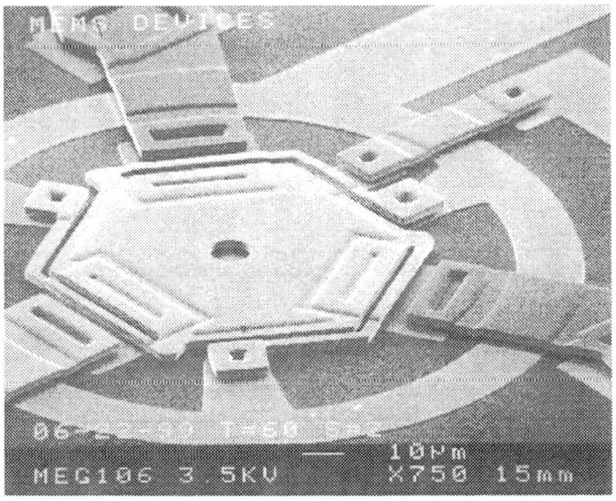

Figure 2. SEM of a surface-micromachined switch
(Photograph courtesy of The Aerospace Corporation)

MEMS devices range in size from a few μm to a few mm with minimum feature sizes between 0.5 μm-to-20 μm. Future nanoelectromechanical system (NEMS) versions could be fabricated using sub-0.1 micron feature and layer thickness to reduce actuation voltage requirements substantially. Reference 16 provides a good review of MEMS devices for space applications.

3. Radiation Effects

Commercial CMOS electronics can usually tolerate total radiation doses up to 1-to-10 kilorads, where a *rad* is the amount of particle radiation that deposits 100 ergs of energy per gram of target material. Most commercial CMOS devices will survive for several years in an 800-km altitude orbit if they have greater than ~5-mm aluminum shielding. Specialized radiation-hard devices are available, but they are expensive and about 7 years behind their commercial counterparts [Reference 17]. Fortunately, commercial CMOS foundry processes, in general, have led to increasing total dose hardness as device features have decreased in size. The increased hardness apparently results from decreased gate oxide thickness [Reference 18]. One commercially-available 0.25-μm process has an apparent total dose limit of greater than 100 krads without design changes and greater than 500 krads with the addition of guard bands, etc. [Reference 17]. MEMS devices are inherently radiation hard because they do not use p-n junctions, while the radiation hardness of NEMS and deep-nanoscale electronic devices such as SETs still needs investigation.

4. Micro/Nano/Pico/Femto/Atto/Zeptosatellites

Microelectronics have created a new generation of microsatellites and nanosatellites. MEMS and NEMS will enable ever-more capable microsatellites and nanostellites and even smaller spacecraft. Since spacecraft with masses between 10-and-100 kg are now classified as microsatellites, and spacecraft between 1-and-10 kg are classified as nanosatellites, one can extend this naming convention down to 0.1-gram mass spacecraft as shown in Table 2.

Spacecraft Class	Mass Range
Microsatellite	10-100 kg
Nanosatellite	1-10 kg
Picosatellite	0.1-1 kg
Femtosatellite	10-100 grams
Attosatellite	1-10 grams
Zeptosatellite	0.1-1 gram

Table 2. Small satellite classification, by mass

Fig. 3 shows a photograph of the smallest active satellites ever launched, the DARPA-Aerospace Corporation PICOSATs [Reference 19]. These 260 gram, 2.5 x 7.5 x 10-cm spacecraft were put into low Earth orbit in February 2000 and were tethered to each other by a 35-meter long fiber [Reference 20]. The square panel on the top is a patch antenna for the 915 MHz uplink/downlink. A spread-spectrum downlink transmitted internal temperature data along with status information on a MEMS RF switch experiment. These spacecraft had no attitude control.

Figure 3. Two DARPA-Aerospace Corporation PICOSATs
(Photograph courtesy of The Aerospace Corporation)

5. Examples of Micro/Nanomachined CMOS Spacecraft Systems

An example of a prototype CMOS Sun sensor "on a chip" is shown in Fig. 4. This 2.2-mm square CMOS die was fabricated using a 1.5-μm process and includes a 10x10 array of active pixel sensors with 82.5-μm pitch. Each pixel includes a 46x46 μm detector and three transistors for row and column selection plus reset. This prototype sensor has a resolution of 4° and a field-of-view of 90° when coupled to 2-mm thick fused silica block with appropriate surface coatings to create a pinhole lens; the fused silica acts as a radiation shield. Reducing the minimum feature size to 0.18-μm allows significantly smaller transistors and hence smaller active pixels while maintaining a reasonable fill factor. Angular resolutions of 0.5° will be possible using a 10-μm pitch. Enough silicon area exists around the sensor array to implement on-chip analog-to-digital conversion and a bi-directional serial interface. Use of radiation hardness-by-design methodologies outlined in Reference 17 should enable a mm-scale "smart" Sun sensor that can survive for more than a decade over a wide variety of orbits.

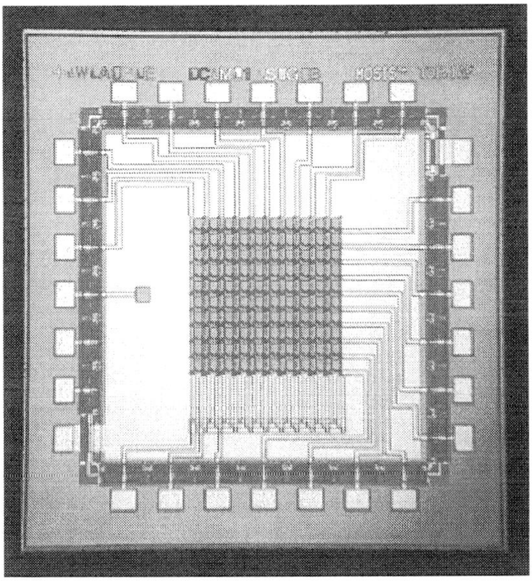

Figure 4. A two-dimensional active pixel sensor array suitable for use in a Sun sensor *(Photograph courtesy of The Aerospace Corporation)*

Earth sensors typically operate at 14-to-16 μm to distinguish the edge of the warm Earth against cold space. Microbolometer arrays, such as the one shown in Fig. 5, can provide infrared imaging of the Earth at these wavelengths. Each pixel consists of 32 serial thermocouple junctions with *hot* junctions in the middle and *cold* junctions anchored to the substrate. The die was fabricated using a 1.5-μm CMOS process where thermocouple junctions are formed using available *poly1* and *metal1* layers. Seebeck thermoelectric coefficients are -1.7 μV/K for *metal1* and -110 μV/K for *poly1* [Reference 21], so 10 or more junction pairs in series can generate mV/K outputs. A post-CMOS etch is used to undercut the "hot" junctions and thermally isolate them. The Earth sensor was fabricated on a 2.2-mm-square die; Fig. 5 shows only 12 elements of a 10x10 array. Fabrication using a 0.18-μm or smaller process would allow more thermocouple junctions per pixel and higher sensitivity.

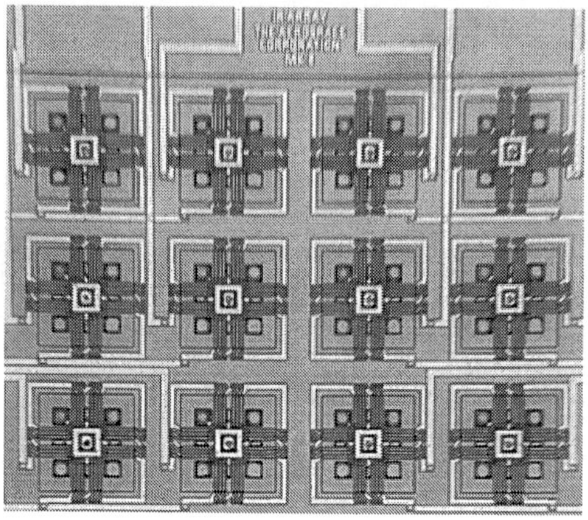

Figure 5. Section of a micromachined bolometer array suitable for use in an imaging Earth sensor
(Photograph courtesy of The Aerospace Corporation)

Fig. 6 shows part of a CMOS die that contains field-emission electron sources and field-ionization test structures for evaluating a particular micromachined ion engine concept [Reference 22]. Electric fields of ~2 V/nm are required for field emission while fields of ~20 V/nm are required for field ionization. This CMOS die was fabricated using a 2-μm CMOS process and relies on 10-nm-scale surface texture to create the very high fields necessary for field ionization. Nanoelectronic CMOS will allow features more than an order-of-magnitude smaller and increased surface area for higher ionization efficiency.

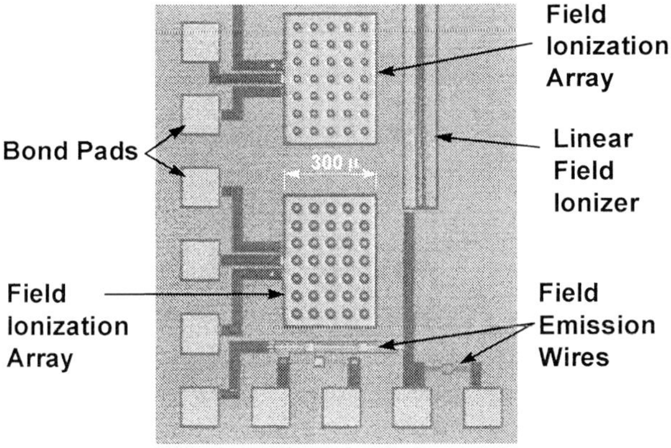

Figure 6. Photograph of one corner of a CMOS die that contains several field-ionization grid structures and thermionic filaments
(Photograph courtesy of The Aerospace Corporation)

6. Summary

Nanotechnology will be commercially-available by 2004 in the form of CMOS circuits. Reduction of the minimum feature size well below 100-nm will then occur, but the future of CMOS is uncertain beyond the next decade. This reduction in minimum feature size will be reflected in MEMS technology which could be used to create mm-to-cm scale attitude determination and propulsion systems for ultra-small spacecraft.

Acknowledgements
I gratefully acknowledge support provided by the Aerospace Corporation through the Corporate Strategic Initiatives. I also gratefully acknowledge DARPA-MTO for supporting the PICOSAT project.

References

1. Intel Microprocessor Quick Reference Guide, available online at <http://www.intel.com/pressroom/kits/quickreffam.htm>
2. Brunner, T.: *Pushing the Limits of Lithography for IC Production*, International Electron Devices Meeting Technical Digest, pp. 9-13, Washington DC, December 1997
3. Silverman, J. P.: X-ray Lithography: Status, Challenges, and Outlook for 0.13 μm, *J. Vac. Sci. Technol. B, Vol. 15,* pp. 2117-2124, 1997
4. Bindra, A.: Advanced 0.13-mm CMOS Process Is Ready to Produce 2-GHz Processors, *Electronic Design, Vol. 49,* p. 27, February 5, 2001

5. Tyagi, S. et al.: *A 130 nm Generation Logic Technology Featuring 70nm Transistors, Dual Vt Transistors, and 6 Layers of Cu Interconnects,* Proceedings of IEEE International Electron Devices Meeting (IEDM), December 11, 2000, available online at <http://www.intel.com/research/silicon/micron.htm>
6. International SEMATECH: *1999 International Technology Roadmap for Semiconductors,* <http://public.itrs.net/files/1999_SIA_Roadmap/ORTC.pdf> *Year 2000 Update; Overall Roadmap Technology Characteristics,* available online at <http://public.itrs.net/Files/2000UpdateFinal/ORTC2000final.pdf>
7. Geppert, L.: Technology 1996: Solid State, *IEEE Spectrum, Vol. 33,* pp. 51-55, January 1996
8. International SEMATECH: *1999 International Technology Roadmap for Semiconductors: 2000 Update; Front End Processes,* available online at <http://public.itrs.net/Files/2000UpdateFinal/FrontEndProcesses2000final.pdf>
9. Philip, H.-S. et al., *Nanoscale CMOS,* Proceedings of the IEEE, Vol. 87, pp. 537-570, April 1999
10. Chau, R. et al.: *30nm Physical Gate Length CMOS Transistors with 1.0 ps n-MOS and 1.7 ps p-MOS Gate Delays,* Proceedings of IEEE International Electron Devices Meeting (IEDM), December 11, 2000, available online at <http://www.intel.com/research/silicon/micron.htm>
11. Ahmed, H.: Single Electron Electronics: Challenge for Nanofabrication, *J.Vac. Sci. Technol. B, Vol. 15,* pp. 2101-2108, 1997
12. Likharev, K.K.: Single-Electron Devices and Their Applications, *Proceedings of the IEEE, Vol. 87,* pp. 606-632, April 1999, available online at <http://rsfq1.physics.sunysb.edu/~likharev/nano/PIEE99.pdf>
13. Likharev, K.K. and Claeson, T: Single Electronics, *Scientific American, Vol. 266,* pp. 80-85, June 1992
14. Madou, M.: *Fundamentals of Microfabrication,* CRC Press, New York, 1997
15. Special Issue on Integrated Sensors, Microactuators, & Microsystems (MEMS), Proceedings of the IEEE, Vol. 86, August 1998
16. Helvajian, H. and Janson, S.W.: Microengineering Space Systems, Chapter 2 in *Microengineering Aerospace Systems*, Helvajian, H. editor, pp. 29-72, AIAA, Reston, VA, 1999
17. Lacoe, R.C. et al.: Application of Hardness-By-Design Methodology to Radiation-Tolerant ASIC Technologies, *IEEE Trans. On Nuclear Science, Vol. 47,* pp. 2334-2341, 2000
18. Osborn, J.V. et al.: Total Dose Hardness of Three Commercial CMOS Foundries, *IEEE Trans. On Nuclear Science, Vol. 45,* pp. 1458-1463, 1998
19. DARPA: Current Efforts: DARPA-MEMS and Microtechnology for Space Applications project summary, available online at <http://www.darpa.mil/MTO/MEMS/frameset.html?summaries.html>
20. Hinkley, D. et al.: A Mechanical Deployment Structure for the PowerSphere Concept, *Proceedings of the 35th IECEC Conference*, 24-28 July 2000
21. Socher, E. et al., Optimal Performance of CMOS Compatible IR Thermoelectric Sensors, *Journal of Microelectromechanical Systems, Vol. 9,* pp. 38-46, March 2000
22. Janson, S.W. et al., *Microthrusters for Nanosatellites,* The Second International Conference on Integrated Micro Nanotechnology for Space Applications (MNT99), Pasadena, CA, April 11-15, 1999

Session 1

New Advances of Nanotechnolgy

Session Chair:

R. Zee, University of Toronto Institute for Aerospace Studies, Canada

The Promises of Nanotechnology: Will They Be Kept?

C. M. Satter, S. E. Matousek, California Institute of Technology, Jet Propulsion Laboratory, 4800 Oak Grove Drive, Pasadena, California, 91109, USA

e-mail: Celeste.M.Satter@jpl.nasa.gov, Steven.E.Matousek@jpl.nasa.gov

Abstract

What are the mission applications of microspacecraft and nanospacecraft? What are the technology needs to enable these missions? What work is currently being done, and what is the state-of-the-art? What are the challenges?

This paper provides an overview of National Aeronautics and Space Administration's (NASA) microspacecraft and nanospacecraft mission applications. A survey of the state-of-the-art and an overview of current development activities are provided, in addition to identifying the technologies needed to enable these mission applications. Technical challenges particular to NASA micro/nano-spacecraft missions are discussed.

1. Introduction

Why is NASA interested in micro/nano-spacecraft? It is because micro/nano-spacecraft are expected to be cheaper to build and fly than conventional spacecraft, and because they can provide greatly reduced launch costs through the use of smaller launch vehicles, shared launches, and/or secondary launch opportunities [Reference 1]. For NASA, these benefits make it possible to fly more frequent, smaller missions instead of one-per-decade "flagship" missions like Galileo or Cassini. And constellations or formations of these small spacecraft can be robust to the loss of any individual element, suffering small degradations to overall mission performance rather than failure of the entire mission.

2. Overview of NASA Mission Applications of Micro/Nano-spacecraft

Applications and concepts for micro/nano-spacecraft can be found throughout all areas of NASA's endeavor (even in relation to human space flight). Applications are particularly driven where mass is severely constrained (i.e., missions to the outer planets), risk is high (in the case of uncertain or hostile environments), or where simultaneous and/or distributed measurements are needed or desirable. The following paragraphs contain descriptions (listed by Directorate) of NASA's micro/nano-spacecraft applications.

2.1 Earth Science (Code Y)

Concepts of interest to NASA's Earth Science Enterprise (ESE) include constellations or formations of micro/nano-spacecraft for distributed or high-

M. Rycroft and N. Crosby (eds.), Smaller Satellites: Bigger Business?, 33–42.

temporal resolution measurements [Reference 2]. Current ESE mission concepts that utilize microspacecraft include:

- Radio Occultation Global Positioning System: 6–100 spacecraft, 30-kg class
- Leonardo (bi-directional reflectance distribution function measurement): 6–12 spacecraft, 30- to 100-kg class
- Global Precipitation Mission: 3–9 spacecraft (plus core), 50-kg class.

For micro/nano-spacecraft constellations, formation flight using micropropulsion is a key enabling technology. Several autonomy technologies are also enhancing or enabling for constellations, including autonomous navigation, relative navigation and collision avoidance, collective planning and scheduling, collective pointing, and fault diagnosis and recovery.

2.2 *Space Flight (Code M)*

In recent years, NASA's Johnson Space Center (JSC) and Ames Research Center (ARC) have been developing robotic astronaut assistants and environmental monitors for the Space Shuttle and the International Space Station (ISS). The Autonomous EVA Robotic Camera (AERCam) Sprint is a 35-cm diameter, 16-kg microspacecraft, which was flown on STS-87 in 1997 [Reference 3]. Another concept under development at ARC is the Personal Satellite Assistant (PSA) [Reference 4]. Intended as a "robotic assistant for astronauts working in space", the PSA (see Fig. 1) is approximately 15-cm diameter.

Key enabling technologies for systems like the PSA and AERCam include autonomous navigation/maneuvering, and mobility or micropropulsion. In the future, evolving micro/nano-sensor technologies could reduce the size and expand the capabilities of such robotic assistants.

2.3 *Aerospace Technology (Code R)*

Code R's interest in micro/nano-spacecraft is more from the perspective of developing the necessary technology than for actual mission applications. As part of their Cross-Enterprise Technology Development Program, Code R's Micro/Nano Sciencecraft Thrust Area aims at producing an orders-of-magnitude reduction of the size and mass of NASA's spacecraft [Reference 5].

2.4 *Space Science (Code S)*

Code S is interested in micro/nano-spacecraft for missions to the outer planets, distributed and/or simultaneous sensing, *in situ* measurements and sensing, and auxiliary spacecraft. Joint microspacecraft mission concept developments have been carried out with the Air Force and with the National Oceanic and Atmospheric Administration (NOAA) [References 6, 7]. Code S mission concepts that utilize micro/nano-spacecraft include:

- The Magnetospheric Constellation mission [Reference 8]
- The Starlight mission [formerly Space Technology 3 (ST3), Reference 9].

For solar system exploration, two of the biggest technology challenges for micro/nano-spacecraft are data collection and data return. Micro/nano-spacecraft typically have size limitations on the instruments they can carry. This limits achievable sensitivity and resolution for science observations. Constellations and formation flight may provide one possible solution to this challenge, but a more fundamental solution would be provided by lightweight or deployable apertures. Lightweight or deployable apertures would also address the second challenge—data return. High-speed downlinks for large volumes of science data require high frequencies and/or high power and/or large antenna apertures. This is particularly true for missions to the outer solar system [Reference 10].

3. Current Development Activities

Major microspacecraft technology development programs in the US include:

- NASA's New Millennium Program (NMP) [ST3 (Starlight) and Space Technology 5 (ST5) missions] [Reference 9, 11]
- Air Force's TechSat 21 Program [Reference 12]
- Air Force Office of Scientific Research (AFOSR) and the Defense Advanced Research Projects Agency (DARPA) University Nanosatellite Program [Reference 13].

A summary of current development activities by technology area follows.

3.1 *Formation Flight*

The NMP ST3 (Starlight) Mission, scheduled for launch in 2005, will demonstrate formation flight for the purposes of long-baseline optical interferometry [Reference 14]. The Nanosat Constellation Trailblazer, or ST5

Mission, will attempt to fly three miniature spacecraft high above the Earth to test methods for operating a constellation of spacecraft as a single system (see Fig. 2) [Reference 15]. The Air Force's TechSat 21 Program will demonstrate formation flight for synthetic aperture radar imaging [Reference 12]. The MIT Space Systems Laboratory is developing the SPHERES (Synchronized Position Hold Engage and Reorient Experimental Satellites) to provide the Air Force and NASA with a testbed for the validation of metrology, control, and autonomy technologies [Reference 16].

3.2 *Autonomy Technologies*

Many autonomy technologies will be demonstrated by the programs described in Section 4.1. Others that can reduce the need for uplink or downlink of large volumes of data include autonomous navigation, on-board data processing, and autonomous image recognition [Reference 17]. Autonomous optical navigation was demonstrated on the NMP Deep Space 1 (DS1) mission [Reference 18].

3.3 *Micropropulsion*

Micropropulsion research has focused on systems which utilize vaporized liquid or solid propellants which are easily stored and do not have the leakage problems associated with pressurized systems [References 19, 20]. Two examples are the subliming solid microthruster and vaporizing liquid microthruster. In addition, many missions require fine pointing control, leading to a requirement for microthruster minimum impulse bit sizes $<10^{-4}$ Ns .

3.4 *Microgyros*

In recent years, teams at JPL, UCLA, and the Charles Stark Draper Laboratory have developed microgyros suitable for use in inertial guidance systems for micro/nano-spacecraft (see Fig. 3) [Reference 21]. These devices enable several orders-of-magnitude reduction in attitude control system mass.

3.5 *Micro- and Nano- Instruments and Sensors*

Reductions in the mass, volume, and power requirements for the science instrument payloads are fundamentally enabling or enhancing for micro/nano-spacecraft. JPL's Center for Integrated Space Microsystems (CISM) and Center for Space Microelectronics Technology (CSMT) have been developing microsensors, instruments, and other integrated "systems-on-a-chip" [References 22, 23].

3.6 *Data Compression Technologies*

Data compression technologies provide another option for reducing downlink requirements. The current state-of-the-art for "lossless" data compression is about 2:1; however, research is ongoing into techniques that can achieve 1–2 orders-of-magnitude compression with some loss [Reference 24].

3.7 *Deployable Apertures*

Currently, mesh deployable antennas, such as the 12-m antennas built by Harris Corporation and flown on the ACES communications satellite [Reference 25] and the 12.25 m x 16 m antennas built by TRW Astro Aerospace for the Thuraya Satellite [Reference 26], are the state-of-the-art for deployable RF communications antennas. However, the emerging technology of inflatable antennas offers potential for the compact stowage of low-mass deployable spacecraft reflectors that are compatible with micro/nano-spacecraft [References 27, 28].

3.8 *Optical Communications*

In some applications, optical communications systems can provide indirect savings in overall spacecraft mass via a reduction in power system requirements as compared with an equivalent data-rate RF system [Reference 29]. Optical communications also present the spacecraft designer with the option of using the same front-end optics for both sensing and communications ("dual-use"), potentially resulting in significant mass savings. This approach was used in the design of the Microspacecraft Technology Development (MTD) demonstration hardware (see Fig. 4) [Reference 30].

3.9 *Aerocapture Technologies*

Aerocapture is a technique in which a spacecraft uses atmospheric drag (instead of chemical propulsion) to decelerate into orbit at another planet. The current state-of-the-art utilizes aeroshells, which are conventional ablative thermal protection surfaces. An alternate technology which has been under investigation in recent years is the ballute. Ballutes are inflatable drag devices which may, in theory, provide more gentle and controllable deceleration at higher atmospheric altitudes. The use of ballutes or aeroshells can significantly reduce the mass required to achieve orbit capture. A chemical propulsion system capable of performing orbit insertion typically constitutes up to 50-85% of the total mass of the space system arriving at another planet. In comparison, the equivalent mass fraction for an aeroshell is typically on the order of 30-35%, and for a ballute may be as low as 20% for some applications [Reference 31].

4. Challenges, Issues, and Evolving Paradigms

4.1 Technical Challenges

Radiation, Micrometeoroids and Dust. Conventional (large) spacecraft often rely on a few millimeters of structural aluminum and burying electronics deep within a spacecraft bus to provide radiation shielding to sensitive electronic components. Micro/nano-spacecraft will not have millimeters of structural material to provide this protection, and may require selective and/or local shielding of radiation-sensitive elements. Similarly, micro/nano-spacecraft will also be more vulnerable to damage by dust or micrometeoroids.

Thermal Control. Thermal time constants are short for small spacecraft, making thermal control more challenging. Adaptive thermal control systems are needed to accommodate widely varying thermal environments or large variations in internal heat dissipation. The dissipation of heat in compact electronics is a particular challenge for micro/nano-spacecraft. New technologies providing solutions to these challenges include miniature active cooling loops, thermoelectric cooling devices, micro heat pipes, and electrochromic materials [Reference 32].

4.2 New Paradigms for Micro/Nano-spacecraft Design

As we gain experience in the design of micro/nano-spacecraft, new "paradigms" are emerging.

Minimize the Payload Requirements. In addition to utilizing new developments in micro- and nano-instruments, approaches such as utilizing shared front-end optics among instruments can provide major benefits.

Minimize the Telecommunications Requirements. Data compression technologies and onboard processing of data are two approaches to reducing downlink demands. Significant savings were also achieved in the Second Generation Micro-spacecraft concepts through the total elimination of uplink capability [Reference 33].

Take Advantage of Multiple-Use Technologies. Very large mass, volume, and power savings have been demonstrated in concepts that utilize the same hardware to perform multiple functions. Examples include the use of minimum gage propellant tanks or high pressure tubing as load carrying structure, and dual use of optical apertures for both communications and imaging. Active Pixel Sensor (APS) arrays [Reference 34] can be utilized for multiple sensing functions ranging from star tracking (sensing dim sources) to imaging of bright targets, thus eliminating the need for multiple detectors. APS arrays can also be fabricated with on-chip processing functions such as centroiding (which reduces image processing requirements) and portions of the array can be selectively read-out (which reduces data output, especially for sparse scenes).

5. Conclusion

While it is difficult to provide a detailed survey of current U.S. micro/ nano-spacecraft technology and development activities in a paper of this limited length, the authors have attempted to provide a fairly comprehensive overview and sufficient references to allow the reader to seek out further details on his or her own. We hope that you find this paper useful.

Acknowledgements

The writing and publication of this paper was performed at the Jet Propulsion Laboratory, California Institute of Technology, through an agreement with the National Aeronautics and Space Administration.

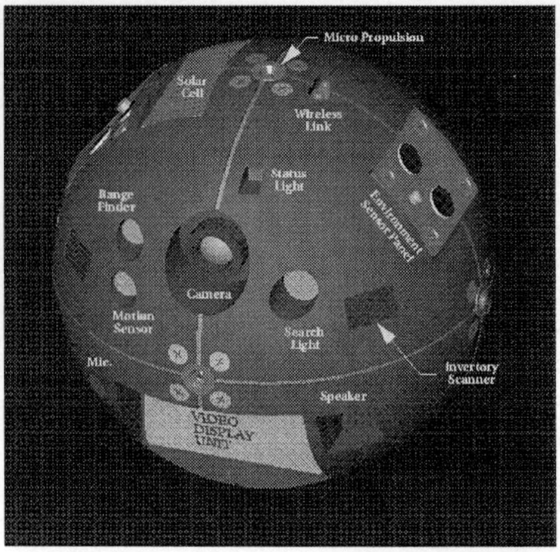

Figure 1. The Personal Satellite Assistant

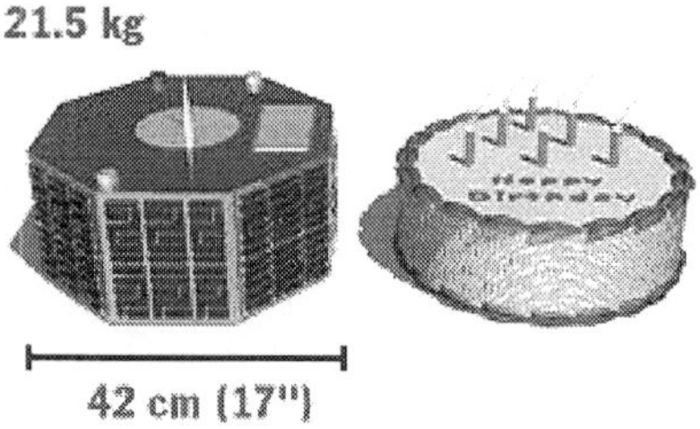

Figure 2. The ST5 spacecraft

Figure 3. Vibratory Microgyro

Figure 4. Microspacecraft Technology Development Hardware

References

1. Rademacher, J. and Leschly, K.: *Microspacecraft Secondary Payload Launch Capabilities & Mission Possibilities,* SSC97-IX-3, 11th AIAA/USU Conference on Small Satellites, Logan, Utah, USA, September 1997
2. NASA: *NASA Earth Science Enterprise Technology Planning Workshop,* <http://nmp.jpl.nasa.gov/workshop-eo4/proceedings.html>. May 6, 2001
3. NASA/JSC: *Autonomous EVA Robotic Camera (AERCam) Sprint,* <http://tommy.jsc.nasa.gov/projects/Sprint>. May 6, 2001
4. NASA Ames Research Center: *Personal Satellite Assistant,* <http://ic.arc.nasa.gov/psa/>. May 6, 2001
5. NASA: Aerospace *Technology Enterprise /Cross-Enterprise Space Technology,* <http://www.aero-space.nasa.gov /programs/mns.htm>. May 6, 2001
6. Collins, D., Moser, R., Ferber, R., Das, A., Jaivin, G., Stallard, M., Smith, J. and Madison, R.: *Novel Missions for Next Generation Microsatellites: The Results of a Joint AFRL-JPL Study,* AIAA/USU Small Satellites Conference, Logan Utah, USA, August 23-26, 1999
7. Office of Systems Development (OSD): *NASA/NOAA/DOD GEOSTORM WARNING MISSION,* <http://www.osd.noaa.gov/sats/geostorms.htm>. May 6, 2001
8. NASA GSFC: *Magnetospheric Constellation Mission/Mission Hardware Concepts,* <http://sec.gsfc.nasa.gov/magcon.htm>. May 6, 2001
9. JPL: *StarLight Website,* < http://starlight.jpl.nasa.gov/ >. May 6, 2001
10. Price, H. W., Carraway, J. B., Matousek, S. E., Staehle, R. L., Terrile, R. J. and Wyatt, E. J.: *Pluto Express Sciencecraft System Design,* IAA-L-0603, 1996
11. NASA JPL: *New Millennium Program Home Page,* <http://nmp.jpl.nasa.gov /index_flash.html>. May 6, 2001
12. AFRL Space Vehicles Directorate: *TechSat 21 Space Missions Using Satellite Clusters,* <http://www.vs.afrl.af.mil/Factsheets/techsat21.html>. May 6, 2001
13. Utah State University Space Dynamics Laboratory: *AFOSR and DARPA University Nanosatellite Program,* <http://www.nanosat.usu.edu/>. May 6, 2001
14. NASA JPL: *Space Technology 3 Project,* <http://nmp.jpl.nasa.gov/st3/proj/form.html>. May 6, 2001
15. NASA JPL: *Nasa's New Millennium Program/Space Technology 5 (ST5)* <http://nmp.jpl.nasa.gov/st5/>. May 6, 2001
16. Massachusetts Institute of Technology (MIT) Department of Aeronautics and Astronautics Space Systems Laboratory (SSL): *MIT Space Systems Laboratory/ SPHERES Formation Flight Testbed,* <http://ssl.mit.edu/spheres/>. May 6, 2001
17. Udomkesmalee, S., Thankoor, A., Padgett, C., Daud, T. and Fang, W. C.: *VIGILANTE: An Advanced Sensing/Processing Testbed for ATR Applications,* In Proceedings of The International Society for Optical Engineering (SPIE), Orlando, Florida, USA, April 22-24, 1997
18. NASA JPL: *NMP/Deep Space 1/Advanced Technologies/Autonomous Navigation,* <http://nmp.jpl.nasa.gov/ds1/tech/autonav.html>. May 6, 2001
19. Mueller, J.: *Thruster Options for Microspacecraft: A Review and Evaluation of Existing Hardware and Emerging Technologies,* AIAA 97-3058 Seattle, Washington, USA, July 7-9, 1997
20. Micci, M. M. and Ketsdever, A. D. (editors): *Micropropulsion for Small Spacecraft,* AIAA Progress in Astronautics and Aeronautics Vol. 187, 2000

21. JPL Commercial Technology Program Office: *JPL – Commercial Technology/HIGH-Performance Gyroscope On A Chip,* <http://techtransfer.jpl.nasa.gov/success/stories/microgyro.html>. May 6, 2001
22. NASA JPL: *Center for Integrated Space Microsystems,* <cism.jpl.nasa.gov>. May 6, 2001
23. JPL: *Center for Space Microelectronics Technology,* <http://csmt.jpl.nasa.gov/csmtpages/index.html>. May 6, 2001
24. Kiely, A. and Klimesh, M.: *Data Compression for Deep Space Missions,* presented at the DESCANSO Symposium, Pasadena, California, USA, September 22, 1999
25. Harris Corporation: *Harris Corporation Announces Successful On-Orbit Deployment of Largest-Ever Commercial Satellite Antennas,* <http://www.harris.com/harris/whats_new/acesdeploy.html>. May 6, 2001
26. Boeing: *Boeing/Thuraya,* <http://www.hughespace.com/factsheets/geomobile/thuraya/thuraya.html>. May 6, 2001
27. ILC Dover, Inc.: *ILC Dover Online /Inflatable Antennas,* <http://www.ilcdover.com/ Products/SpaceInf/Antennas.htm>. May 6, 2001
28. L'Garde, Inc.: *Inflatable Reflector Development,* <http://www.lgarde.com/programs/ird.html>. May 6, 2001
29. Lemmerman, L. A. and Wercinski, P. F.: *Small Neptune Orbiter Using Aerocapture,* STAIF '97, Albuquerque, New Mexico, January 26-30, 1997
30. JPL/Jones, R. M.: *JPL Microspacecraft Technology Development (MTD) Program* <http://techreports.jpl.nasa.gov/1997/97-0180.pdf>. May 6, 2001
31. Hall, J.L.: *A Review of Ballute Technology For Planetary Aerocapture,* Presented at the 4th IAA Conference on Low Cost Planetary Missions, Laurel, MD, May 2–5, 2000
32. Birur, G. C. and Bruno, R. J.: *Thermal and Mechanical Microspacecraft Technologies for Deep Space Systems Program X2000 Future Deliveries,* 2nd International Conference on Integrated Micro/Nanotechnology for Space Applications, Pasadena, California, USA, April 12-15, 1999
33. Collins, D.: *Miniature, Low-Cost, Highly-Autonomous Spacecraft – A Focus for the New Millennium,* IAF Conference on Benefits of Space for Humanity, Oslo, Norway, July 1995
34. JPL/CMST: CSMT/ Active Pixel Sensors (APS) at JPL, <http://csmt.jpl.nasa.gov/csmtpages/APS/index.html>. May 6, 2001

Canada's First Microsatellite - An Enabling Low-Cost Technology for Future Space Science and Technology Missions

R. E. Zee, Space Flight Laboratory (SFL), University of Toronto Institute for Aerospace Studies (UTIAS), 4925 Dufferin Street, Toronto, Ontario, Canada M3H 5T6

e-mail: rzee@utias-sfl.net

P. Stibrany, Microsatellite Systems, Dynacon Enterprises Limited, 3565 Nashua Drive, 2nd Floor, Mississauga, Ontario, Canada L4V 1R1

e-mail: ps@dynacon.ca

Abstract

The MOST (Microvariability and Oscillations of STars) astronomy mission has been chosen by the Canadian Space Agency's Small Payloads Program to be Canada's first space science microsatellite. It is currently planned for launch in early 2002 to conduct long-duration stellar photometry observations in space. A major science goal is to set a lower limit on the age of several nearby "metal-poor sub-dwarf" stars, which may in turn allow a lower limit to be set on the age of the Universe. To make these measurements, MOST will incorporate a small (15 cm aperture), high-photometric-precision optical telescope developed by the University of British Columbia. A direct benefit of meeting the formidable challenges of the MOST mission is the creation of a versatile spacecraft bus that can be used for a multitude of science and technology missions that could not be performed previously using a spacecraft of similar size and cost. Moreover, the expertise now exists for collaboration with the European community on low-cost space technology projects.

1. Introduction to the MOST Mission

Canada's first microsatellite and space telescope, MOST (Microvariability and Oscillations of STars), is expected to break new ground both scientifically and technologically. The benefits include the potential for collaboration with the European community in future joint low-cost missions or technology development projects. The MOST satellite bus and ground stations are currently being developed by Dynacon Enterprises Limited and the Space Flight Laboratory at the University of Toronto Institute for Aerospace Studies (UTIAS/SFL) with help from AMSAT. The modular spacecraft design, and more importantly, the expertise, may be applied to a wide range of future missions with varying requirements and degrees of complexity.

The MOST mission [Reference 1] entails placing a low-cost satellite equipped with a high photometric precision telescope into Earth orbit (Fig. 1),

M. Rycroft and N. Crosby (eds.), Smaller Satellites: Bigger Business?, 43–51.

and using it to conduct observations on a series of stellar targets. The baseline orbit is a dawn-dusk Sun-synchronous orbit, with 98° inclination at an altitude of about 785 km. For long-duration stellar photometry, the satellite must stay inertially pointed to better than half an arcminute accuracy and stability. The extremely demanding attitude control requirements, the challenging data management tasks, and the implications that these have for the thermal design of the satellite and the development of suitable electronics, have created a versatile microsatellite bus capable of accommodating the most challenging science objectives. The expertise developed at Dynacon and UTIAS/SFL can also be exploited in adapting the spacecraft design to the needs of the most ambitious space scientist. Moreover, the ground stations installed for the MOST mission can be used for many different missions, with customized interfaces tailored to specific mission requirements.

Figure 1. Depiction of MOST in orbit

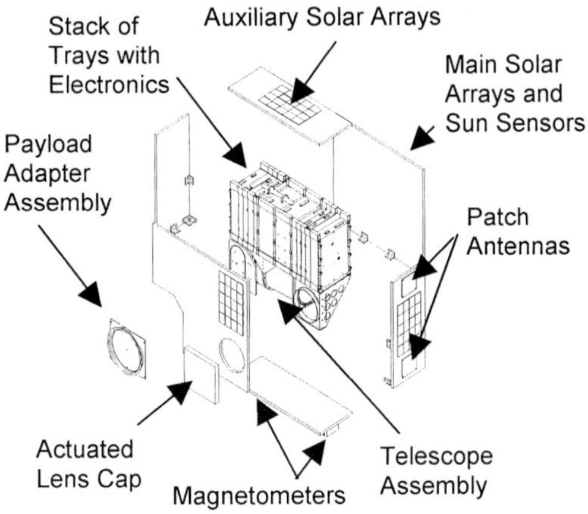

Figure 2. General layout of MOST

2. Multi-Mission Microsatellite Design

The guiding philosophy for the design of this system is to employ proven microsatellite designs from some of the most experienced microsatellite designers in the world as a starting point, making adaptations to these in order to meet the core science objectives of the mission. The resulting system design allows the mission to be achieved while re-using structural, power, on-board computing, telemetry and command, and ground station subsystem designs from previous successful microsatellite missions. This in turn can be used for future microsatellite missions with ease.

The satellite's general layout is shown in Fig. 2. It is driven by the secondary payload volume envelope provided by the Delta II launch vehicle, and incorporates a stacked set of Bus functional modules, with an adjacent Payload bay defined by a set of panels carrying solar arrays. The overall dimensions, including an allowance for the launch vehicle interface hardware, but excluding appendages for antennas and magnetometers, are approximately 63 by 58 by 25 cm.

2.1 *Bus Overview*

The main features of the MOST Bus design are:

- It is based on the stacked-tray modular concept that has been used successfully in the AMSAT Microsat series of satellites, as well as in the University of Surrey/SSTL UoSATs [Reference 2]. The main functional elements of each subsystem (electronics boards, reaction wheels, batteries) are housed within a set of trays of identical cross-section, which are bolted together to form a "box-beam" structural back-bone for the satellite
- The total mass of the microsatellite is approximately 50 kg, with the fundamental frequency of vibration above 100 Hz. The satellite structure has been optimized to withstand all launch loads with comfortable margins
- The On-Board Computer Subsystem is based upon a multiprocessor architecture involving an off-the-shelf V53 housekeeping computer communicating with custom DSP boards using the RS-422 standard. The Motorola DSP56303 is used in the Attitude Control Computers in addition to the CCD Drive Computers. A Simple Serial Protocol has been designed that is available as a digital communications standard for microsatellite developers
- The Telemetry and Command Subsystem consists of custom dual S-band receivers and transmitters, using GMSK modulation, and communicating with the AX.25 and PACSAT protocols. The transmitters have an efficiency of about 30 %, much better than most off-the-shelf products available. A data rate of 9.6 kbps is available for the MOST uplink, while 38.4 kbps is used on the downlink
- The Power Subsystem uses a centralized switching, decentralized regulation topology for maximum flexibility and control. The power supply to the satellite uses Peak Power Tracking to maximize the energy conversion efficiency of the solar arrays. The total available power from the solar panels during nominal operations is approximately 30 W
- The Attitude Control Subsystem (ACS) uses as its foundation a design that has been used successfully in other microsatellites. Magnetometers and

magnetorquers provide a simple but low-accuracy basic attitude control capability for detumbling

- To create the high-accuracy attitude control capability needed to achieve the science goals, a set of small reaction wheels provides three-axis actuation. These are commanded based on pointing error signals generated by the science instrument, when operating in fine-pointing mode. On MOST, the instrument also acts as a star sensor. This tightly couples telescope pointing with the performance of the Attitude Control Subsystem
- The Thermal Control Subsystem is entirely passive, except for trim heaters that regulate the temperature of the CCD focal plane. This is no small feat considering that most previous microsatellites are spinning spacecraft with much less difficult thermal trade-offs. In addition, deployable solar arrays are not used, creating an even greater challenge.

2.2 *Payload Overview*

The MOST satellite's payload comprises the science instrument, along with the electronics required to operate it and interface it with the on-board computers. The main requirements on the instrument are:

- To collect science photometry data from the selected target star. This is required to be photometrically very precise in order to maximize the signal to noise ratio, to be photometrically stable over sample runs of several weeks duration when subject to up to 25 arc-seconds of instrument boresight wander, and to have a wide dynamic range in order to maximize the number of usable target stars
- To collect star-field image data with which the ACS will measure the instrument's pointing error. The baseline requirement here is to produce star field image data once per second, using variable exposure times of as little as a tenth of a second.

The main requirements on the payload data processing electronics are to operate the instrument's camera, to process photometry measurements from the target star using variable exposure durations between 1 and 60 seconds, and to process the star-field image data at 1 Hz, with a maximum latency time of one second, to produce boresight pointing error measurements with an accuracy of 1 arc-second.

The science instrument features a dual-CCD-based visible-light camera that receives light through the largest optical telescope that can fit within the available volume and mass allocations. The baseline design has a 15-cm aperture, and is about 55 cm long. A Maksutov optics design is used. The CCDs are passively cooled, but actively heated to control their temperatures and to maximize photometric precision.

2.3 *Ground Station Overview*

A pair of ground control stations is planned, to issue commands to operate the satellite's payload, attitude control subsystem, and other housekeeping functions. The stations are to be located at UTIAS/SFL and the University of British Columbia.

Both ground stations are based on the same design – commercial UHF/VHF transceivers with programmable Doppler correction, L- and S-band up- and down-converters, and AMSAT-developed satellite tracking and radio-operation software running on commercial PC-class computers. These are inexpensive to develop and reliable in operation, and have the data throughput capacity needed to downlink the data generated by the satellite's payload. The antenna system consists of a collection of Yagi antennas for VHF, UHF, and L-band communications. Two and three meter reflector dish antennas are used to communicate at S-band.

The inexpensive ground station architecture and system design is augmented by custom software used for remote ground station control over the Internet. Specialized command interfaces have been written that talk to a coordinating program that organizes all requests into a logical command sequence for the satellite. Each command interface has a virtual link to the relevant hardware on board the satellite. To the user, it is as if the computer on the ground is talking directly to that hardware in space. These programs may be tailored to the specific attitude control and scientific payload requirements of any microsatellite mission.

3. Enabling Technologies for Future Missions

Several key technologies have been developed for future microspace missions. These are in addition to the expertise developed in creating optimized structures, thermal control systems, and high-efficiency power systems available at Dynacon Enterprises Limited and UTIAS/SFL. The technologies, together with the design expertise created by the MOST mission afford Canadians continued low-cost access to space for increasingly ambitious space science and technology missions.

3.1 On-Board Computers

The On-Board Computer Subsystem for MOST is shown in Fig. 3. An off-the-shelf V53 computer manages housekeeping tasks (including power and telemetry management) and the satellite radios. The housekeeping computer is also connected to redundant Attitude Control Computers, or "Supernodes", across RS-422 buses. A custom simple serial protocol ensures bus contention issues are avoided. The Supernode is a general-purpose attitude control computer based on the Motorola DSP 56303 processor. It is built to run tightly timed control laws, extended Kalman filters, and ephemeris, orbit, and magnetic field models. The multipurpose ACS Supernode can be tailored to interface with and process data from commercial off-the-shelf Sun sensors and magnetometers. It also has the ability to drive up to three independent magnetorquers for magnetically based pointing control. Several serial communications ports allow networking with other on-board computers (telemetry and command channels) while also allowing the direct control of Dynacon's miniature reaction wheels. Each Supernode routes data to other processors like the CCD Drive Computers and the microcontrollers in the reaction wheels, and also watches for fault conditions necessitating the closure of the telescope lens cap to protect the instrument. The ACS Supernode is comprised of two parts: a standard compute core, and an application specific nodule that can be tailored to the needs of a particular mission. Like other custom boards, the Supernodes monitor and report their own health, and the health of the actuators and sensors under their control.

The CCD Drive Computers are instrument computers responsible for gathering science data and providing star sensing measurements to the ACS Supernodes. Based on the latest digital signal processors, with timing accuracies in the millisecond range and timing precision in the microsecond range, the CCD drive electronics for MOST must acquire, process and store continuous streams of photometric data, while protecting the data from radiation upsets using Error Detection And Correction (EDAC) software. These electronics also serve to generate precise attitude error estimates using highly optimized star tracker routines. One CCD Drive Computer is in charge of science data readout, processing, and managing the tight timing requirements between processing and taking images. The computer is responsible for clocking the Science CCD and setting the length of exposures. CCD temperature control is also managed by this computer. The requirements of high-precision photometry require very low noise levels (down to 1-2 bits RMS out of 14-bit wide channels). These requirements are met by a very carefully designed power system, grounding and shielding strategy. The CCD Drive Computer is ideally suited to manage instruments requiring very low noise in the measurement signal chains. The second computer is used to collect star field images for precise attitude determination. Using highly optimized software developed by Dynacon Enterprises Limited, star field images are correlated against local star maps for precise attitude error sensing under strict timing requirements. Both CCD Drive Computers regulate the temperatures of their respective CCDs, and report telemetry from the instrument and the electronics.

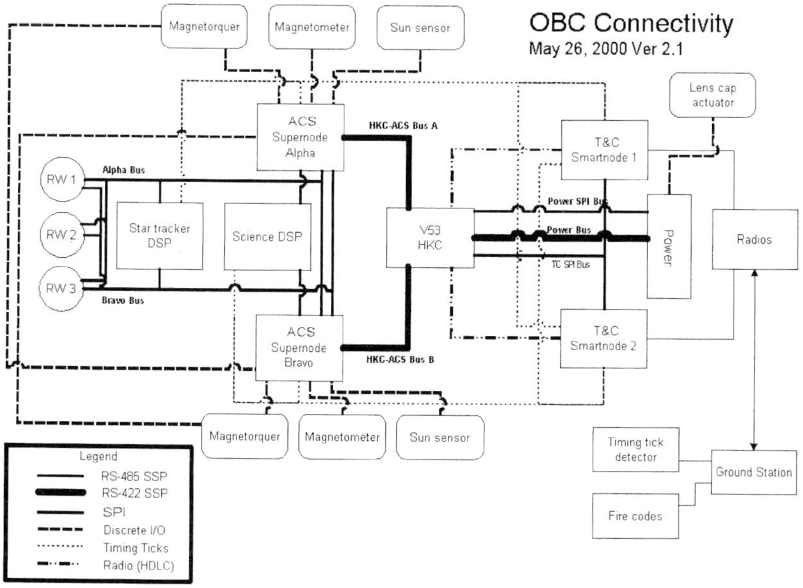

Figure 3. MOST on-board computer architecture

3.2 *Telemetry and Command*

Contact with the ground is established using high efficiency receivers and transmitters designed and developed by the UTIAS Space Flight Laboratory in collaboration with the Electrical and Computer Engineering Department of the University of Toronto. Transmitter efficiencies upwards of 30 % are possible for missions where available power is at a premium. Switchable uplink and downlink rates, and the use of standard amateur radio protocols allow the inexpensive operation of the satellite using affordable ground stations that can be controlled from locations defined by the scientist. Easy-to-use terminal programs allow seamless communications with the satellite payload, as if the scientist's computer were directly connected to the instrument.

The communications subsystem for MOST consists of telemetry and command "Smartnodes", receivers, transmitters and antennas. The Smartnodes developed by UTIAS/SFL, provide the modem interface to the receivers and transmitters and also gather 24 channels of analog telemetry. Variable modem speeds are commandable from the ground and various modulation schemes may be adopted. The scheme chosen for MOST is Gaussian Minimum Shift Keying (GMSK), but others are possible. The Smartnodes also provide telemetry collection and conditioning services to the rest of the satellite. "Firecode" ROMs provide emergency and/or immediate control of key satellite features, and sense special bit sequences sent from the ground which instantaneously get translated into the appropriate actions on the satellite.

3.3 Attitude Control

The MOST attitude control system is unusual for a microsatellite, requiring highly-accurate three-axis inertially-fixed stabilization (to less than half an arcminute), far better than can be achieved using the gravity-gradient boom stabilization approach typical of many past microsatellites. Dynacon is providing the MOST Attitude Control System, based on its miniature reaction wheel (called "MicroWheel") and High Performance Attitude Control (HPAC) products. MOST's HPAC capability will enable it to be one of the first operational space science microsatellites.

From Dynacon, a complete attitude control software suite is available. Dynacon's High Performance Attitude Control System Library consists of a number of modules that can be used to tailor the attitude control system to the needs of the particular micro-satellite mission of interest. The hardware elements of the HPAC System include the ACS Supernode (available from Dynacon and UTIAS/SFL) and Dynacon's MicroWheels. Dynacon's MicroWheels have on-board processing and rate sensing. They may be operated in various modes, and the electronics can be tailored to speak the required communications protocol including CAN and UTIAS/SFL's RS-422 Simple Serial Protocol. The reaction wheels, together with the ACS Supernode and software, are being used by other programs such as CHIPSat (SpaceDev and University of California at Berkeley), and the Australian FEDSat.

The MicroWheel is a "smart" sensor/actuator pair. It includes both a reaction wheel, and a solid-state angular rate sensor, both of which are operated locally by an embedded micro-controller. The rotor dimensions are specified by the customer, to allow angular momentum and mass to be traded. The micro-controller implements local speed and torque control loops for the reaction wheel, so that the reaction wheel accurately tracks speed or torque commands from a satellite's attitude control computer. Commands and telemetry (measured wheel speed, acceleration, torque, rate sensor reading, temperatures, voltages and internal pressure) are communicated via a serial interface.

4. Conclusion - Canadian Expertise and Facilities Available for Partnering

The various components of MOST represent modular technologies that may be modified and rearranged to suit a wide range of microspace missions. After achieving such high precision, three-axis pointing requirements as demanded by MOST, the opportunities to apply the technologies abound. Other proposed missions that use MOST technology include the Near Earth Space Surveillance (NESS) and the Canadian Robotic Asteroid Flyby and Tentatively Impact (CRAFTI) missions.

The MOST mission has enriched the Space Flight Laboratory, a unique university lab in Canada involved in low-cost space projects and microsatellite missions, and Dynacon Enterprises Limited, expert in systems engineering, quality assurance, attitude control, and power systems. A staff of engineers trained by AMSAT stands ready to meet the challenges of tomorrow's demand for low-cost access to space. As a university lab currently working with Dynacon, Canada's leading enterprise in low-cost space systems solutions, the Space Flight Laboratory is able to draw upon funding from both the public and private sectors. Both Dynacon and UTIAS/SFL encourage scientists and engineers from the European community to join them in defining collaborative space projects of mutual interest.

Acknowledgments

Dynacon Enterprises Limited gratefully acknowledges the support of the Canadian Space Agency, Space Science Branch. The UTIAS Space Flight Laboratory would like to acknowledge the support of Dynacon Enterprises Limited, the Ontario Research and Development Challenge Fund, the Center for Research in Earth and Space Technologies (CRESTech), the University of Toronto, and the Natural Sciences and Engineering Research Council (NSERC) for financially supporting the MOST project and the installation of lab facilities. The authors would also like to thank Dr. Jaymie Matthews and Dr. Rainer Kuschnig at UBC for their contribution to the smooth execution of the MOST program.

References

1. Carroll, K.A., Zee, R.E. and Matthews, J.: The MOST Microsatellite Mission: Canada's First Space Telescope, *Proceedings of the Twelfth AIAA/USU Conference on Small Satellites,* Logan, Utah, 1998
2. Davidoff, M.: *The Satellite Experimenter's Handbook,* The American Radio Relay League, Newington, CT, 1994

PROBA
(Project for On-board Autonomy)

D. Bernaerts, J. Bermyn, Verhaert Design and Development nv, Hogenakkerhoekstraat 9, B-9150 Kruibeke, Belgium

F. Teston, ESA/ESTEC, Keplerlaan 1, 2201 AZ Noordwijk, the Netherlands

Abstract

PROBA is an ESA mission dedicated to in-orbit technology demonstration, near-Earth environment monitoring and preparatory Earth observations. PROBA is a small spacecraft equipped with a selected set of technologies providing advanced on-board functions to support mission operations with minimum ground involvement. This spacecraft autonomy is exercised and demonstrated in realistic scenarios through the utilization of the payload instruments.

1. Introduction

PROBA is a mission included in the ESA's General Support Technology Programme [Reference 2], as part of an overall effort to promote technological missions using small spacecraft. The next step is a follow-on project (PROBA 2) which is included in the next phase of the GSTP and is due to start in 2001. An industrial team led by Verhaert Design and Development, Belgium, is responsible for the project. It is supported by several European subcontractors and suppliers. The payload instruments are provided to the industrial team under ESA's responsibility. This paper describes the PROBA mission, the spacecraft and ground segment design, and the payload instruments.

1.1 PROBA Mission Summary

PROBA is planned for launch in August 2001 on a PSLV from Antrix, India. It will be injected directly into its final polar, Sun-synchronous orbit slightly elliptic. The orbital drift (away from Sun-synchronism) is compatible with the PROBA mission requirements. There is thus no need for on-board propulsion. Navigation of the spacecraft is performed autonomously on-board by the Attitude Control and Navigation Subsystem (ACNS) with a combination of GPS measurements and orbit propagation. The spacecraft is kept three-axis stabilised by means of attitude measurements provided by an autonomous star tracker and by on-board control through a set of reaction wheels and magneto-torquers.

The payload of PROBA is mainly composed of a spectrometer (CHRIS), two Earth environment monitors (DEBIE and SREM) and two imagers (WAC and HRC). These instruments have been selected because they put severe

M. Rycroft and N. Crosby (eds.), Smaller Satellites: Bigger Business?, 53–68.

requirements on the spacecraft technology in terms of ACNS, data handling and resources management, in addition to their scientific interest. For example, the spectrometer uses the spacecraft high accuracy slewing capabilities to perform multiple images of the same scene on Earth from different viewing angles. Also, the planning and the execution of the spectrometer and the imagers observation requests use the on-board flight dynamics function of PROBA.

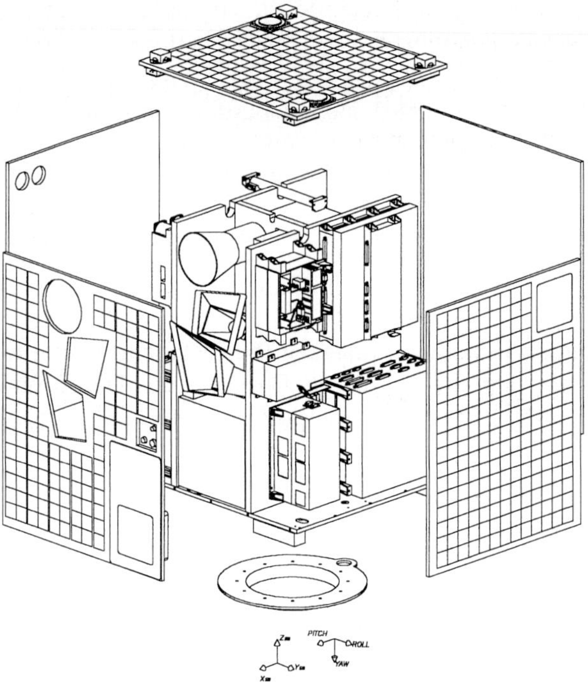

Figure 1. Exploded view of PROBA demonstrating the internal H-assembly with the units and the outer panels with the body mounted solar cells.

2. PROBA System Description

PROBA a micro-satellite of about 100 kg, (see Fig. 1) is built in a classical manner using aluminium honeycomb panels. Body-mounted Gallium Arsenide solar panels provide power to the spacecraft and a Li-ion battery is used for energy storage. A centrally switched 28 V regulated bus distributes the power to the units and instruments. A high performance computer provides the computing power to the platform, and a Digital Signal Processor (DSP) based computer with a solid state recorder provides the processing power to the imaging payload. The telecommunications subsystem is omni-directional using CCSDS-compatible up-link and down-link for S-band communications with the ground. The set of ACNS units support Earth and inertial 3-axis attitude pointing as well as on-board navigation and manoeuvring computations. The spacecraft platform provides full redundancy.

Whereas commonly available units and well-proven concepts are used for the communications subsystems and part of the power subsystem, the system design of PROBA is innovative in many respects, especially in the areas of attitude control and avionics. A core of technologies to demonstrate spacecraft autonomy is accommodated in the attitude control and the avionics subsystems and forms an integral part of the PROBA system design. They are, for example, a GPS receiver for navigation and attitude determination, an autonomous star tracker for attitude determination, a high-performance computer, a DSP for on-board scientific data processing and analysis, and a solid state mass memory, as indicated in Table 1.

2.1 *Mechanical and Thermal*

The PROBA structure was designed to meet the following requirements:

- to provide a carrying structure compatible with the ASAP5 and PSLV launcher requirements
- to accommodate mainly off-the-shelf units and payloads with frozen mechanical design
- to provide easy unit access and modularity compatible with a flexible integration and checkout approach
- to be reusable to maximum extent for other technology demonstration missions.

The carrying part of the structure is composed of three aluminium honeycomb panels mounted in an H-structure and a bottom board. Almost all units are mounted on these inner panels. The nadir pointing bottom board of the spacecraft acts as the interface with the launcher. The outside panels have body-mounted Gallium Arsenide solar cells. The thermal control of the spacecraft is completely passive, by an appropriate choice of the different paints, by the application of Multi-Layer-Insulation and by control of the conductive links between the different units and the carrying structure.

System		Description
Orbit	LEO Sun-synchronous (right ascension node at 345 degrees) Near-polar	Optimisation for the requirements: Best orbit for main imager Good orbit for other payloads Selection of low cost launcher with fligh opportunity in a short time
	Operational Lifetime	2 years for consumables (battery and solar cells)
Mechanical	Dimensions	600 x 600 x 800 mm
	Mass	< 100 kg
Thermal	Passive thermal control	
ACNS	Attitude control	3-axis stabilized, providing high accuracy nadir and off-nadir pointing capabilities
	Sensors	Cold redundant dual head advanced sta trackers, redundant 3-axis magnetometers, GPS receiver
	Actuators	4 magnetorquers, 4 reaction wheels
	Absolute Pointing Accuracy	Better than 360 arcsec
	Absolute Pointing Knowledge	Better than 125 arcsec
Avionics	Processor	Cold redundant radiation tolerant ERC3 RISC processor
	Memory	8 Mbyte RAM, 2 Mbyte FLASH
	Interfaces	RS422, TTC-B-01, analog and digital status lines, direct high speed interface t telemetry
	Uplink Communications	Hot redundant S-band receivers, 4kbps
	Downlink Communications	Cold redundant S-band transmitters, 1 Mbps
	Communications Packet Standard	CCSDS
Power	Solar Panels	5 body mounted GaAs panels, 90W pea power
	Battery	36 Li-ion cells, 9Ah, 25V
	Power Conditioning System	28V regulated power bus, redundant battery charge and discharge regulators, power distribution system and shunt regulators
Software	Operating System	VxWorks
	Data Handling and Application Software	Newly developed for PROBA

Table 1. PROBA platform specifications overview

2.2 *Attitude Control and Navigation System (ACNS)*

The main requirements on the PROBA ACNS are:

- 3-axis attitude control including high accuracy nadir and off-nadir pointing and manoeuvring capabilities in accordance with the selected Earth observation instrument requirements
- ACNS software control of the spacecraft based only on target oriented ground commands (i.e., commands specifying the targets longitude, latitude and altitude); the spacecraft sensors shall acquire all required information autonomously
- technology demonstrations of GPS attitude, and the use of Computer-Aided Software Engineering tools to develop the ACNS software.

To meet these requirements, PROBA has been fitted with a high-accuracy double head star tracker, a GPS receiver and a set of reaction wheels for the nominal ACNS operation. This set of sensors and actuators is complemented with the magnetotorquers and 3-axis magnetometers to be mainly used for momentum dumping and during the initial attitude acquisition operations after separation or non-nominal events. Finally, the core of the ACNS subsystem consists of the ACNS software.

The autonomous star tracker is the main attitude determination sensor during nominal mission phases. It provides full-sky coverage and achieves the high accuracy required in Earth observation. The sensor can autonomously reconstruct the spacecraft's inertial attitude starting from a "lost in space" attitude without any prior estimates of the spacecraft orientation. This is done with a typical performance of a few arc-seconds up to an arc-minute. The attitude can be reconstructed at relatively high inertial rates, which allows the ACNS software to perform gyro-less rate measurements which are sufficiently accurate to control large-angle precise and stable manoeuvres. The star tracker selected for PROBA is provided by the Technical University of Denmark.

Knowledge of the PROBA orbit is acquired autonomously with a GPS receiver, supplied by SSTL (UK). The provision of range data from GPS satellites (or pre-processed position, velocity and time measurements from the GPS receiver software) will allow on-board determination of the osculating orbital elements of the spacecraft and a correlation of on-board time with Universal Time Coordinated (UTC) required in the various on-board ephemeris generators. Knowledge of the orbit will allow pointing the spacecraft to any orbit-referenced attitude (including the normal-mode nadir pointing) without the need for an Earth sensor. In addition, using an on-board Earth-rotation ephemeris calculator, pointing to any user-selected Earth coordinates is also possible. The GPS receiver thus forms a crucial component in the on-board autonomy demonstration. In case of GPS failure, the ACNS software obtains the navigation data from NORAD two-line elements automatically uplinked by the ground. Finally, the GPS receiver will be used for the GPS-based attitude determination demonstration.

During nominal operations, the generation of control torques is ensured by four Teldix (Germany) reaction wheels mounted in a tetrahedron configuration. Their storage capacity is 0.04 Nms and their maximum torque capability is 5 mNm at 1500 RPM. As already indicated, momentum dumping is ensured by two redundant, three-axis magnetometers and by four magneto-torquers.

All ACNS sensors and actuators are controlled by the ACNS software (developed by Université de Sherbrouck) running on the central ERC 32 computer, and provides complete flight dynamics calculation functions, including:

- Navigation function: autonomous estimation of the orbit using GPS measurements and the autonomous determination of attitude using the advanced star sensor
- Guidance functions: prediction of orbital events (eclipses, next target passages, next station passage etc) and the on-board generation of the reference attitude profile during imaging; these functions provide essential support to the on-board mission planning system
- Control function: execution of the attitude control commands for attitude acquisition and hold.

In performing these functions, the ACNS software has to deal with further complications such as latency of the detectors and synchronisation between the different on-board clocks.

Pointing to geographical Earth references will either be to a fixed target (e.g. during ground station pass for antenna pointing or during utilisation of one of the imagers) or in a scanning motion over a 19 km user-selected target area during the CHRIS imaging mode. Each of the five scans over the target area are executed back and forth at 1/3 the nominal nadir push-broom velocity in order to increase the radiometric resolution. A detailed description of the ACNS is given in [Reference 1].

2.3 *Avionics*

The avionics is composed of:

- a high-performance redundant central computer (DHS) responsible for spacecraft telecommand and telemetry, all spacecraft computing tasks and interfaces to every unit of the spacecraft
- a Payload Processor (PPU) with a solid-state recorder and a DSP for payload processing and data storage
- a redundant set of S-band receivers and transmitters.

DHS The DHS unit was designed to integrate in a single redundant unit all the core functions of the spacecraft avionics and to provide sufficiently high-performance computing to support not only the traditional attitude control and data handling tasks but also spacecraft autonomy (i.e. the processing normally performed on-ground has been migrated onboard in the case of PROBA).

To this end a high-performance RISC processor, the ERC 32, has been used. The ERC 32 is a radiation tolerant (> 80 krad) SPARC V7 processor providing 10 MIPS and 2 MFLOPS with a floating-point unit. A memory controller includes all the peripheral functions needed by the processor, such as the address decoders, the bus arbiter, the EDAC, 2 UARTS, 3 timers and a watchdog. The chip set is manufactured with the MHS 0.8 micron CMOS/EPI radiation tolerant technology.

The DHS includes 2 of these processors. They are normally used as cold redundant hardware. However, in order to cope with potentially higher processing demands, it is also possible to run the DHS in dual processor mode, where both processors are running concurrently and exchanging data with high speed serial links.

The other functions of the DHS are:

- 2 hot redundant telecommand decoders supporting COP-1 packet telecommanding and direct (MAP-0) ground commands
- cold redundant spacecraft interfaces, RS-422 or TTC-B-01 for data exchange, pulse, analog, digital, ... for commanding, house-keeping and time distribution
- 2 cold redundant telemetry generators, each supporting 3 Virtual Channels
- a reconfiguration unit performing the reconfiguration of the DHS in case of software and hardware failure or transient (e.g. Latch-up).

PPU The PPU is the computer controlling the imaging instruments (provided by Astrium UK). It includes a DSP, the TCS21020 for high capability data processing and improved usage of the on-board mass memory through compression. The TCS21020 is a radiation tolerant (> 100 krad) 32 bit DSP. It is fully compatible with the ADSP 21020 from Analog Devices. It provides 15 MIPS and 45 MFLOPS. The chip is manufactured by the TEMIC/MHS 0.6 micron SCMOS/2RT+ process. The mass memory is 1.2 Gbit and uses three-dimensional packaging technology for high-density storage. The PPU interfaces to a Miniaturised Radiation Monitor (MRM), an interface to an extra star tracker (PASS), and an interface to a house-keeping bus.

This latter will be used on PROBA to measure temperatures and radiation dose in remote locations of the spacecraft. It uses a collection of devices called SIP (for Smart Instrumentation Point) which are small modules of 15 mm/7,5 mm/5 mm, weighing less than 3 g. They use advanced packaging and include the temperature and total dose sensors, the Analog to Digital converter, and the bus interface. The SIPs were developed by Xensor Integration (NL). The PPU controls also the imagers covered in the payload section of this paper.

TT&C The S-band link capacities are 4 kbit/s for the packet telecommanding and a maximum of 1 Mb/s for the packet telemetry. A standard ESA modulation scheme (PSK/PM) is used for the uplink and BPSK for the downlink. Off the shelf units have been used for TT&C support. Two hot redundant receivers are connected through a combiner/diplexer to Zenith and Nadir antennas provide omni-coverage commanding of the spacecraft. Two cold redundant transmitters are connected through a switching unit to Zenith and Nadir antennas also provide omni-coverage telemetry from the spacecraft.

2.4 *Power*

The basic power consumption of the platform is 35 W. The excess of power will be allocated to the payload. The duty cycle of the instruments will be calculated onboard taking into account the available power and energy but also the scientific requests and the available data storage.

The body mounted solar arrays will provide a worst case (end of life, summer solstice) peak power generation ranging from 45 to 67 W per panel. Averaged along the sunlit part of the orbit, the power generation at bus level will be 85 W. The solar arrays are built with 22 strings of 39 to 41 cells grouped in 6 sections. The cells are of a standard size of 3.8x3.5 cm with integrated diode. The panels are provided by Officine Galileo (Italy).

The 9 Ah Li-ion battery will be used mainly in eclipse as the peak power demands in day phase will be almost all covered by power from the arrays. The duty cycle of the instruments will not create a Depth of Discharge higher than 20 %. The battery weights 2.2 kg and is provided by AEA (UK).

The functions of the PCS are:

- conditioning and distribution of power to users by means of a 28 Volt regulated bus, four non-switchable power lines to the essential sub-systems and twelve switched power lines to payloads and non-essential sub-systems (PDU)
- sourcing of the power from six solar array sections and control by a Sequential Switching Shunt Regulator (S3R)
- sourcing of the power during eclipse or peak power loads, from one battery through two Battery Discharge Regulators (BDR)
- battery monitoring and management (BME)
- battery charging by means of two Battery Charge Regulators (BCR).

Failure tolerance of the PCS is provided by:

- redundant PCS interfaces, BDRs and BCRs
- majority voting of 3 independent voltages in the Main Error Amplifier controlling the BCR, the BDR and the S3R
- each of the 6 sections being made of one shunt transistor and two output diodes in series and a seventh "section" connecting a resistor across the power bus
- redundant End of Charge detection in the BME
- over-current output protection in the PDU.

The PCS also protects the spacecraft against a power bus under-voltage by turning off all switched power outputs in case of bus or battery under-voltage. The PCS is controlled by the on-board computer through a redundant 16-bit memory load interface (TTC-B-01) and discrete ground commands

2.5 Payloads

CHRIS The larger instrument is a Compact High Resolution Imaging Spectrometer (CHRIS) provided by Sira Electro-Optics Ltd (UK) (see Fig. 2). The scientific objective is to provide multi-spectral data (up to 62 bands) on Earth surface reflectance in the visible/near-infrared (VNIR) spectral band (415 to 1050 nm), with a spectral sampling interval ranging between 2 and 10 nm at high spatial resolution (25 m at nadir). The instrument will use the PROBA platform pointing capabilities to provide Bidirectional Reflectance Distribution Function (BRDF) data (variation of reflectance with view angle) for selected scenes on the Earth's surface, mainly of land areas, vegetation and for aerosols.

Figure 2. CHRIS instrument

The objective is also to validate techniques for future imaging spectrometer missions possibly on agile small satellite platforms, particularly with respect to precision farming observations, regional yield forecasting and forest inventory.

The instrument is an imaging spectrometer of basically conventional form, with a "telescope" forming an image of Earth onto the entrance slit of a spectrometer, and an area-array detector at the spectrometer focal plane. The instrument will operate in a push-broom mode during Earth imaging. The detector is a thinned, back-illuminated, frame-transfer CCD. CCD rows are assigned to separate wavelengths, and CCD columns to separate resolved points in the Earth image.

The platform will provide slow pitch during imaging in order to increase the integration time of the instrument. This increase in integration time is needed to achieve the target radiometric resolution, at the baseline spatial and spectral sampling interval.

The platform will process imaging demands from ground control specifying, target location (requiring roll manoeuvres to point across-track for off Nadir targets), viewing directions for each target in one orbit (requiring pitch manoeuvres to point along-track), spectral bands and spectral sampling interval in each band, and spatial sampling interval.

The platform will perform the required pitch and roll manoeuvres and transmit control signals to CHRIS to initiate and terminate imaging, with the required spectral and spatial characteristics. In-flight calibration for radiometric response, using a dark scene on Earth or the calibration device, will also be supported by CHRIS and the platform. The digitised data from CHRIS will be stored in a mass memory unit, processed and compressed with a DSP and transmit to ground on command.

SREM The primary objectives of ESA's Standard Radiation Environment Monitor (SREM) are for space environmental technology research, which is a hazard to future missions. However, the SREM data are also available for scientific studies such as particle sources, transport and loss, energisation of radiation belt particles.

The path of PROBA will cover the "polar horns", where energetic electrons of the outer radiation belt are transported to low altitudes, as well as the South Atlantic Anomaly (SAA) with its enhanced proton fluxes (the inner part of the inner radiation belt). PROBA will also be exposed to energetic particles from the Sun during energetic events, and cosmic rays. These latter environments are modulated by the Earth's magnetic field. With SREM, mapping (1), temporal variations (2) and possibly directional measurements (3) of these particles populations will be carried out. The measurements of SREM will also be correlated with on-board degradations caused by radiation on the electronic parts, the CCDs, and the solar cells.

(1) Through continuous operation of SREM, the models of the radiation belt positions, the particle fluxes and the geomagnetic shielding will be compared, updated or renewed for various electron and proton energies, and for cosmic rays.

(2) Through long-term operation of SREM, the radiation belt flux variations (storm injections, radiation belt motions), the solar particle event variations, the variation in geomagnetic shielding in response to storms, and the long-term variations in the SAA and Cosmic Rays (CR) fluxes will be studied. The measurements will be correlated with other spacecraft STRV, SOHO, GOES, etc.

(3) Using the manoeuvring capability of PROBA, the anisotropy of the radiation environment at low altitude can be measured. PROBA will be rotated such that the SREM points locally vertically (up and down), and collects

particular angles at different locations. In the SAA, PROBA will be slewed up to ±45°. In order to obtain good statistics, this procedure should be performed repeatedly and over a period of time to see long-term flux variations, and during high solar activity to observe possible short-term atmospheric influences. Finally, since the SREM has measurement channels for high-energy protons and heavy ions, asymmetries in cosmic ray fluxes may be observed.

SREM was developed by Contraves (Switzerland).

Figure 3. The SREM (left) and DEBIE (right) payloads. Both the processing unit and the 2 sensors of DEBIE are displayed.

DEBIE The DEBris In-orbit Evaluator (DEBIE) detector (Fig. 3) on PROBA will measure, for the first time, the small size particulate fluxes in a polar orbit. Information on the sub-centimetre meteoroid and space debris population in space can only be gained by the analysis of returned material (for lower orbits) or by such *in situ* measurements. DEBIE, developed by Patria Finnavitec (Finland), uses a combination of impact ionisation, momentum transfer and foil penetration to detect impacting particles. The mass and velocity of the particle can be deduced from the recorded signals. The lower detection threshold is about 10^{-14} g.

Two separate sensors facing in different directions (velocity direction and normal to the orbit plan) are mounted on the external walls of PROBA. According to the present models a few impacts per day are expected. The impact measurements will be sent to ground at each contact together with the time and the sensors attitude and location at the time of the impact.

Imagers PROBA accommodates two imagers from OIP (Belgium), a Wide Angle Camera (WAC) and a High Resolution Camera (HRC) (see Fig. 4). Imaging requests will be scheduled using the prediction and planning functions of the spacecraft. The spacecraft can store and compress hundreds of these images, which will be telemetred to ground and then distributed using the Web.

WAC is a miniaturised (7x7x6 cm) black and white camera using a 640x480 CMOS Active Pixel Sensor with a field of view of 40 by 31 degrees. Images are digitised on 8 bits before transmission to the spacecraft.

HRC is a miniaturised black and white imager with 10 m ground resolution. The telescope is of the Cassegrain type with an aperture size of 115 mm and a focal length of 2296 mm. The detector is based on a CCD and uses 3D packaging technology. It contains 1024 x 1024 pixels of 14 μm size. The field of view (along the diagonal of the detector) is 0.504°. Images are digitised to 10 bits before transmission to the spacecraft.

2.6 On-board Software

On-board software, running on the central Data Handling System is a new development for PROBA. It uses VxWorks as operating system and is implemented in C by Spacebel (Belgium). Apart from the classical functionality of performing the spacecraft control, housekeeping and monitoring tasks, it contains also the autonomy related functionality (such as failure detection), ACNS software and the on-board mission planner. The latter function plans ground requests for payload operation tasks taking into account the available on-board resources and target visibility as predicted by the ACNS software. The ACNS software, described previously, is produced by the autocoding tool of Xmath and integrated in the rest of the on-board software. The on-board software follows the ESA Packet Utilisation Standard for the communications with the ground. PROBA and its software are designed such that it can be completely reprogrammed in flight.

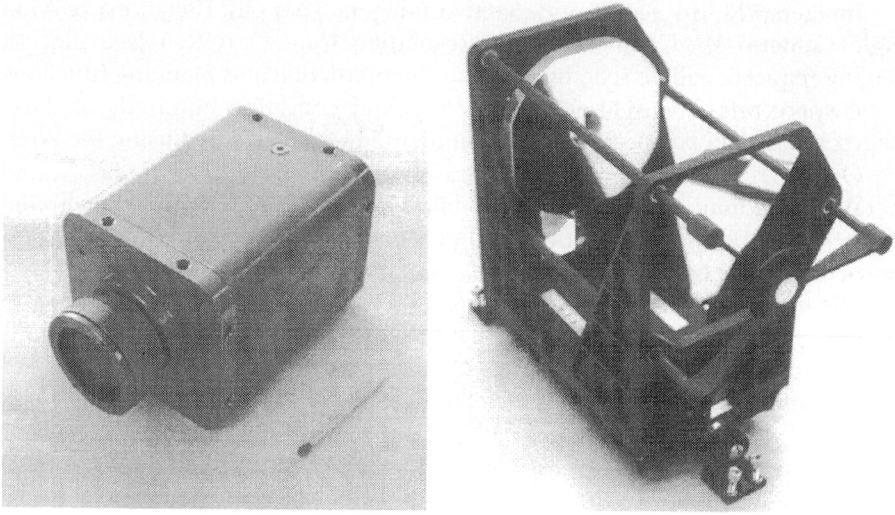

Figure 4. The two imagers to be accommodated on PROBA: the Wide Angle Camera (left) and the High Resolution Camera (right) providing 10 m resolution.

3. Development

The classical ESA development approach has been adapted to the objectives and the constraints of PROBA. The project life cycle has been split in 3 main phases, the System Design Phase, the Production and Qualification phase and the Integration and Acceptance phase. Between each phase a peer review is performed by Estec.

The model philosophy adapted for PROBA is based on a Structural and Thermal model and the Proto-Flight model of the spacecraft. The PFM approach at spacecraft level was further supported by partial electrical models of most of the bus units.

In the area of software validation, spacecraft testing and operations preparation, the project has tried to optimise the available resources. This has been translated into the usage of a common environment for spacecraft testing and operations and the production of a Software Validation Facility (provided by SSF Finland) in the early phase of the project. This latter tool simulates the entire spacecraft and allows that the on-board software executable code is executed in this facility as if it was running in the real hardware while enhanced testing and debugging capabilities are available. This facility is also connected to the common spacecraft test and spacecraft operations environment to provide a spacecraft simulator for the operation teams.

4. Ground Station and Operations

One of the benefits of on-board autonomy is the reduced need for ground operators and hence lower costs. With the ground station at Redu (Belgium), about 4 times 10 minutes of visibility per day will be available in average. The station, provided by SAS (Belgium), consists of a portable 2.4-m dish with S-band RF front-end and a control centre with limited facilities. The ground station provides the following functions:

- Automatic link acquisition based on Norad elements and spacecraft navigation data
- Communications set-up protocol for the types of data (and their bit-rates) to be received
- Automatic uplink of previously screened observation requests and spacecraft planning commands
- Automated call of ground staff in case of detection of on-board anomalies
- Automated science data filing, notification of scientists and data distribution.

It is the intention, that with the combination of on-board autonomy and automation in the ground station, the involvement of ground operators during nominal and routine mission phases is limited to routine maintenance tasks on (typically) a weekly basis. To implement the described automations, the ground segment provides besides classical telecommand and telemetry also operation procedures support. These procedures implement the automatic link acquisition, setup procedures, routine spacecraft operation procedures and (limited) telemetry analysis. The same operations environment will also be used during subsystem and system check-out tests performed during the spacecraft integration and validation phase.

5. Conclusions

The PROBA mission design fulfils the ESA objectives of in-orbit technology demonstration, near-Earth space environment monitoring and preparatory Earth observations. Its technological units represent a significant part of the spacecraft and provide the advanced capabilities required by the instruments. PROBA should demonstrate that a micro-satellite can efficiently combine in-orbit technology demonstration and operational missions, and yet remain sufficiently flexible to be reconfigured for other missions.

Acknowledgements
The authors would like to thank E. Daly, P. Nieminen, G. Drolshagen, D. Olsson (ESA/ESTEC), and SIRA for their valuable inputs to this paper.

References:

1. J. de Lafontaine, J. Buijs, P. Vuilleumier, P. Van den Braembussche, K. Mellab "Development of the PROBA Attitude Control and Navigation Software", *ESA-SP-425, Proceedings of the 4th ESA International Conference on Spacecraft Guidance, Navigation and Control Systems,* ESTEC Noordwijk, 18-21/10/1999, p. 427-441.
2. F. Teston, R. Creasey, J. Bermyn, K. Mellab "PROBA: ESA's Autonomy and Technology Demonstration Mission", *Proceedings of the 13th Annual AIAA/USU Conference on Small Satellites,* August 23 - 26, 1999
3. D. Bernaerts, F. Teston, J. Bermijn "PROBA (Project for on-board autonomy), *Proceedings of the 5th International Symposium on Small Satellites Systems and Services, La Baule,* June 19 - 23, 2000

SNAP-1: Design, Construction, Launch and Early Operations Phase Results of a Modular COTS-Based Nano-satellite

C. Underwood, V. Lappas, Surrey Space Centre, University of Surrey, Guildford, Surrey GU2 7XH, UK

e-mail: c.underwood@surrey.ac.uk; v.lappas@surrey.ac.uk

G. Richardson, J. Salvignol, Surrey Satellite Technology Ltd. (SSTL), University of Surrey Campus, Guildford, Surrey GU2 7XH, UK

e-mail: g.richardson@surrey.ac.uk

Abstract

On 28 June 2000 the UK's first nano-satellite, SNAP-1, was launched from Plesetsk on a Cosmos launch vehicle, together with a Russian COSPAS-SARSAT satellite called Nadezhda. An SSTL-built Chinese micro-satellite, Tsinghua-1, was also flown. The main mission objective of SNAP-1 was to demonstrate that a sophisticated, fully agile nano-satellite could be constructed rapidly, and at very low cost, using an extension of the modular-COTS-based design philosophy pioneered by SSTL for its micro-satellites.

The SNAP-1 mission has been highly successful, becoming the first nano-satellite (i.e. sub-10kg satellite) to have demonstrated full attitude and orbit control via its miniature momentum-wheel-based attitude control system and its butane-propellant-based propulsion system. This paper discusses Surrey's design philosophy for nano-satellites, and describes the architecture of the SNAP-1 vehicle in this context. The initial results of SNAP-1's mission are reviewed, and future exploitation plans discussed.

1. Why Nano-Satellites?

Over the past decade interest has grown in nano-satellites (i.e. sub-10 kg satellites). Advances in microelectronics/mechanics mean that the capabilities of nano-satellites can easily match or exceed those of typical micro-satellites of the early 1990's. The smaller size and mass of nano-satellites makes it feasible to launch several together effectively reducing the launch cost per vehicle, and opening up the possibilities of new mission scenarios, using clusters or constellations of spacecraft. When combined with a simplified architecture, and a Commercial Off-The-Shelf (COTS) based approach it is possible to decrease the time-to-completion significantly, and substantially reduce costs for some mission scenarios.

2. The Surrey Nanosatellite Applications Platform (SNAP) Programme

The SNAP programme was conceived in the mid-1990's in the Surrey Space Centre as a means of demonstrating a low cost COTS-based nano-satellite platform for both technological and educational use [Reference 1]. The

M. Rycroft and N. Crosby (eds.), Smaller Satellites: Bigger Business?, 69–77.

programme was adopted by SSTL, and the SNAP-1 mission itself was formally started in October 1999, when a launch opportunity (initially in April 2000) was identified alongside the Chinese Tsinghua-1 micro-satellite - itself built by SSTL. In practice, the launch slipped to June, and the fully tested SNAP-1 spacecraft was delivered in early May - i.e. design-to-orbit (for a brand new spacecraft design) took just nine months!

The objectives set for the mission were:

- to develop and prove a modular COTS-based nano-satellite bus
- to evaluate new manufacturing techniques and new technologies
- to obtain images of Tsinghua-1 during deployment
- to demonstrate the systems required for nano-satellite constellations, i.e. 3-axis attitude control, precise GPS orbit determination, and automatic orbit manoeuvres
- to rendezvous if possible with Tsinghua-1 later.

3. Surrey Nano-Satellite Design Philosophy and the SNAP-1 Architecture

The success of nano-satellites relies on lowering mission costs and reducing the time to completion. Based on this premise, Surrey's nano-satellite design philosophy is an extension of the COTS-based microsatellite design approach that has been applied so successfully over the past 20 years. This is to:

- facilitate concurrent design
- make it modular
- standardise both the electrical and mechanical interfaces
- make it easy to assemble and test
- use COTS technologies, but make the design robust
- above all, keep it simple!

For SNAP-1 a simple standard electrical interface for each module was prescribed, consisting of regulated 5 V and raw battery (V_{batt}) power connections, with a single bi-directional Controller Area Network (CAN) bus for data transfer. These connections were provided by a 9-way connector which was standard to all modules. This simplified testing and harness design. In addition to the 9-way connector, a 44-way connector was allowed in the modules to provide for specific point-to-point connections where absolutely necessary. All SNAP-1 modules (except the power system) could be powered and tested using just the 9-way connector. A standard module box form was also defined at the beginning of the SNAP-1 programme; thus, every module on SNAP-1 had the same external dimensions. This approach allowed mechanics, avionics and payload design to occur in parallel and largely in isolation. It also allowed procurement to start very early in the programme.

4. SNAP-1 Avionics

4.1 *Power system*

The SNAP-1 spacecraft uses four body mounted solar panels populated with 20 % efficient GaAs cells each producing nominally 12 V at 0.5 A. This gives an orbit-average power of approximately 6 W. The minimum required bus power is only 650 mW (which represents only the receiver and power system being on). Thus, under nominal power conditions the on-board computer (OBC) and attitude control system (ACS) can be in continuous use as well, with the GPS navigation system, payloads and transmitter being activated periodically as required. Solar array temperatures, voltages and currents are measured by the power system. Each solar panel has it own independent battery charge regulator (BCR) which implements maximum-power-point tracking in hardware. This approach gives maximum power transfer efficiency and flexibility for future panel configurations. The BCRs charge a 10 Whr battery consisting of six 'A' sized SANYO KR-1400AE Cadnica cells in series giving a nominal V_{batt} =7.2 V. The total mass of the battery pack is 270 g (equivalent to an energy density of 37 Whr/kg). Overcharging of the battery is prevented using a temperature compensated end-of-charge voltage trigger to switch to trickle charging.

The power conditioning module (PCM) provides a regulated 5 V supply, and an unregulated 12 V supply for the spacecraft systems. The maximum total current that can be safely drawn from the PCM is 3 A. In total, the battery can sustain up to 10A (i.e. ~60 W) output for a few minutes. The Power Distribution Module (PDM) uses miniaturised low R_{DS}-on FET (Field Effect Transistor) power switches, which are based on the design used in current SSTL micro-satellites. There are 8 switched 5 V lines and 8 switched V_{batt} lines, and a single switched 12 V line. All switches provide over-current protection. Essential systems (e.g., the receiver) are connected directly via a fuse (for ground testing).

4.2 *VHF Receiver*

The VHF receiver is a single frequency crystal-controlled device, operating at a selected frequency in the 140-150 MHz band. The uplink modulation scheme is frequency-shift-keying (FSK) which ensures that the receiver is compatible existing SSTL micro-satellite ground-station uplink facilities. The nominal uplink data rate is 9600 bps, which ensures that the signal bandwidth remains within the limits specified for amateur radio communications. For non-amateur use, the receiver can be operated up to a maximum of 76.8 kbps. The fixed frequency design minimizes the tuning requires and simplifies system set-up and testing.

In the primary mode of operation, digital data are up-linked to the spacecraft asynchronously using a CAN-based packet communications protocol. The receiver's CAN-microprocessor passes these packets on to the internal CAN bus and directly on to the OBC. This direct link to the OBC allows program code and data to be uploaded efficiently. The receiver also sends "received data" and "received data-clock" signals to the OBC, which allow synchronous communications to be supported. This provides compatibility with existing spacecraft operating systems at Surrey's Mission Operations Centre (using AX.25) and also provides basic compatibility with industry-standard Consultative Committee on Space Data Systems (CCSDS) telemetry/telecommand (TTC) formats. Other formats such as TCP/IP may soon be supported via the OBC running LINUX.

4.3 S-Band Transmitter

The S-band transmitter is designed to operate between 2.4 and 2.5 GHz. It delivers between 100~200 mW of RF Power from ~3.3 W DC. The downlink modulation scheme is nominally binary-phase-shift-keying (BPSK), but quadrature-phase-shift-keying and offset quadrature phase-shift-keying are also supported. The nominal data rate on SNAP-1 was fixed at 38.4 kbps (BPSK); however, the transmitter is capable of operating at data rates between 10 kbps and 10 Mbps. Two selectable scramblers are provided, CCITT V35 and Intelsat, and differential / Viterbi encoders are implemented in a Field Programmable Gate Array (FPGA). The transmitter is also CCSDS compatible. The S-band transmitter contains a virtual firmware-based TTC system, which is capable of poling the other spacecraft systems via the CAN bus to acquire telemetry data, before formatting and sending the data to the ground for "quick-look" health-check purposes.

4.4 AOCS

The Attitude and Orbit Control System (AOCS) is designed to allow 3-axis momentum-biased operation with autonomous orbit manoeuvres. Attitude knowledge is provided via a compact internal 3-axis flux-gate magnetometer and a Kalman-filter attitude-estimator. Attitude stabilisation and control is provided by a Surrey-designed miniature pitch-axis momentum wheel, running nominally at 2000 rpm, together with three miniature magnetorquer rods used for momentum dumping and magnetic attitude control (if required). A 12-channel GPS navigation system, based on the GEC-Plessey "ORION" GPS receiver, provides accurate orbit knowledge (~15 m position uncertainty). A cold-gas propulsion system, using liquefied butane as a propellant through a single 50 mN thruster, allows orbit manoeuvres to be carried out [Reference 2]. A coiled titanium pipe was used to store 32 g of propellant, enough for a total velocity change of approximately 3.5 m/s. Full 3-axis stabilization (to better than 1° in roll and yaw and 0.2° in pitch) and orbit manoeuvres are controlled via the OBC.

4.5 OBC

The On-Board Computer (OBC) uses a 220 MHz StrongARM 32-bit SA1100 RISC processor (running nominally at 88 MHz). The 4M x 8-bit program memory is protected by a double-bit-correcting (16,8) code implemented in hardware. Low-level programs and a bootloader are stored in a 1M x 16-bit flash memory. The OBC runs a multi-tasking executive, and is programmed in "C". It greatly enhances the spacecraft's capabilities providing automatic control of the spacecraft but, as with all of our spacecraft, it is not relied upon for basic operations.

5. Mechanics

Similar to Surrey's micro-satellites, the SNAP-1 primary structure consists mainly of the aluminium-alloy electronics module boxes. SNAP-1 uses three stacks of three modules arranged in a triangle (see Fig. 1). The structure is closed with aluminium honeycomb panels and the four solar panels are body mounted. The centre of the spacecraft provides a volume suitable for housing larger systems, which cannot be accommodated inside a module box. On SNAP-1 the propulsion system, magnetometer, pitch momentum wheel and two of the three magnetorquer rods were located in this central space.

By design, the spacecraft geometry results in each module box having three sides with a good field-of-view (FoV) to space or to the Earth. This simplifies the accommodation and integration issues of payloads which require external sensors. Additionally it simplifies the RF harnessing, as the RF module boxes effectively act as bulkheads. Thermal control is passive via thermo-optical surfaces applied to the boxes and honeycomb panels.

A simple low-shock three-point separation system was designed for SNAP-1. This employed three tensioned cables to latch the spacecraft down during launch with two (redundant) pyrotechnic guillotine-cutters to release the cables for separation.

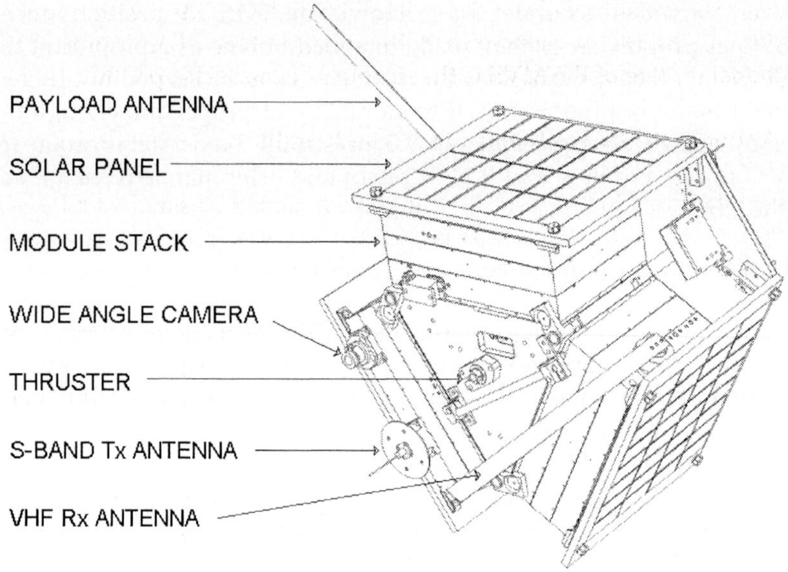

Figure 1. The SNAP-1 spacecraft

Standard aerospace metal alloys and aluminium skinned honeycomb panels (in preference to composites) were used throughout the spacecraft. Although this results in some mass penalty, it greatly simplifies design, reduces manufacture time, and maximises the ability to deal with design changes late in the programme.

6. Payloads

Three payloads were flown on SNAP-1:

- a VHF spread-spectrum communications payload for a commercial customer
- a UHF Inter-Satellite Link aimed at deriving relative position data via differential GPS to aid rendezvous manoeuvres between SNAP-1 and Tsinghua-1
- a Machine Vision System (MVS) for remote inspection of Tsinghua-1, imaging deployment and Earth imaging.

The MVS consisted of four ultra-miniature CMOS active pixel array video cameras - three wide-angle cameras (WAC) each with 90° Field of View (FoV) and one narrow-angle camera (NAC) with 20° FoV. Processing and storage of images was performed by a 220 MHz StrongARM SA-1100 processor with 2 MBytes of FLASH memory and 8 MBytes of SRAM [Reference 3]. A useful additional feature of the MVS is that it could act as back-up OBC.

7. Launch and Early Operations Phase Results

On 28 June 2000 SNAP-1, Tsinghua-1, and a Russian COSPAS-SARSAT satellite, Nadezhda, were launched from Plesetsk on a Cosmos launch vehicle into an 700 km Sun-synchronous (near polar) orbit.

The launch and early operations phase proceeded smoothly and very successfully. The spacecraft was acquired on the first pass, and data received showed that the machine vision system had successfully imaged the deployment of Tsinghua-1 from Nadezhda (see Fig. 2).

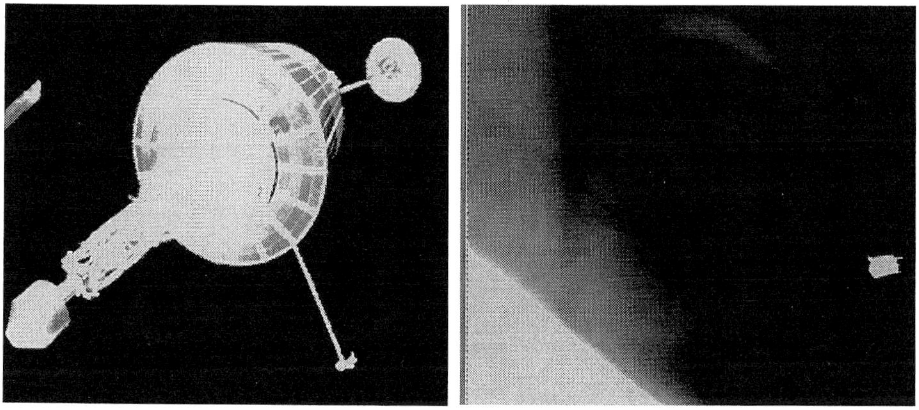

Figure 2. (Left) Image of Nadezhda taken by SNAP-1 two seconds after separation and (Right) Image of Tsinghua-1 and the limb of the Earth taken by SNAP-1 ten seconds after separation

Initial data from the AOCS showed that SNAP-1 was tumbling at 26 °/s. Two days after launch the automatic AOCS magnetorquer rate damping controller (built into the AOCS module firmware) was activated, and within a day the high initial tumbling was completely damped and the spacecraft was placed into a slow stable, rotation about the pitch axis. It was then discovered that, left to its own devices, SNAP-1's space facet would track the Earth's magnetic field vector almost perfectly. It was soon determined that this was the result of small amount of residual magnetism in the propulsion system solenoid valves. Over the course of the following months the attitude control algorithms were refined and new software uploaded to the spacecraft. In late November 2000, full nadir-pointing momentum-biased operation was achieved [Reference 4].

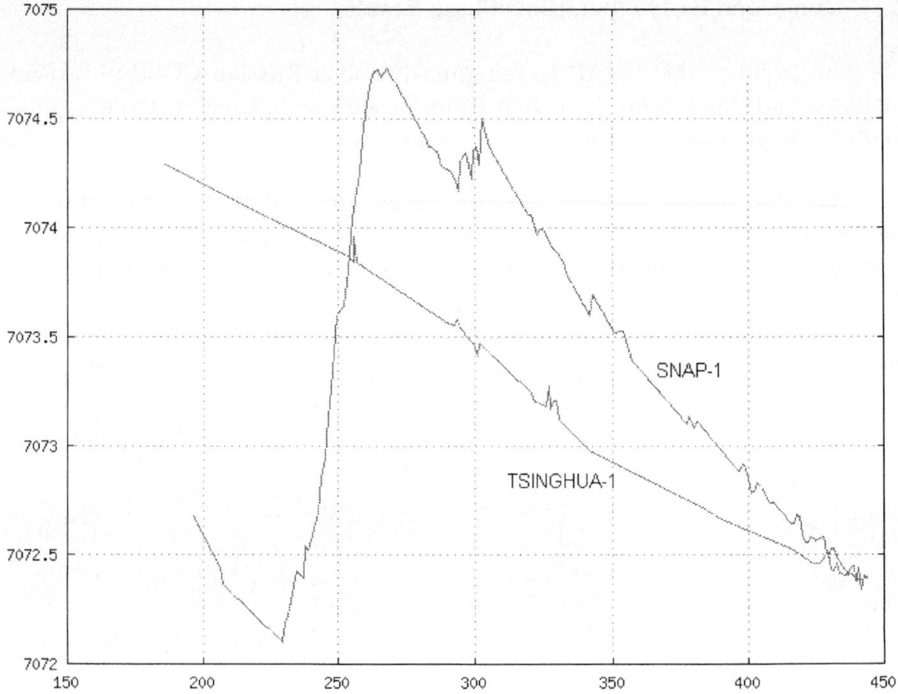

Figure 3. Variation of SNAP-1 and Tsinghua-1 semi-major axes (in km) from the launch date to March 2001 (in days)

During this period, all the onboard systems and payloads were checked and found to be working satisfactorily. On 28 August 2000 the propulsion system was used for the first time, and shortly afterwards orbital manoeuvres were started to try to bring SNAP-1 and Tsinghua-1 back together. This was made difficult by the differential effects of atmospheric drag which meant that, without firing the thrusters, SNAP-1 would fall ~10 m per day with respect to Tsinghua-1. This, coupled with SNAP-1's initial orbital insertion ~2 km below Tsinghua-1 meant that the cold-gas thruster had to be used extensively to regain altitude. Thus, a long sequence of firings was initiated under the automatic control of the OBC which instructed the AOCS system to change attitude as appropriate, and used the GPS navigation system to keep track of the orbital changes. Over the following 30 days, the thruster was fired ~4 times per day (giving a Δv change of ~10 cm/s per day), by which time SNAP-1 had climbed to ~1 km above Tsinghua-1 (see Fig. 3). High solar activity (and hence marked differential drag) meant that a further sequence of firings was necessary to have any chance of completing the rendezvous, and unfortunately the propellant ran out before this could be completed. SNAP-1 passed through Tsinghua-1's orbital altitude on 18 March 2001, at an along-track separation distance of ~2000 km. Thus, whilst rendezvous was not achieved, the agility and manoeuvrability of SNAP-1 under automatic control was amply demonstrated.

8. Conclusions

The SNAP-1 mission has been highly successful. It has met its purpose of demonstrating that low-cost (< £ 1 million), modular, COTS-based nano-satellites can be constructed rapidly (< 9 months) to achieve sophisticated mission objectives. Indeed, SNAP-1 has demonstrated, in-orbit, the navigation, attitude, and orbital control technologies needed to achieve agile nano-satellites capable of "formation flying". The SNAP concept has also played a role in spacecraft engineering education, with the USAF Academy becoming the first to adopt the core SNAP systems for use in its own nano-satellite programme.

References

1. Underwood, C.I., Crawford, M and Ward, J.W.: A Low Cost Modular Nanosatellite Based on Commercial Technology. *Proceedings of the 12th Annual AIAA/USU Conference on Small Satellites*, SSC98-V-4, AIAA/USU, 1998
2. Gibbon, D and Underwood, C.I.: Low Cost Butane Propulsion Systems for Small Spacecraft, Paper to be presented at the 15th AIAA/USU Conference on Small Satellites, SSC01-XI-1, 2001
3. Lancaster, R. and Underwood, C.I.: "The SNAP-1 Machine Vision System", *Proceedings of the 14th Annual AIAA/USU Conference on Small Satellites*, SSC00-III-6, AIAA/ USU, 2000
4. Steyn, W.H. and Hashida, Y.: In-Orbit Attitude Performance of the 3-Axis Stabilised SNAP-1 Nanosatellite, Paper to be presented at the 15th AIAA/USU Conference on Small Satellites, SSC01-V-1b, 2000

Report on Panel Discussion 1:

New Advances of Nanotechnology

S. Janosik, H. Spitzl, A. Volp, International Space University, Strasbourg Central Campus, Parc d'Innovation, Boulevard Gonthier d'Andernach, 67400 Illkirch-Graffenstaden, France

e-mail: Janosik@mss.isunet.edu; spitzl@mss.isunet.edu; volp@mss.isunet.edu

Panel Chair: R. Zee, University of Toronto Institute for Aerospace Studies, Canada

Panel Members:

D. Bernaerts, Verhaert Design and Development nv, Belgium
L. French, NASA Jet Propulsion Laboratory, California, USA
V. Lappas, Surrey Space Centre, UK
A. Lewin, Orbital Sciences Corporation, USA

This panel discussion involved questions and answers on topics ranging from the impacts of increasing the number of smaller satellites in orbit to the role of miniaturization technology in the development of the small satellite market.

The discussion raised two different issues concerning usage of the radio frequency spectrum by small satellites - the greater number of small satellites decreases the amount of frequency spectrum available to any one, and there may be technical difficulties with inter-satellite interference in the case of constellation systems. **R. Zee** suggested devoting one satellite in a system to communications, while **A. Lewin** explained that Orbcomm uses a dynamic way to allocate downlink channels. **V. Lappas** suggested options such as unique receiver design and sharing the spectrum.

The next topic concerned justifying small satellites with regard to the International Space Station (ISS). Eventually, satellites in higher orbits will cross the ISS orbit as they can carry only a limited amount of fuel, thus requiring the ISS to use its precious resources in avoiding maneuvers. Is there a policy concerning what should be done at the end of a small satellite's life? **R. Zee** commented that different orbits could be used for the small satellites, and, together with **V. Lappas** and **L. French**, suggested the use of microthrusters to propel the small satellites around the ISS. **A Lewin** also cited effective

M. Rycroft and N. Crosby (eds.), Smaller Satellites: Bigger Business?, 79–80.

engineering design with adequate margins as solutions to the issue, as did **L. French**, who further mentioned that NASA GSFC has conducted studies in this area.

Safety questions were specifically raised on the liquid propulsion system used on SNAP-1. **V. Lappas** responded that the butane used is not harmful, and that the titanium tube used as a tank is as safe as possible. He added that safe engineering and testing is indeed an issue, especially in the academic environment.

In line with the overall Symposium theme, the panel was questioned as to the viability of the small satellite market. **D. Bernaerts** explained that, at a small company such as Verhaert, creative cost-cutting approaches are necessary. He noted that the first satellite produced is clearly an investment rather than a profit generator and that the numbers produced need to be increased to sustain the company. The numbers of countries realizing their own satellites is on the rise, creating a small market, but that will not be large enough without an increase in the number of payloads.

Finally **L. French** described a role for smaller satellites as a telecommunications relay, especially for interplanetary missions as these missions become more focused. Longer times and technology demonstrations, as tested in such projects as the NASA New Millennium program, are required before the MEMS technology is used for interplanetary missions where reliability is crucial.

Session 2

Micro/Nanosatellites - Benefits, Services and Applications

Session Chair:

R. Blott, DERA, Farnborough, UK

Nanosatellite Science Applications

John E. Oberright, NASA-GSFC, Greenbelt Maryland, 20771 USA

e-mail: John.E.Oberright.1@gsfc.nasa.gov

Abstract:

The notion of a "nanosat" or satellite in the range of 1 to 10 kg is not new. In fact, some of the very earliest satellites were technically "nanosats". These were barely functional but did demonstrate that artificial satellites were feasible. Driven by the belief that bigger is better, spacecraft grew larger and larger for the next thirty years. Mission designers took full advantage of the large launch vehicles developed for human flight and military endeavors. This approach has proven to be very successful. Complex, sophisticated spacecraft now perform incredible missions on a regular basis.

So if we have built small spacecraft a long time ago, and now deploy large spacecraft to provide services and conduct space exploration, why is there interest in "nanosats"? The answer is simple. It is economics. Designs that provide today's complexity and yesterday's size are critical for tomorrow's missions.

There are mission concepts, which if done with today's technology, would not be feasible because of cost. Constellations of satellites using "off the shelf" technology to provide simultaneous collection of data over a wide area of geospace could be built today. However, the cost of this hardware and the massive multiple launch vehicles to deploy such systems would be prohibitive. Yet, these distributed sensor webs are key elements in the future plans of NASA's Earth and Space Enterprises. Highly capable, inexpensive and reliable nanosats must be made a reality

The same technology and design techniques that are necessary for nanosats can also be applied to large single-purpose missions. This will increase mission design flexibility and allow a balance of performance, reliability, launch vehicle selection and total cost.

This paper is intended to stimulate a new thinking about future mission possibilities enabled by the development of nanosat systems and the underlying technologies they employ. Examples are provided to initiate the conversation; applications are left to visionary planners.

1. Nanosat History

In the process of understanding where we are going, it is often useful to identify where we have been and why we were there. In Oct. 1957 the first artificial satellite placed in orbit and successfully transmitted information to the USSR creators on the ground. Its mass was 84 kg exclusive of the final rocket stage. The first successful "nanosat" (less than 10 kg) was launched by the United States a few months later and was followed by many others over the next several years. These "nanosats" had masses that ranged from 1.5 to 9.8 kg. The reason these early satellites were "nano" sized was due to launch vehicle constraints. In fact, the size of these satellites was somewhat of an embarrassment to the US since the USSR was launching bigger, more complex satellites and finally orbited the first human in April of 1961.

M. Rycroft and N. Crosby (eds.), Smaller Satellites: Bigger Business?, 83–91.

As the "space race" gathered momentum the emphasis was placed on the size and the reliability necessary to support human space flight. There was an implication that if one could routinely fly humans into space, one could also routinely look down on another's country and drop things on them whenever they wanted. And so, the designers of unmanned science, weather, and communication satellites took full advantage of the large launch vehicle capabilities and built spacecraft sized to fill them. This is not to deny that there are some specific applications where the economy of scale dictates massive spacecraft and heavy lift launch vehicles. A good example of this reality is communication satellites in geosynchronous orbit. The general belief that "bigger is better" spilled over into evaluating the worth of science missions.

2. Renewed Interest in Nanosats

This historical reality is mentioned to make the point that nanosats are not necessarily a novel idea. Why then is there such an interest in nanosats? The cost of a spacecraft is historically tied to its mass. A low mass spacecraft will be a low cost spacecraft. Also, there continues to be an expectation that small launch vehicles with nanosat capability and very low cost will soon be available. Designers have started to respond to these promises with new technology and more efficient designs to meet the low cost model projecting great capabilities at bargain prices. Seeing the potential for missions enabled by pushing this trend to the limit, the science community has already identified the need for multiple small, inexpensive spacecraft.

The important word here is inexpensive. There must be a significant reduction in mission level cost through the development and application of small spacecraft techniques. No one is going to be successful in proposing smaller spacecraft if they will be more expensive. The mission level requirements must be satisfied while the life cycle cost must be reduced. Only if the use of small spacecraft supports these objectives will they be acceptable. A pledge to use nanosats in a mission regardless of cost will not find much support.

Is the Nanosat concept a matter of size or design approach? The use of "less than 10 kgs" to define a "nanosat" probably does us a disservice. Except for providing a scale of reference for size it does nothing to indicate the new thinking in spacecraft design necessary to realize the potential. The challenge is to develop a new or "Small Satellite" approach to spacecraft systems architecture. The approach to designing a small satellite when applied to any spacecraft should produce a better system and a lower cost mission. The emphasis should be on top down integrated system designs, rather than designs that start with subsystem building blocks. The attempt should be to use size as a driver to achieve an integrated design approach and then apply this thinking to all future missions.

3. Status of Nanosat Developments

There are activities underway around the world focused on developing the small components and elements needed to make small lightweight spacecraft. These efforts are expected to continue as better materials and techniques are brought together to make ever smaller items. Unfortunately, these small component technologies have largely delivered an assortment of disjointed elements with the overall system, and interfaces left as an exercise for the users. Spacecraft designers have successfully integrated a variety of these technologies and combined them with existing and modified hardware to produce some remarkable flight units. However, the system solutions have been done one mission at a time and often required inefficient use of resources.

The overhead of developing custom mechanical and electrical interfaces and unique software for every mission does not lead to the lowest mass or the lowest cost. There is not a common vision of what a small spacecraft functional system design should look like. What seems to be missing is an architecture for small scale spacecraft that uses higher levels of functionality than the traditional subsystem approach. So far, the most successful organizations in the development of smallsats are universities and small businesses. This may be because these organizations have not developed a structure that is compartmentalized according to a traditional subsystem architecture. The small organization can therefore think and act in a more systems oriented manner.

4. Science Applications for Nanosats

The projected use of nanosats for science applications within NASA is best viewed through the agency's enterprise structure. These applications are therefore presented as Earth Science Enterprise missions, Space Science Enterprise Missions, and the Technology Demonstrators which support these future concepts. Other applications, such as military and commercial missions are not addressed in this paper.

5. Earth Science

NASA's Earth Science Enterprise (ESE) attempts to develop the capability for measuring and understanding earth processes. The eventual results will drive operational system requirements to provide the integrated information that will improve life on Earth now and for future generations. In general, the space-based measurements for Earth science, are extensive. The ideal data set would provide global coverage with high spatial resolution in three dimensions, delivered in real time. The measurements would have high spectral resolution of all reflected and emitted energy, plus active soundings using multiple wavelength sources.

The ESE is in the process of identifying a sub-set of this ideal that will support understanding and monitoring of earth processes that have the greatest benefit to humanity. A progression of development flights will be used to demonstrate and validate the recommendations for possible implementation as future operational missions. The eventual space-based sensors that will be recommended will undoubtedly be extensive. Mission architects are working with the orbits, apertures and numbers of spacecraft necessary to meet projected requirements. A major trade is between a few large multiple sensor platforms in high orbits with large apertures, and many small single sensor spacecraft in lower orbits with smaller apertures. One distinct advantage of going to the smaller more numerous approach, is the flexibility and resilience of the deployed system. The loss of a few orbiting units may not have a significant effect on the overall mission data and objectives. The constellation can be directed to respond to a unique event and increase temporal cover age of a specific location. Also data delivery to the ground would be distributed over multiple spacecraft rather than squeezed through a few high rate down links. Generically, this type of data collection system has come to be called a "Sensor Web". A graphic example of a sensor web is shown in Fig. 1.

ESE has identified three categories of missions in their long range planning. These are:

- exploratory missions of discovery and process research
- systematic measurement missions to continue the collection of key Earth system parameter data
- operational Instrument Development missions to provide new or enhanced sensors for operational observing systems.

In addition to this core program structure, there are small missions under the University Earth System science program, and technology demonstrations under the New Millennium Program (NMP) - EO series.

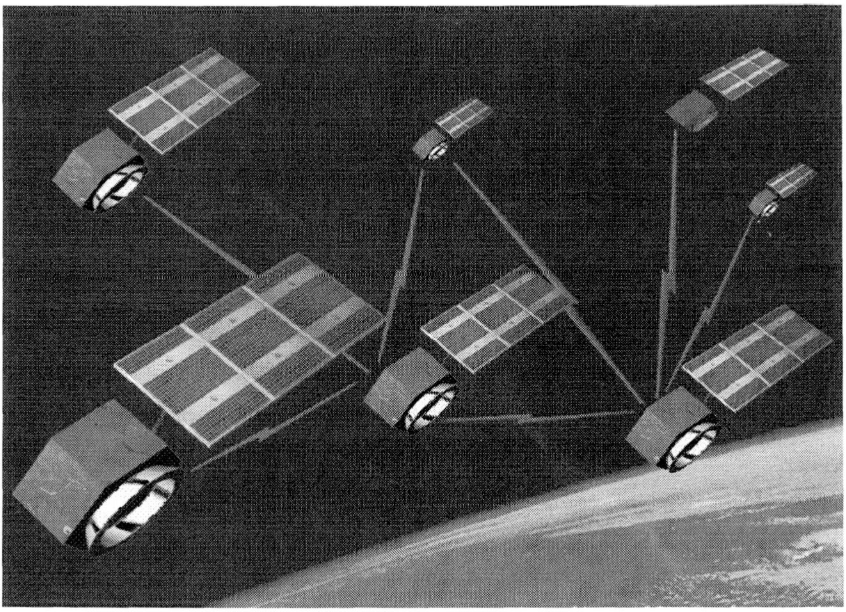

Figure 1. Sensor web concept for Earth Science Enterprise Missions

While any missions that emerge from these categories can benefit from a nanosat design approach, several being discussed are dependent on the use of nanosat class spacecraft. Some examples are:

- Leonardo, a mission to measure the angular distribution of Earth reflected solar energy, could consist of 12 s/c in a formation flying constellation
- Global Precipitation Monitor will be a follow on mission to TRMM; It will be in the systematic measurement category and consist of up to 8 small spacecraft in various orbits coordinated with a larger spacecraft
- A GPS Occultation Constellation will explore the potential for using GPS signals sensed through the atmosphere to provide three dimensional profiles of temperature, pressure and humidity. Eight nanosat sized spacecraft will be used in this exploratory mission. If the results show significant benefits, hundreds of nanosats may be deployed to provide a global stream of real time data.

6. Space Science

Space Science Enterprise (SSE) missions tend to look for very specific data. This may be extensive observation of relatively few targets, a long distance probe traveling to a specific location, a whole sky survey for a specific phenomenon, or multipoint simultaneous in-situ measurements. All of these mission types can benefit from nanosats or a nanosat systems approach.

Once the practical limit of large single surface reflectors is passed, highly cooperative multiple spacecraft can be used to improve focused measurements. Using many spacecraft operating as a single instrument, virtual apertures can be formed in space providing a nearly limitless set of opportunities for science.

Long distance probes will always benefit from lower mass which reduces propulsion requirements and/or delivers more sensing capability to the target location.

Measuring the immediate local environment at many points at the same time, simply requires many sensing units distributed throughout the volume of space being investigated.

SSE has organized their future plans into four themes, which together, seek to understand the solar system and the universe, past, present and future. "Roadmaps" for these themes have been developed which identify future missions to support this understanding. Some of these missions call for very large spacecraft, but many are looking to small and very small spacecraft working collaboratively to make very difficult measurements. Outside the main theme structure of SS other missions are supported through programs such as University Explorers and the New Millennium Program (NMP) - Space Technology (ST) series.

The Sun Earth Connection (SEC) theme, which explores the relationship between the dynamic solar processes and their effects on Earth's changing environment. Several constellation missions are on the on the SEC roadmap. These include:

Global Ionosphere-Thermosphere-Mesosphere (ITM) Constellation will measure the dynamic physical and chemical processes that occur where the solar wind interacts with the Earth's upper atmosphere. Time and spatial resolution requirements dictate a network of 24 satellites in circular and elliptical orbits with equal longitudinal spacing.

Magnetospheric Constellation will consist of up to one hundred spacecraft distributed throughout the anti-solar portion of the Earth's magnetosphere. The simultaneous field, plasma and particle measurements available from these platforms will allow the understanding of geomagnetic storm and sub storm development in response to violent coronal mass ejections from the Sun. A visualization of this mission in its deployed configuration is shown in Fig. 2.

In preparation for this type mission, the NMP ST-5 flight opportunity selected the Nanosat Constellation Trailblazer proposal for implementation. This proposal called for the integration of multiple new technology elements into a nanosat sized spacecraft and flying three of these spacecraft together as a constellation. This mission is discussed in further detailed under the Technology Demonstrators section of this paper.

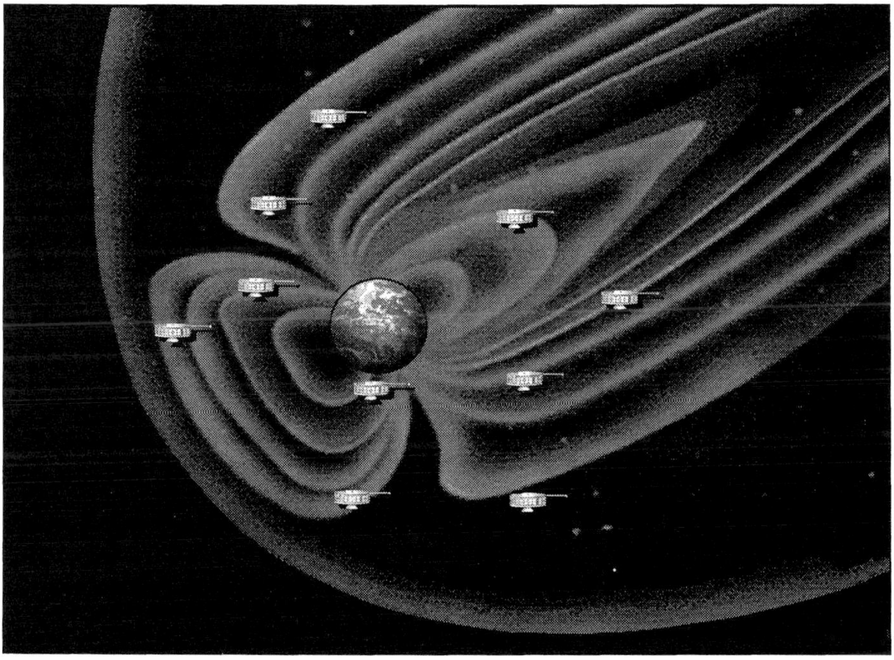

Figure 2. Space Science Enterprise - Magnetospheric Constellation

7. Technology Demonstrators

There are small satellite missions that are being flown and will be flown in order to demonstrate a new technology development or design concept. This provides an opportunity top reduce risk before a full science campaign is funded. To make this effective, the demonstration mission must be as inexpensive as possible and still maintain a high probability of success. NASA's New Millennium Program has supported several technology demonstration missions. The ST-5 was specifically selected to demonstrate a number of technologies that enable and support spacecraft in the range of tens of kilograms.

The Space Technology 5 (ST-5) project being implemented under NASA's New Millennium Program (NMP) is typical of the exploratory efforts to develop and validate technologies for future mission concepts based on nano-sized spacecraft. There are eight subsystem technologies selected through open competition and integrated into a spacecraft design that supports research quality science instruments. The system design, interfaces and software were all unique and required some compromises to accommodate the technologies. Three identical spacecraft will be sent into orbits which provide a required separation range at apogee. The spacecraft will be capable of communicating with each other and providing intra constellation ranging. Also the mission will demonstrate the 3 real spacecraft in orbit operating with 97 virtual spacecraft in a simulator all under common ground control. The project is looking for a launch in 2003. An expanded view of the spacecraft layout showing the new technology features is contained in Fig. 3.

Many other technology demonstration type missions are in planning and implementation by various organizations around the world. Notable among these are the University Explorers supported by NASA and the TechSats supported by the US Air Force.

8. The Launch Vehicle Dilemma

Today's launch vehicle capabilities are not a good match for single small satellites. Any launch vehicle capable of reaching orbit, carries a fixed burden of reliability, safety, facilities and approval. The cost of these burdens alone may be many times more than the cost of the spacecraft being launched. Finding a ride as a secondary payload may provide an acceptable price, but the orbit desired by the secondary and an adequate schedule to develop the payload may be compromised. So there are three options for launch under reasonable cost constraints. Constellation type missions that make full use of a launch vehicle capability appear to be the best fit with existing systems for unique delivery to orbit. The "shared ride" concept, which accommodates a number of different spacecraft each with similar investments and all agreeing to a common orbit, may be acceptable. The alignment of a single small spacecraft with a vehicle having secondary payload capability, and going to the right orbit at the right time is a solution, but difficult to find.

9. Conclusions

The background and current status of nanosatellite developments should lend confidence in the ability to produce very capable small spacecraft in the future. There is a very real intention to utilize quantities of small spacecraft in future Earth and Space Science application missions. Significant investments are being made in the development and flight validation of the technologies for very small components. There is a need for systems level functional architecture development that supports nanosat optimization and design approach. There is not a launch vehicle that meets the low cost requirements for very small payloads. Delivery systems for whole constellations and secondary payload slots are currently the only practical options for small satellites.

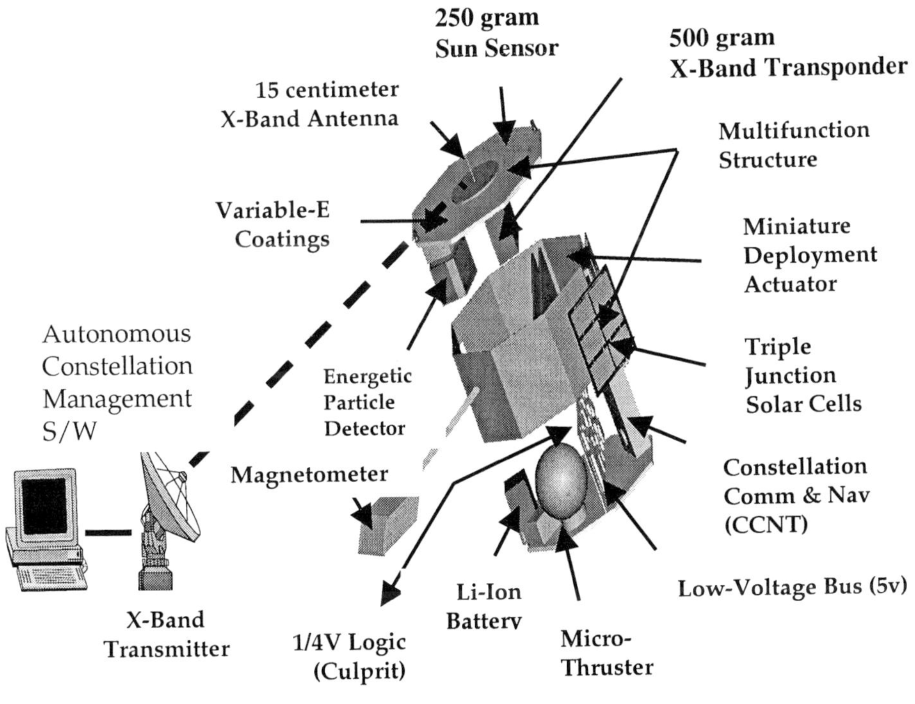

Figure 3. New Millennium Program: ST-5 Technology Demonstrator

Creation of an Information Society by Satellite Communications R&D: the Role of Small Satellites

T. Iida, Communications Research Laboratory, 4-2-1 Nukui-Kita, Koganei, Tokyo, Japan 184-8795

e-mail: iida@crl.go.jp

Abstract

This paper has two objectives: to discuss the role of small satellites in the satellite communications R&D (research and development) in the IT (Information Technology) era, and to show communications technology topics in small satellite applications. The need for small satellite communications is first discussed. The future (next three decades) vision of satellite communications R&D is then described, corresponding to the recent rapid increase of capacity of terrestrial links. Third, the role of small satellites in future satellite communications R&D is discussed. Finally, small satellite related communications technology topics are described to show that there are many interesting research issues.

1. Introduction

In this paper, there are two objectives. One is to discuss the role of small satellites in satellite communications R&D (research and development) in the IT (Information Technology) era. The other is to show the communications technology topics in small satellite applications.

1.1 Necessity for Satellite Communications in the IT Era [Reference 1]

Recently the terrestrial communications system of both fiber optic cables and cellular phones has grown tremendously. In this situation, there are some who question the role of satellite communications in the future. We give three reasons for satellite communications systems below.

First, both satellite and terrestrial communications systems are needed because there are some areas that terrestrial systems cannot cover., e.g., in developing countries. Satellite communications have wide coverage. Needless to say, the satellite can contribute to bridging the digital divide (G8-Okinawa charter [Reference 2]). Satellite communications are required for anti-disaster measures. This is clear from the Great Hanshin/Awaji earthquake disaster of 1995 when over 6000 people died in Japan.

Second, satellite communications are still the only example of success in space commercialization. Internet demand is increasing rapidly; more than 11% of world ISP (Internet Service Providers) use satellites. The market for the

M. Rycroft and N. Crosby (eds.), Smaller Satellites: Bigger Business?, 93–100.

satellite Internet is now 8 times greater than it was 3 years ago [Reference 3]. The demand for satellite communications will continue to grow in the future.

Third, the author has proposed a gigabit satellite [Reference 4] considering that the time will come when the demand can not be satisfied even if both satellite and optical fibers are used. In addition, it has been clarified that satellite links have no problems for high data rate transmissions and seamless connections with optical fibers via Asynchronous Transfer Mode (ATM) [Reference 5].

1.2 Future Vision of Satellite Communications R&D [Reference 1]

In order to develop a harmonious infrastructure with terrestrial communications systems, the speed of satellite communications must be increased to meet the speed of terrestrial communications systems. It is useful to consider a so-called Internet satellite and its development. Then, the communications satellite R&D for the next 30 years can be proposed as shown in Table 1 [Reference 1].

Time Frame	Generation of Satellite	Capacity	Characteristic
2000-2010	1st generation Internet satellite	5-50 Gbps	Gigabit satellite
2010-2020	2nd generation Internet satellite	50-500 Gbps	Global reach
2020-2030	3rd generation Internet satellite	0.5-5 Tbps	GEO platform

Table 1. Proposed communication satellite

The generation change of a time frame of ten years, as shown in Table 1, seems to be too long (in recent times called "Dog year"). One generation of ten years may be long, but it must be reasonable because it may be about a generation of infrastructure. By an Internet satellite is meant one that provides fixed, broadcast and mobile satellite communications.

Now we consider the capacity of proposed future generation satellites. The capacity of the terrestrial backbone network increases tremendously year by year; a prediction is shown in Fig. 1 [Reference 6]. According to Fig. 1, the capacity of a single fiber as a backbone network, shown in the upper part of Fig. 1, is about 400 Gbps-5 Tbps in the 2000-2005 time frame and will increase to 2.5 Tbps-40 Tbps around 2010. The dotted thick bold line shows the extrapolation, considering that there could be a limitation in the capacity of a single optical fiber cable, around one Peta bit per second.

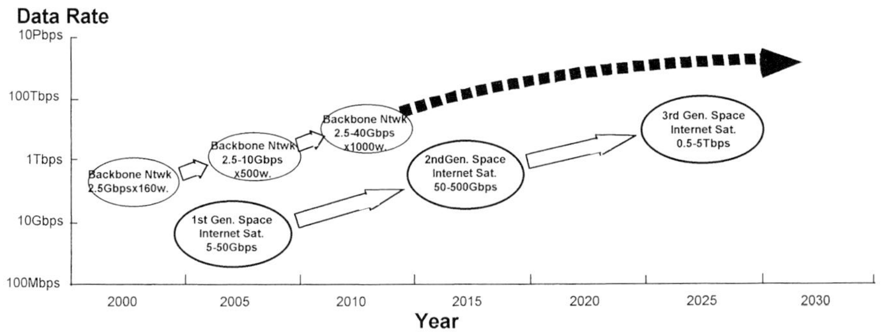

Figure 1. Prediction of required data rate which is detained by extrapolating the data of 2000-2015

The satellite capacity should be increased along with the terrestrial capacity, although the complete follow up is impossible. The satellite can be used not only as a backbone network but also as a fixed or mobile access network. We assume here that the satellite shares at least 1% of the backbone network capacity.

In the first generation Internet satellite, the gigabit satellite has 1.2 Gbps/channel and the total capacity is estimated to be about 6 Gbps [Reference 7]. On the other hand, the iPSTAR is planned to have total 50 Gbps capacity, to be launched in 2005 [Reference 8]. Therefore, it is reasonable to assume that several to several tens of Giga bit per second (5-50 Gbps) is the total capacity of the first generation Internet satellite.

As far as the second generation system is concerned, we believe that the capacity will increase by an order of magnitude (10 times). Therefore, the total capacity per satellite will be several tens to several hundreds Giga bit per second (50-500 Gbps) for the second generation Internet satellite. Several hundreds to several thousands of Giga bit per second capacity (500 Gbps-5 Tbps) is taken to be the total capacity of the third generation satellite, in order to survive in the future IT environment. The capacity of the third generation Internet satellite could not be realized by a conventional satellite. Therefore, a geostationary platform will become necessary and a technical breakthrough is needed.

As far as the mass of the satellite is concerned, the first generation Internet satellite will be around 3 tonnes, the second generation 10-30 tonnes and the third generation several hundred tonnes as shown in Fig. 2 [Reference 4]. How the satellite mass might be decreased will depend heavily on new miniaturized devices.

1.3 Role of Small Satellites in Future Satellite Communications R&D

It has been considered that a constellation of small satellites could provide some of the communications infrastructure. However, mainly due to economics, these systems may be difficult to implement except for a special purpose system such as a military application. The economical difficulty stems from the vast development of terrestrial systems, especially cellular phones. If so, what is the role of small satellites in the future satellite communications R&D?

The Internet satellites mentioned earlier will have to be launched every 5-7 years, because of their large size. Therefore, to accelerate technology development, launching small satellites every few years at any time between launches of the large satellites will be needed. In this respect, the small satellite will be useful as a test-bed for space proof, space infrastructure such as formation flight, miniaturization of equipment, and characteristics of rapid development. The rapid development characteristics of small satellites is very important to link the communications satellite developments with the rapid changes of the IT related technologies.

An example of the space infrastructure suggested, that is a step to constructing a future space infrastructure repairing satellite, is the OMS (Orbital Maintenance System) concept as shown in Fig. 3 [Reference 9]. In order to conduct safe and reliable operations of future communications and broadcasting satellites, there has to be greater recognition for the necessity of the OMS, which manages inspection, reorbit, and simple repair of failed satellites in space. In order to develop the fundamental technology for the OMS, the micro-OLIVe (OMS Light Inspector Vehicle) mission is planned [Reference 10] for partial verification of the image-processing computer and inspection-monitoring camera, which are important and critical items for the OMS.

2. Communications Technology Topics on Small Satellite Applications

Finally the small satellite related communication topics are described to show that there are many interesting research items in the communications systems. In Chapter 1, it was described that small satellites might not be the main stream for the future satellite communications infrastructure. However, the special purpose other than communications can be considered, such as observation on a planet. In this conjunction, many small satellites might be used or many micro robots might land on the planet. In this case, how to establish the communication will be of interest. Related topics are described below.

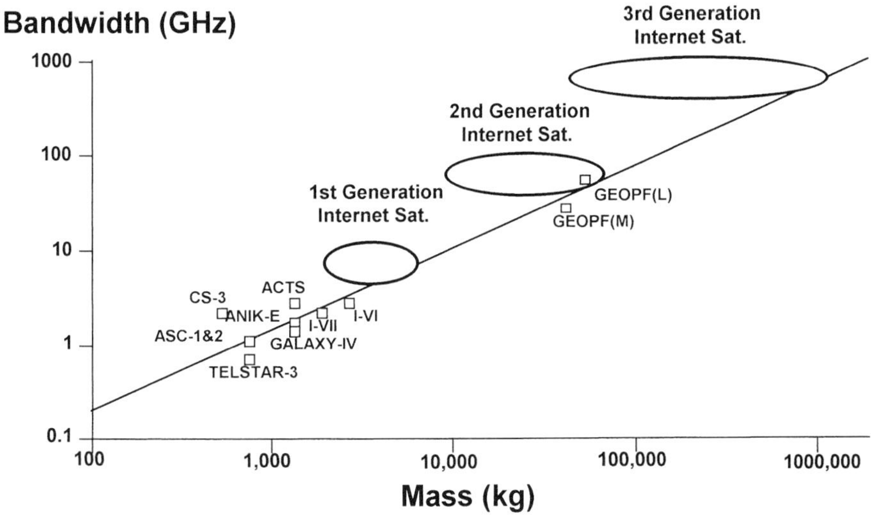

Figure 2. Bandwidth vs. mass of communications satellite

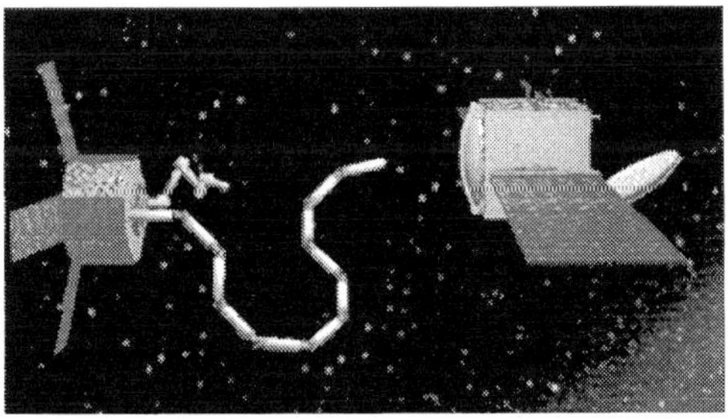

Figure 3. Concept of OMS

2.1 *Ad-Hoc Network System [Reference 11, 12]*

There are many mobile terminals (called "nodes"), when a message is sent from one node as a sender to another node as a receiver, the message can be received though there are many node relays as shown in Fig. 4. This ad-hoc network has such characteristics as:

- the network is independent of the base station and is not the wire network, which are indispensable for the traditional network
- disperse node has equal terms and performs autonomously
- the information is exchanged directly among nodes through wireless communication
- the information can be exchanged even between nodes which could not communicate directly through the relay by other nodes (multi-hop wireless communication).

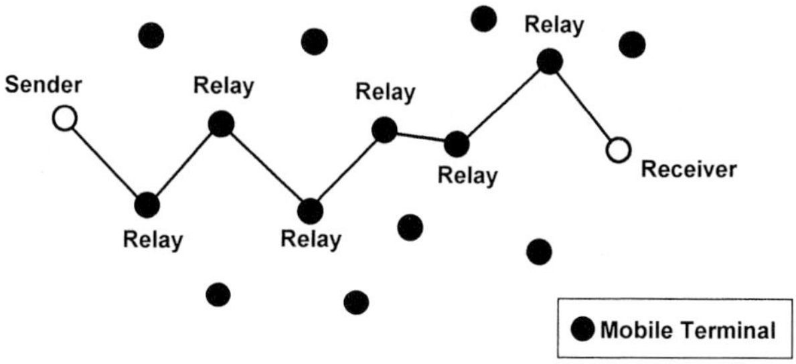

Figure 4. Ad-hoc network system

The name "ad-hoc" network stems from the instantaneous installation of the network in any area as long as the node is available. One of the protocols is a "flooding" protocol. However, this protocol is not efficient. Therefore, issues remain to be solved. Although the detail of the ad-hoc network cannot be described here, if the node corresponds to a small satellite, micro-robot or "bluetooth" device which is a small wireless local area network terminal, a very interesting system configuration could be considered.

2.2 *Vehicle-to-Vehicle Communications System*

A vehicle-to-vehicle communications system is studied in the ITS (Intelligent Transport System). The development of road transport, while bringing greater convenience to all, has led to significant problems such as traffic accidents and environmental pollution. The Intelligent Transport System

(ITS) addresses many of the problems inherent in existing transport systems. In the CRL (Communications Research Laboratory), R&D of millimeter wave band communications between traveling front and back vehicles is conducted [Reference 13]. A 60 GHz band street propagation experiment and a 10 Mbps transmission experiment by FSK (Frequency Shift Keying) on a highway have been conducted, and multimedia information transmission of moving images has been carried out experimentally.

2.3 *Other Applications*

Micro-Robot-to-Micro-Robot Communications System

In the case of using multiple robots for cooperative work, the micro-robot-to-micro-robot communication is indispensable. However, such communications are very difficult to establish, because the communication device must be made very small. Therefore, case-by-case studies have been conducted, but there does not seem to be any systematic study made so far. But, as required for increasing the data capacity to transmit any image data, more studies will be required in the near future.

Micro Air Vehicle Communications System [Reference 14]

A small engine and a small airplane, micro air vehicle (MAV), have been researched and developed as military technology. In the future, it may enter a room if it becomes the size of an insect, and it may be recognized only as a fly on the wall. There are many technical hurdles to overcome to realize an MAV. They are an actuator, a motor, a camera, other very small parts. There is a big problem storing energy and converting it to enough power to move a small plane. In addition, in a small device, there is sky dynamic difficulty due to the small Reynolds number. There seems to be no systematic study for communications. Further study will be needed in this field.

Smart Dust [Reference 15]

The "Smart Dust" is a device for autonomous sensing and communications in a cubic millimeter, prepared by the University of California at Berkeley. Size reduction is paramount, to make the nodes as inexpensive and easy-to-deploy as possible. The target of the Smart Dust is millimeter-scale nodes. It is certainly within the realms of possibility that future prototypes of Smart Dust could be small enough to remain suspended in the air, buoyed up by air currents, sensing and communicating for hours, or even days, on end.

3. Conclusions

In this paper, future satellite communications R&D have been discussed taking into account the development status of terrestrial communications systems. Some topics related to small satellite communications systems have also been described. Through the discussion, it is concluded that satellite communications will still be important in the future, and that small satellites will be useful for accelerating the construction of, and for conducting research on, future satellite communications systems.

Acknowledgements

The author would like to thank Dr. Shin'ichi Kimura, Dr. Masayuki Fujise, and Dr. Koichi Gyoda, all with CRL, for their contributions on new communications technology topics.

References

1. Iida, T. and Suzuki, Y.: *Communications Satellite R&D for Next 30 Years*, 19th AIAA ICSSC, Paper No. 233, Toulouse, France, April 17-20, 2001
2. <http://www.g8kyushu-okinawa.go.jp/e/documents/it1.html/>
3. Bates, J.: Satellite Internet Market Expanding Rapidly, *Space News, Vol. 11*, No. 16, p. 4, April 24, 2000
4. Iida, T. and Pelton, J. N.: Conceptual Study of Advanced Asia-Pacific Telecommunications Satellite for Future Gigabit Transmission, *Space Communications, IOS Press, Vol. 11*, No. 3, pp. 193-203, 1993
5. <http://www.nren.nasa.gov/tpd/>
6. MPT Telecommunication Technology Council: Basic Plan of Information Communication R&D –Policy for R&D of Information Communication Technology in the 21st Century, February 2000
7. Kadowaki, N., Nakao, M., Yoshimura, N., Ogawa, Y. and Nishinaga, N.: *Space-Based High Data Rate Internet with On-Board Processing Technology,* 19th AIAA ICSSC, Paper No. 187, Toulouse, France, April 17-20, 2001
8. <http://www.thaicom.net/ipstar/introduction.html/>
9. Kimura, S., Tsuchiya, S., Araki, K., Suzuki, Y. and Suzuki, R.: *OMS for NELS – A Concept for a Robot-Assisted Service for Removing Satellites from a LEO Constellation,* 51st IAF Congress, Rio De Janeiro, Brazil, IAA-00-IAA.6.6.02, 2000
10. Kimura, S., Tsuchiya, S. and Okamoto, O.: *Micro-OLIVe: Preliminary Experiment on Remote Inspection Using Small Satellite,* 50th IAF Congress, Amsterdam, The Netherlands, IAA-99-IAA.11.1.05, October 4-8, 1999
11. Mase, K., Nakano, K., Sengoku, M. and Shinoda, S.: Ad Hoc Networks, *Journal of Institute of Electronics, Information and Communication Engineers, Japan, Vol. 84,* No. 2, pp. 127-134, February 2001 (in Japanese)
12. <http://www.ietf.org/html.charters/manet-charter.html/>
13. Kato, A., Horimatsu, T. and Fujise, M.: *Experimental Investigation for Inter-Vehicle Communication System Using Millimeter Wave,* 7th World Congress on Intelligent Transport Systems, No. 3231S, Torino, Italy, November 2000
14. Wilson, J. R.: Mini Technologies for Major Impact, *Aerospace America, Vol.36,* No.5, pp.36-42, May 1998
15. <http://robotics.eecs.berkeley.edu/~pister/SmartDust/>

MYRIADE, the CNES Micro-satellite Product Line for Science and Innovation

C. Bouzat, Micro-satellite Division, Centre National d'Etudes Spatiales, 18, Avenue Edouard Belin, 31401 Toulouse Cedex 04, France

e-mail: charles.bouzat@cnes.fr

Abstract

With the low cost launch opportunities, initially offered by the Ariane ASAP structure and now largely developed by new launch companies using converted missiles, and with strong support from the national scientific community, CNES has decided to extend the PROTEUS product to develop a new micro-satellite product line called MYRIADE.

This new product is compatible with a secondary payload launch configuration on Ariane 5, PSLV (Indian launcher) or DNEPR and, in the mid-term, with other launchers like STARSEM/SOYUZ, ROCKOT, or LEO-LINK. The goal is to achieve scientific missions at a cost of 10 million Euro per mission (payload development excluded, but launch and operations included), on average at two missions per year. Four missions and a cooperation with Brazil have already been initiated. These are:

DEMETER, for launch early in 2003 to study electromagnetic effects and ionospheric disturbances potentially linked to volcanic or seismic activity.

PARASOL, using an instrument derived from POLDER, for atmospheric aerosol and cloud parameters.

MICROSCOPE, based on the ONERA high accuracy accelerometer, which aims to verify the equivalence principle between inertial mass and gravitational mass with an accuracy of 10^{-15} instead of 10^{-12} corresponding to ground measurements today; this drag-free micro-satellite uses electric propulsion and is planned for launch at the end of 2004.

PICARD, to be launched around mid-2006, concerns precise measurements of the solar diameter, the solar constant and of their variations and variability. These measurements will be analyzed to improve our knowledge of solar factors influencing the Earth's climate.

This MYRIADE product line is also a means for supporting cooperation with other countries wishing either to develop a joint scientific mission or to access space activities. This paper describes the rationale of the CNES micro-satellite MYRIADE in terms of these various objectives, the organization and the methods used for different missions, keeping in mind performance, quality and cost.

1. Introduction

Over the past five years CNES has studied the possibility of developing a micro-satellite product line for scientific, technology demonstration, radio communications and Earth observation missions. The success of micro-satellite development in Surrey Satellite Technology Ltd., and in Germany, and various experiments undertaken by other countries wishing to develop space activities have shown a growing interest for low cost missions to benefit from opportunities like the Ariane ASAP, PSLV or other launchers demonstrating the importance of micro-satellites of up to 120 kg. After the development of PROTEUS in partnership with ALCATEL SPACE, CNES decided on a micro-

M. Rycroft and N. Crosby (eds.), Smaller Satellites: Bigger Business?, 101–108.

satellite project organization, with a prime role for controlling the industrial organization and setting up the methods and tools, with the opportunity of adapting the product to each mission in a low cost way and on short time scales. This program, with a Euro 15 million per year funding from 2003, is the opportunity for CNES to test innovations in programmatic, system, technological and methodology areas.

The main goals of the MYRIADE program are to develop a product line:

- At low cost (~ Euro 5 million per satellite)
- A short (18 to 30 months) development program
- Strict product assurance.

These characteristics allow a new programmatic approach such as:

- To decide on a short time scale program offering the best phasing with technology evolution
- To take higher risks for programs with uncertain scientific returns
- To create opportunities for international cooperation
- To offer access to space for new partners like universities.

Micro-satellites are a key to technological progress:

- To achieve an efficient micro-satellite, miniaturized technologies are essential
- Non-space-qualified technologies, developed for large markets (for example, automotive or mobile phone) can bring benefits from the electronic industry.

2. Low Costs

For micro-satellites, low costs have to be met in the following context:

- Develop innovation
- Use in-house design
- Use technical knowledge spread through a large organization
- Try to find the performance-cost optimum with targeted investment
- Ensure adaptability, with no mission-specific designs.

To deal with that context and offer maximum cost reduction, the following principles have been considered:

- Try to analyze in depth, and freeze mission requirements during the feasibility phase as soon as possible
- Design robustness with high performance avionics and powerful configurations (e.g., deployable and steerable solar array, 3-axis stabilization for fine pointing, CCSDS communications)
- Use functional redundancy as much as possible
- Reliability is not sacrificed but obtained by quality control processes
- Minimize radiation effects (e.g., upset effect and latch up) by selection of parts
- Reuse designs whenever possible
- Use commercial standard interfaces (e.g., RS422/RS485 data handling interfaces)
- Use COTS devices at part level and for standard equipment
- Make strategic procurements to prevent obsolescence
- Have on board autonomy to simplify ground
- Ensure full qualification at equipment level, with no satellite engineering model, but thoroughly tested protoflight model
- Focus management on key points, minimize paperwork, and use PC-based software tools
- Have colocated and innovative teams communicating well.

3. Organization and Methods

CNES is responsible for the design and development of generic items. CNES technical departments are deeply involved in the design and in R&D developments.

CNES is consequently responsible for the selection and procurement of all equipment items from small companies. Preparing for the industrialization phase, CNES has selected the LATECOERE Company for satellite integration. CNES will remain in charge of the system and satellite engineering work performed in cooperation with scientific laboratories responsible for the payload and data analysis.

The success of such a program depends on the clear identification of what is considered as the right level of acceptable risks. So at the beginning of the program, these must be assessed through program system reviews of accepted critical points and associated control methods.

Low cost and short time scales mean:

- Rapid mission assessment and feasibility

- Improved understanding of quality and performance.

In order not to over-specify the system, the project team must interact with the principal investigator and the payload team to adjust the mission requirements to ensure the goals of the mission while taking the finite resources into account.

4. Description of Micro-Satellite System

Development of the CNES micro-satellite product line is characterized by a high performance design in terms of the capability offered to the payload (mass, power, pointing accuracy and agility, telemetry and processing), with low costs given by the use of COTS (parts and equipment), together with the use of innovative methods of engineering and management (risk management, quality management, internal or external customer/supplier processes, design and manufacturing tool standardization).

The CNES micro-satellite program is based on:

- A reference subset of functional chains fully developed and qualified on the occasion of the first mission, DEMETER, to study the effects of earthquakes and volcanoes on the ionosphere
- Options fully studied and available for new missions, most of them flying on DEMETER even if not required by that mission
- Models and simulations, mostly but not exclusively software, allowing a rapid study of any new mission in the frame of the Engineering Center
- A ground system for multi-mission control.

In addition, this system has the capability to use CNES multi-mission resources (the CNES S-band stations network and the Orbitography Operational Center) for station acquisition and in case of any anomaly.

The satellite architecture is mainly based on the following product line functional chains:

1- An attitude and orbit control function, which includes:

- An attitude determination system based on 3 coarse Sun sensors (only used for the acquisition and the safe mode), a 3-axis magnetometer and a stellar sensor when precise attitude measurement is needed or when the pointing accuracy must be better than few degrees. Three raw gyros (1-

axis gyros) are added to the design to give attitude information during orbit maneuvers

- An attitude control system based on the use of up to 3 reaction wheels (fine attitude control and maneuverability) and 3 magnetotorquers, for reaction wheels unloading or coarse attitude control mode
- An optional orbit control system using a hydrazine propulsion subsystem with four thrusters (in the minimum configuration) to have maneuver capabilities and 3 axis attitude control during maneuvers.

2- Electrical power generation, regulation and distribution, including :

- One GaAs solar steerable generator (with 2 deployable panels)
- One Li-Ion battery of 11 Ahr
- A power regulation and distribution box.

The solar generator can be fixed (Sun pointing mission) or oriented with a 1-axis mechanism around the -Y axis (normal to the orbit plane). The cant angle of the deployed solar array will depend on the mission.

3 - A control and command function which is implemented on:

- A central On Board Computer (OBC), including 1 Gbit memory, based on a T805 transputer and several micro-controllers (PIC 16C73) for interfacing with the equipment
- Two S-band transmitter and receiver chains, with two antennas (Tx /Rx) used quasi-omnidirectional coverage. The telemetry data rate is 400 kbits/s in the operational modes and 25 kbits/s in safe modes.

The structure and the satellite thermal control will be customized for each mission, except when the basic structure can accommodate the payload identified for the mission.

The mass budget of the DEMETER satellite is approximately 115 kg divided between avionics (19 kg including payload management/mass memory and X-band telemetry), structure and thermal (34 kg), power supply (19 kg), propulsion (8 kg) and payload (35 kg).

Concerning the power budget, the solar generator should give 160W maximum power at 70 °C after one year in orbit. The solar generator configuration (with a cant angle of 14° for DEMETER) and the 10: 30 local time orbit (ascending node) lead to a maximum permanent available power higher than 80 W, and up to 90 W during the X band TM transmission.

The raw attitude control mode has an attitude control accuracy less than 5°. With the use of the stellar sensor the attitude control accuracy is better than 0.1° (3 sigma) worst case and the measurement accuracy is better than 0.01°.

For the OBC computer, the goal was to fulfill the low cost/low mass/low power optimum while having a computing and communications performance similar to the state-of-the-art computers used in current projects. A 3 kg/5W/2 liters/2 Mips/ 1 Gbits/ 15 krads goal was assigned. Even if in the conservative space domain it is not so difficult to get performance using current technology (SPARC X processor, memory, radiation-hardened parts), the main challenge was to fulfill all the constraints together. New architectural solutions were necessary.

The conventional printed circuit boards are populated on both sides with plastic surface mounted parts (except for FPGAs and processor using ceramic packages). Internal serial buses connect the four baseline modules (Power, TMTC, CPU and I/O). For external interfaces direct and dedicated RS485 links are used. As the processor is not latch-up immune, specific circuits protect and allow a reboot of the computer when a non recoverable error occurs. With regards to COTS parts use, some "system hidden" redundancy scheme was implemented for each critical internal node (watchdog and over-current protection against heavy ion effects, coded command against failure propagation).

For the communications equipment, low cost was achieved by miniaturization and technology from commercial markets. To achieve low cost/low power/low volume, the driver was to find and use macro-parts for the heart of the equipment which belong to numeric TV or mobile phone electronic production methods (for example, ASICs for decoding and demodulation in TV receivers). These parts are both innovative and reliable. Other functions are achieved in more conventional hardware, but always using more advanced technologies (multi-layer circuits, CMS parts, automatic report as for radiotelephone devices).

Today, the results are encouraging. With respect to a standard, full space development program, savings by a factor of 2 have been obtained for the power consumption and on volume; the recurrent costs have been cut by a factor of almost 10.

The first mission DEMETER will also include some technological experiments, such as on board autonomous orbit control system, and pyrotechnic priming by a laser system.

The communication between the ground segment and the satellite is performed through two links, S-band X-band. The S-band link is full compliant to the CCSDS standard, whereas the X-band link complies with this standard at the packet level.

The X-band is devoted to scientific high data rate telemetry, whereas the S-band is used for the housekeeping and technological telemetry or for scientific missions with limited data transmission needs (up to 600 Mbytes/day). All the TC are transmitted via S-band, using a protocol which guarantees automatic re-emission of TC frames detected as erroneous or lost by the board peer.

For S-band, the specified performance objectives must be guaranteed for a minimum elevation of 10° with a system margin of 3 dB for the TM and 10 dB for the TC. For X-band, the specified performance objectives must be guaranteed for a minimum elevation of 15° with the X-band ground station. The maximum data coded rate is 18 Mb/s. The modulation and coding used is a Multidimensional Trellis Coded Modulation Concatenated with Reed-Solomon bloc code (MCTMCRS).

The Command Control Ground Segment (MIGS) is based upon the Proteus Generic Ground System. As far as the CCC is concerned, it :

- Prepares the programming messages, taking into account the payload from the Mission Centers
- Prepares and monitors communications with the satellite using a TTCET
- Reconciles the TM for quick look data and for alarm generation
- Evaluates the performance of the platform
- Performs orbit and attitude monitoring.

The TTCET are automatic S-band stations, having a 3.1 meter diameter antenna to:

- Establish and maintain the satellite to ground radio frequency link for all programmed visibility passes (transits)
- Receive and temporarily store the received telemetry during a transit
- Receive and transmit to CCC the housekeeping data
- Accept the connection with CCC or a Mission Center to transmit payload data and part of the received platform data
- Transmit telecommands to the satellite
- Perform Doppler measurements during the transit, for orbit calculation and antenna positioning
- Compensate for the Doppler effect on the telecommand link

- Ensure correspondence between on board and UTC time.

5. Conclusions

Since the decision to start the program in 1998, we can state the following points:

- The CNES product line is now a national French standard ; the industrial organization is settled and both industrial satellite prime contractors, ASTRIUM and ASPI, have been associated with the development through engineering partnerships which allow them to use the design for their own applications
- Four scientific missions have been approved (DEMETER, PICARD, PARASOL and MICROSCOPE). A cooperation with Brazil is agreed to develop jointly a dedicated micro-satellite for scientific and technological purposes, and a mission for the Ministry of Defense based on four micro-satellites will be developed by ASTRIUM, that means a total of 9 micro-satellites based on the same CNES product line up to 2004
- All the industrial organization involved with the manufacture of generic equipment and in their integration are agreed.
- Definition of the reference product has been finalized; in the fall of 2001 the integration of the DEMETER micro-satellite will start
- For such a program large CNES staff (around 70 people) relevant to all the competences developed in CNES is mobilized, to develop the product and to decide upon the engineering workshop for the next missions.

This CNES Micro-satellite program is an opportunity to apply the strategic plan defined by CNES. The micro-satellite product line defined as a subset of tools and functional chains (equipment, software, architectures, simulation test bench, and engineering workshop) will be qualified by mid-2002 and the first application for the DEMETER mission will be flown 2003.

Air Quality Improvement through Micro-Technology

J. D. Burke, 165 Olivera Lane, Sierra Madre, CA 91024, USA

e-mail: jdburke@its.caltech.edu

L. Zhang, Chinese Academy of Space Technology, P.O. Box 9628-35, Beijing 100086, P.R. China

e-mail: zhanglh@public.east.cn.net

Abstract

With dense population and transport, highly-developed industry, intensive agriculture and concentrated energy use, Europe and China have high impacts on their natural environments. For this paper we have chosen to focus on one such impact - air pollution. Pollution sources, topography, wind and weather patterns cause air pollution to degrade the quality of human and other life in both regions, but there are important differences such as those due to rural coal burning in China. In Europe, the creation of the European Union has provided new policy approaches for monitoring and controlling both internal and trans-boundary air, water and waste pollution. For air quality management an initiative called EUROAIRNET exists for the purpose of obtaining and distributing objective, reliable and comparable information through the use and improvement of measurement systems already comprising more than 5000 ground sites. Similar efforts are underway in China, with the aid of National Environment Protection Bureaus and other institutions. At the same time, new technologies are becoming available for measuring and controlling air pollution and its effects.

In this paper we discuss an application of small ground, airborne, and space devices to augment existing and planned European and Chinese air quality monitoring networks. Measuring instruments, data storage and transfer devices, distributed and standardized data analysis systems, air and space platforms and public education and outreach activities all can be made more efficient and economical through the use of new micro-technology. Microsatellites can serve in important roles, complementing the larger atmospheric and land remote sensing satellites already operating or approved in near-term plans. As the economic impact of environmental pollution becomes more widely recognized and as regulatory policies are adjusted, systems architectures including micro-technique on the ground, in the air and in space may provide new business opportunities.

1. Introduction

In considering how emerging micro-technologies may be applied to improving air quality, we focus on two temperate industrial regions - Europe and China. Obvious needs exist in both areas (see Fig. 1 and Fig. 2). However, the main sources of air pollution are quite different, meteorological factors are very different (Europe maritime, China continental), and the social and governmental structures for dealing with air quality, though their goal is the same, have different forms.

M. Rycroft and N. Crosby (eds.), Smaller Satellites: Bigger Business?, 109–118.

Figure 1. A SeaWiFS view of a haze cloud over Europe.

In both Europe and China, efforts are underway to understand the physics and chemistry of air pollution. Some of these scientific experiments are elaborate, involving many different coordinated techniques used over large regions. An example is given in Reference 1, a web site for the multi-sensor ACE-2 campaign conducted in 1997 to compare clean oceanic and polluted European continental air masses.

In this paper, however, we intend to concentrate on applications rather than science, applications that can perhaps be implemented on the basis of science already in hand, but are not yet in service because the required technology is just emerging, and the needed public pressures and governmental responses have not yet happened.

By approaching air-quality applications in this manner we are, of course, making several assumptions. The first is that people and their governments will realize the worth of improved air quality and be willing to pay to get it. While it may be possible to base that decision on strict grounds of economic payback, we do not expect to be able to prove that as a present reality. Rather, we observe increasing evidence that both the real and the intangible benefits of improved air quality are given increasing weight in public policy, both in Europe and in China. As an example of this in Europe, a major effort is in progress to reduce transalpine truck traffic in favor of rail transport.

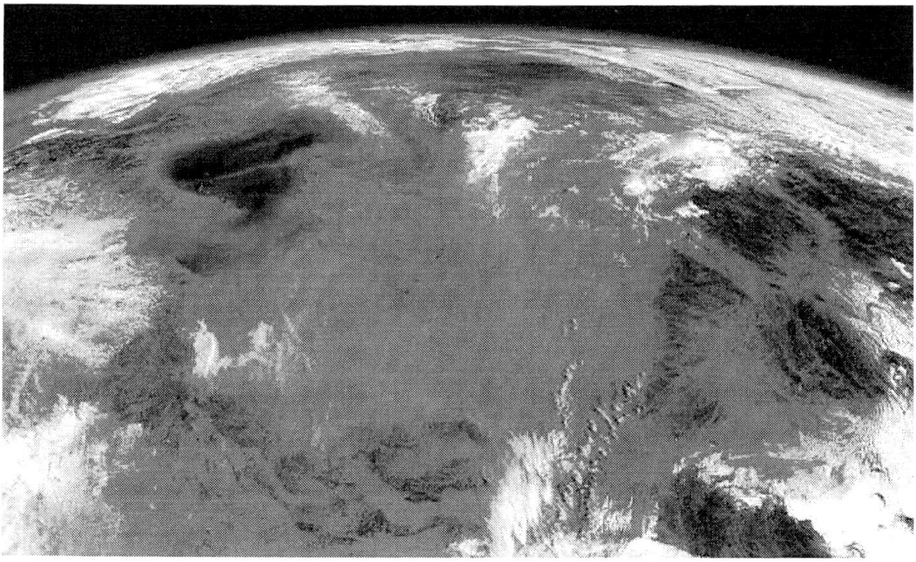

Figure 2. A SeaeWiFS view of a dust-and-haze cloud over China

A second assumption is that micro-technology can indeed make an important difference in the rate of progress. That can probably be proven only in retrospect. It is our goal to suggest some early steps in this direction, using ground-based, airborne, and spaceborne observational systems linked into a unified architecture using cell-phone-derived communications through the Internet. Why the Internet? Our reason is that not only can it serve as the information interconnect for distributed measurements and distributed data management, but also it can be the vehicle for bringing the results to public attention and eventually causing governmental action.

A third assumption is that, as postulated in the next section, is that university student teams can gain access to very advanced micro-technology for designing and building small satellites, the associated ground systems and the required software.

2. Education Component

Because air quality improvement is inherently a slow, long-term process, the system architecture that we envision for Europe and China must have an education component. Giving children an early understanding of the problem, plus real, meaningful activities that they can do during their school years, plus incentives to go on dealing with it as adults, is one way to assure the sustained social and political efforts required for success.

In California's South Coast Air Quality Management District such a program for children is in progress, with as-yet unquantifed results. In both Europe and China, similar efforts could be mounted with but small changes in existing environmental education plans. In Europe the program would be inherently international because air pollution is a trans-boundary problem. In China the program could be synchronized among the many existing, independent local educational authorities through the rapid spread of Chinese Internet services.

Once young people reach university, their air-quality activities can branch into one of two main tracks - technical (monitoring equipment and networks, modeling and data systems) and non-technical (policy, law, education and outreach). With the emergence of micro-technology, progress on both of these tracks can be accelerated. Collegiate groups in many countries have already built and operated small satellites (and their associated ground systems) in the class that we envision as part of an air-quality monitoring network. The new element is to have such projects use the products of advanced technology developments such as the multifunction microsystems being developed in the Center for Integrated Space Microsystems at Caltech/NASA Jet Propulsion Laboratory [Reference 2].

Such student-built and student-operated ground, air and space devices have, up to now, existed mostly as the result of independent local initiatives. However, with the advance of the concept of smallsat constellations, proposals are being put forward for several universities each to each supply one satellite to a collaborative constellation. A smallsat effort of that kind could be a low-cost, low-risk addition to existing national and international air-quality projects.

On the other, non-technical track, new technology fuels the burgeoning information industry that enables rapid proliferation of new ideas. Our envisioned system architecture takes advantage of this by invoking cell-phone-derived techniques with special software as a network element. People to whom cell phones are an essential part of daily life (tens of millions already in both Europe and China) may soon be using them for data as well as voice. Designing and building the required distributed software for such a system would itself be a challenging effort for young people who would then carry important skills into their later careers.

If a suitable output from a ground-air-and-space air quality network can be provided in a form that the cell-phone public will choose to use widely, this can be another component in a continent-wide education and outreach synthesis.

3. Present Observations

Both natural and anthropogenic air pollution phenomena across the often cloudy European plain involve complex physical and chemical interactions that are the subject of interesting research. In this paper we point to some relevant examples and some main findings that point a way toward possible system solutions. Europe's frequent cloud cover establishes some physical limits on what can be observed using the various available instrumental techniques. Satellite observations (Meteosat, NOAA, etc.) routinely give the locations of cloud cover. Therefore, even though smallsats (e.g., Stellenbosch University's Sunsat) can also do cloud imaging, this is not a priority goal for the small (10 - 50 kg) spacecraft that we envisage here.

Measuring the properties of cloud-covered low-altitude hazes is not a task for small satellites, but rather for ground-based and perhaps airborne instruments such as lidars. However, current projects have either a scientific focus or a focus on local environment. Missing at present is a public consensus on how to apply such techniques in a distributed, routinely-operating, continent-wide, air monitoring system. A start in that direction has been made by the European Union with EUROAIRNET [Reference 3]. One purpose of EUROAIRNET is to harmonize the observations of some 5000 existing air monitoring sites in Europe, so that data can be compared on a common basis and conclusions reached about air quality for the whole continent. At present, EUROAIRNET and similar efforts in China depend entirely on ground-based observations. Extending the monitoring with more advanced ground instruments such as Doppler acoustic sounders and lidars, plus airborne and spaceborne instruments is obviously technically feasible, but there is as yet no consensus on how to do that for routine operations. Instead, European initiatives are still largely of a scientific and early-operational nature, as represented for example by ENVISAT, a large, powerful multipurpose orbiting platform that is expected to give coordinated remote-sensing data from a suite of instruments designed to investigate a wide range of surface and atmospheric phenomena worldwide [Reference 4]. In China, a small satellite constellation for environment and disaster monitoring is under development.

To obtain more frequent observations of any given region, and especially to obtain specific supplementary data for air quality monitoring, a different system architecture is needed, namely, several small satellites with specialized instrument payloads integrated with a distributed ground system. This is the observational focus that we define and advocate.

4. The Need for Space-borne Instruments in Europe and China

For Europe, because of the maritime climate and prevalent low clouds, smallsat payloads cannot be expected to yield cloud-free haze and photochemical smog observations over a large fraction of their observing time. China's typically drier air masses offer a more favorable opportunity. However, the worst air-pollution problem in China is not photochemical smog; it is suspended particulates resulting from the widespread use of coal-burning and other unregulated, e.g. transport and agricultural, smoke sources. In addition, windblown sand and dust (resulting over time in great loess deposits) affect large parts of the country.

Can a single constellation of 10 - 50 kg smallsats make a useful contribution to air monitoring over both Europe and China? Our answer is conditional: It is yes, if relatively simple measurements such as near-ultraviolet extinction due to haze can be realized in small payloads and the subcloud haze problem can be handled by non-space observations in Europe, or no, if these conditions cannot be met at low enough cost, low enough risk, and in a short enough time to be economically interesting. One critical need for turning the answer from a no to a yes is to find a way for smallsat projects to move beyond proof-of-concept into continuous operation. This transition may be greatly aided by on-board autonomy of the satellites and unattended operation of the ground systems, both now becoming practical through micro-technology.

5. Potential Users

Existence of a need does not guarantee that it will ever be met. Populations in the industrial world have suffered for generations under increasingly bad air quality. Only recently have large public communities recognized that, first, it is technically and economically possible to make improvements, and second, such improvements have leverage beyond their near-term cost-benefit relations. The obvious user and sponsor of any new air-quality system is government. Private entities such as businesses and academic institutions have motives and resources that prevent the large initial investments needed.

In Europe it may be possible to assemble a multi-government consortium. ESA is a model for space, and EUROAIRNET is a model for ground systems. In China there may be great potential for unified government action, as shown by already-existing initiatives toward coordinating a multitude of independent local air-quality agencies. However, it is still hard to imagine these governmental entities reaching, in the short term, agreements sufficient to launch whole new, integrated ground-air-space observational systems for air quality management.

For that reason we envisage an incremental approach that begins at the university student level, with only modest government support in terms of funding but with a clear commitment to sustain that support and demand reasonably quick results. The initial user is indeed expected to be government (in Europe multiple, in China single), but with a stated intent to move sponsorship and operation soon into the commercial economy. Making that possible depends ultimately on whether or not air-quality applications of micro-technology, the Internet, and cell-phone-based communications with appropriate complex software can really pay dividends.

In the United States, corporations both large and small do sponsor university projects and the dissemination of air-quality data to the public, for example by placing advertising based on inputs from governmental entities such as NOAA, the Environmental Protection Agency, and California's ARB and AQMD. Newspapers publish daily smog reports and forecasts on their weather pages. But at present it is universally expected that the cost of gathering and disseminating the original data will continue to be borne by governments.

6. Possible System Architectures

For our present purpose we divide small satellites into two classes - those using commercial off-the-shelf technology to perform relatively simple missions at low cost and risk, with buses in the 50 - 100 kg mass range, and a new and as yet unproven class using more advanced (and presumably more risky) technology to perform simple missions with 10 - 50 kg mass, or alternatively more complex missions in the 50 -100 kg class.

For the air quality application considered in this paper, we envision a two-step sequence. Initially a consortium of university teams would build and operate a constellation of perhaps eight autonomous miniature satellites in the 10 - 50 kg size range, carrying modest observing instruments such as miniature infrared and near-ultraviolet spectrometers. Though small, with limited effective observing apertures, these spacecraft would be highly capable in other ways due to their use of emerging technology. For example, they might use optical on-board data buses, GPS-derived on-board orbit determination, orbit maintenance via micro-propulsion, and systems-on-a-chip [Reference 2]. Observations from this smallsat constellation would be very limited in ground sampling distance, spectral resolution and data volume in comparison to observations from large instruments such as MISR and MOPITT on TERRA [Reference 5] and MODIS on ENVISAT [Reference 4]. The main useful result would be more frequent coverage of particular regions within a few hundred km of small ground terminals, which could be located either at their host universities or at sites selected for best air-resource coverage. These measurements would have an educational and public-outreach benefit. Whether or not they would significantly accelerate air-quality progress is unknown.

However, even a partial success of such a constellation could energize a first step in integrating space observations with existing ground and airborne air-quality networks. In our vision this would happen as follows: At each university's receiving station, a cell-phone and Internet link would be autonomously activated (using special-purpose dedicated network software built by student teams) for each satellite pass, sending real-time data to an existing site associated with other established air-quality monitoring systems. In Europe this could be a station of EUROAIRNET; in China it could be a station of an Environment Protection Bureau.

An architecture of this kind obviously bypasses some ground data processing that is normal for current scientific and applications missions. Instrument calibrations, removal of atmospheric signals that affect but are not part of the desired measurement, georegistration by ground-based processing, and other such data manipulations would be absent, and the data rates would be limited relative to those from, for example, ENVISAT. But the satellite data could still be useful because they would be merged with the much more accurate but relatively infrequent observations from large satellites, and also with routine ground-truth observations from existing surface sites. On-board autonomous data management and satellite positioning using GPS, both made possible in tiny spacecraft by microelectronics, would obviate the need for some of the conventional ground data processing. With the spacecraft maintaining their orbital constellation phasing autonomously, the unattended ground stations could switch themselves on and off automatically for observing each pass and sending bent-pipe data into the ground net. Also, substantial receiving antenna gain could be used, because the receiving antenna could point autonomously at the satellite. Although store-and-forward capability is also a proven possibility, we see no need for it in this architecture.

If such a simple multi-university system were to be successful in meeting its modest goals it would demonstrate achievements essential for the next step, namely, proof of the supposedly risky miniature and autonomous spacecraft technologies for scale-up from the 10 - 50 kg class to the 50 - 100 kg class, and proof of the cell-phone-and-Internet data networking. Validation of the complex software required to tie together the network would be a permanent asset for the second step. For that next step the program would have to move beyond an academic setting. A model for this is the formation of Surrey Space Technology Ltd. Corporation out of the University of Surrey group.

As in every other proposal for new commercial space applications, the success of step two depends on low-cost access to orbit. Piggyback launches are probably incompatible with the setting up of constellations for specific geographical coverage, so in our vision there would still have to be substantial government budgets for step two. However, there is hope that such budgets will be found in the future, as public concern with deteriorating air quality continues to become more effective in influencing government actions.

7. Airborne Sensors

Airborne *in-situ* and remote-sensing observations, an ancient technique for meteorology, have not been widely used in routine air quality monitoring, probably because of the cost of continuously flying aircraft. Yet airborne techniques both *in-situ* (sampling) and remote (lidar, etc.) are fully proven as research tools, and could be used operationally if the cost of routine operation were radically reduced. However, high-altitude aircraft have the same disadvantage as satellites. They look at the optically-opaque clouds from above, while most of the physical and chemical action is going on in the air mass beneath and inside the clouds.

Microtechnology may enable new methods for low-altitude tracking of polluted subcloud and in-cloud air masses. One way is to use tiny balloons with "soft" instrument payloads, small and frangible enough not to cause any more damage than a bird if ingested by a jet engine. Another is to place micro-payloads on board commercial airliners.

With the advent of GPS in air traffic control it will soon be possible for such piggyback *in-situ* monitoring devices to operate entirely autonomously with no attention by either the aircraft operator or the ground controller. The data would simply be delivered into the air-quality system by a pick-off from the data stream of the existing air traffic control system.

As in the case of the small satellites described above, this parasitic monitoring operation could begin as a university project. However, it might be privately sponsored from the beginning. Collaborating airlines could support the development and integration and carry the tiny autonomous payloads (a few kg at most) in exchange for good publicity as helpers in the air-quality improvement effort.

8. Conclusion

Micro-technology has already advanced to a stage where innovative prospects are available for improved air monitoring using large networks of ground and airborne sensors, plus small constellations of micro-satellites. However, just monitoring the nature, sources and movements of air pollution is not enough to ensure air-quality improvement. Long-term public and governmental action is essential. In this paper we have shown ways in which such action may be stimulated. With the help of progressive programs, beginning in schools and universities, to apply micro-technique for measurement, and with the Internet for dissemination of the results, it is possible that air-quality improvement in Europe and China may be accelerated in coming years.

References

1. Raes, F. and Bates, T. (Guest Eds.) *The Second Characterization Experiment (ACE-2)*, 16 June – 25 July 1997, TELLUS 52B, 2, April 2000, <http://www.ei.jrc.it/ace2/>
2. Alkalai, L. et al., CISM: Center for Integrated Space Microsystems, 1999 Annual Report. Jet Propulsion Laboratory, 2000, <http://cism.jpl.nasa.gov>
3. Larssen, S. et al., Criteria for EUROAIRNET. European Environment Agency, February 1999, <http ://www.binary.eea.eu.int :80/t/tech12.pdf> 20 October 2000
4. European Space Agency: ENVISAT Instruments. <http://envisat.esa.int/instruments:index.html> 13 June 2001
5. NASA Goddard Space Flight Center: TERRA. <http://terra.nasa.gov/> 13 June 2001

Small Satellites in the Future Space Infrastructure

K. Maeda, Satellite Program and Planning Department, Office of Satellite Technology, Research and Applications, National Space Development Agency of Japan (NASDA), World Trade Center Bldg. 2-4-1 Hamamatsu-cho, Minato-ku, Tokyo, 105-8060, Japan

e-mail: maeda.korehiro@nasda.go.jp

Abstract

In the near future, various space infrastructure will be established which include Earth observation system, communications and navigation system, on-orbit service system, space research. Many of such systems will consist of small satellites (1 to 500 kg). In this paper, the value of small satellites in the future space infrastructure and activity of NASDA and Japan in such fields will be introduced. Technological and research issues such as components/parts, constellation and formation "flying" and selection of radio frequencies for communications, and countermeasures for space debris will be presented.

1. Introduction to NASDA Space Activities

In the future space infrastructure, Earth observations, communications and navigation, on-orbit servicing, space research and others, will rely on various smaller satellites. In order to realize such a system, not only smaller parts and components with higher performance but also system technologies such as formation "flying", constellations, and communications links between satellites will need to be developed. However, constellation and formation "flying" produce many debris and some considerations on this are presented. In this paper, the role and uses of small satellites in the future space infrastructure, and the activities of NASDA and other research organizations in Japan are discussed.

NASDA has launched about 40 satellites since 1975. There are larger Earth observation satellites such as ADEOS (1996, 3560 kg), TRMM (1997, 3500 kg), ADEOS-II (2001, 3700kg) and ALOS (2003, 4000 kg). There are larger communications satellites such as ETS-VI(1994, 2000 kg), COMETS (1998, 2000 kg) and ETS-VIII (2003, 3000 kg). Sudden destruction of solar paddle in ADEOS led to sudden stop of operation of all sensors after about one year since the launch. Failure to transfer ETS-VI and COMETS into Geostationary orbit (GSO) led to operation of all equipments in non-GSO. However, if a system has many small satellites, most functions are not lost at the same time. On the other hand, in the case of a large satellite, all functions are lost simultaneously.

NASDA has developed a piggy-back satellite called "μ-Lab Sat". NASDA also opened an announcement of opportunities concerning piggy-back satellites for H-IIA rocket for ADEOS-II in 1997. By organizing selection committee and

M. Rycroft and N. Crosby (eds.), Smaller Satellites: Bigger Business?, 119–126.

examining 7 proposals, NASDA selected WEOSS (Whale Ecology Observation Satellite System) proposed by Chiba Institute of Technology (CIT). Under a cooperation between Australia and NASDA, the FedSAT (Federation Satellite) developed by CRCSS (Cooperative Research Center for Satellite Systems) became one of piggy-back satellites for ADEOS-II.

NASDA mainly has funded a Satellite Design Contest since 1993; it is held jointly by JSME (The Japan Society for Mechanical Engineering), JSASS (The Japan Society for Aeronautical and Space Sciences), IEICE (The Institute of Electronics, Information and Communication Engineers), ISAS (Institute of Space and Astronautical Science), NASDA and JSF (Japan Space Forum), and is supported by other organizations. Only university students can apply for this contest. The above mentioned WEOSS got a top prize for IEICE in the first Satellite Design Contest held in October, 1993. At the eighth Satellite Design Contest held on October 22, 2000, a satellite called JatGun Sat TOPPU proposed by Tokyo Institute of Technology and a satellite called TAKO (Target Collaborativize) Flyer by Tohoku University were selected as top prize for design, and top prize for concept, respectively.

NASDA has planned the MDS (Mission Demonstration test Satellite) series, and MDS-1 will be launched in 2002. The mission objectives of MDS-1 are to verify the functioning of commercial parts in orbit, to verify the miniaturization technology for components, and to measure the space environment data, especially radiation. NASDA issued an announcement of opportunities for MDS research missions in 1998, and 7 proposals were selected from 46 proposals. Joint research between each selected organization and NASDA has been conducted during 1999 and 2000 fiscal years. NASDA opened an announcement of opportunities for candidate MDS satellites in 1999. NASDA selected three proposals from 12 proposals; space experiment for solar array with higher voltage proposed by ISAS, LEO intersatellite optical communications and multi-media communications proposed by TAO, space evaluation for cloud radar proposed by ESTO. Further, NASDA has initiated a study of next generation LEO satellite system which can be applied to various missions.

2. Small Satellites in the Future Space Infrastructure

2.1 Earth Observations

In the field of Earth observations, NASDA proposed small satellites with a superconductor submillimeter limb sounder to observe trace gases in order to study ozone destruction process in the stratosphere [Reference 1]. However, a small 4K cooler system was not available at that time and this was realized as SMILES (Superconductivity SubMIllimeter-wave Limb-Emission Sounder) using a SIS (Superconductor Insulator Superconductor) mixer mounted on the exposed facility of the JEM (Japanese Experiment Module) attached to the International Space Station (ISS). NASDA also proposed a constellation of small Earth observation satellites, each of which has one sensor to observe disaster phenomena with higher observation probability [Reference 2]. Formation "flying" is also important in some Earth observation applications. NASDA has also proposed an interferometry mission with a pair of satellites with SAR to measure terrain height precisely. Small satellite data collection systems are also important to obtain data of the ocean and land including life.

2.2 *Communications and Navigation*

In the field of communications and navigation, NASDA proposed a constellation of small satellites [Reference 3]. In order to realize continuous communications and enhance positioning accuracy, proper constellations are needed. Concerning radio amateur communications, the JARL (the Japan Amateur Radio League) developed JAS 1 (1986, 50 kg), JAS 1b (1990, 50 kg) and JAS 2 (1996, 50 kg) to contribute to amateur service communications. Optical communications equipment has the potential to reduce sizes, and NASDA has developed OICETS (Optical Inter-orbit Communications Engineering Satellites, 2002, 570 kg) which will have an optical link with ARTEMIS (in GEO) developed by ESA.

2.3 *On-orbit Service*

In the field of on-orbit service systems, NASDA developed and launched ETS-VII in 1997. This satellite consists of chaser and target (410 kg). The Main objectives are to demonstrate rendezvous/docking and space robot technologies using manipulators. Target observation capability will be demonstrated by ejecting a very small target and observing the target with a CMOS camera mounted on μ-Lab Sat, in collaboration with CRL (Communications Research Laboratory). As for activities for satellites having some problems on-orbit, there are needs for inspection of targets, fuel supply, power supply, or to exchange battery or other components or to repair. As for countermeasures for space debris, there are needs for capture, change of orbit, burning in the atmosphere, or controlled reentry may be considered. In the near future, these may be realized by using small satellites/segments which have different functions [Reference 4]. Many types of robots have been developed for use on the Earth and demonstrated to show their technological evaluation, entertainment and other purposes. In the near future, various types of space robots will play an important role in an on-orbit servicing system. Astronauts may be replaced by space robots in some dangerous activities under strong solar activity or risky space debris situations.

2.4 Space Research

NASDA and ISAS have developed the lunar observation satellite SELENE (SELenological and Engineering Explorer, 2004) which has a small relay satellite (39 kg). This relay satellite will be separated from the main satellite to observe lunar gravity anomalies using a 4-way Doppler. ISAS is also developing a piggy back satellite called INDEX (Innovative-technology Demonstration Experiment) (50 kg) as shown in Reference 6. This will be launched by a H-IIA rocket; the main mission objectives are engineering technology demonstrations and observations of the aurora.

2.5 Technology Evaluation

In the field of technology evaluation in orbit, small or low cost satellites are essential. The first example for NASDA is ETS-1 (1975, 82.5 kg) to measure the launch environment of the N-I rocket. A payload LDREX (Large Deployment Reflector Experiment) (182 kg) attached to the second stage of Ariane V rocket was launched on December 20, 2000. As a result of this experiment, it was found that design improvements were needed in the unfolding mechanism of the large antenna. Thus, this type of experiment proved to be very important to evaluate a critical function before launch of ETS-VIII which has two large antennas (19m x 17m).

2.6 Summary of Various Future Space Infrastructures

Table 1 shows a summary of various current and future space infrastructures.

3. Technology Items

The following items should be studied: small, efficient and reliable parts and components, proper orbital design of formation "flying" and constellations, proper radio frequency for communications, including up and down and inter-satellite links, and countermeasures for space debris.

3.1 Parts and Components

In order to reduce size and realize higher performance, it is very important to develop proper parts/components. By using MDS-1, μ-Lab Sat and Hypersat,

NASDA will develop various parts and components based upon commercial parts. Concerning batteries, Ni-Cd/Ni-MH is used for μ-Lab Sat , and a Li-ion cell will be a candidate for μ-Lab Sat-2 because its energy density is higher and the cell voltage is three times larger than that of Ni-Cd/Ni-MH. Concerning solar cells, InGaP/GaAs/Ge is promising for higher efficiency (about 30%) , and CIS ($CuInSe_2$) / CIGS($Cu(InGa)Se_2$) is also promising due to its strong resistance to radiation.

Type of Space Infrastructure	Typical Uses	Examples
Earth observations	• Simultaneous observations • Higher probability of observations • Data collection from DCP	• Formation "flying" (Interferometry) • Constellation [Reference 2] • WEOSS (50 kg)
Communications and navigation	• Continuous communication • Amateur communications • Precise positioning • Optical communications	• Constellation [Reference 3] • Amateur satellite JAS-1 (1986, 50 kg) , JAS-1b(1990, 50 kg), JAS-2(1996, 50 kg) • Constellation [Reference 3] • OICETS (570 kg)
Space research	• Lunar gravity anomalies • Aurora observation	• Relay satellite (39 kg) for SELENE (ISAS/NASDA) • INDEX (50 kg, ISAS)
On-orbit servicing	• Function by group • Rendzvous and docking • Inspection • Robotics • Reentry	• TOPPU ,TAKO • Chaser and target (410 kg) of ETS-VII • μ-Lab Sat • Manipulators in ETS-VII and JEM • Dash (92 kg, ISAS)
Technology evaluation	• Measurement of launch environment • Evaluation of parts/components • Evaluation of critical function	• ETS-I (1975, 82.5 kg) • μ-Lab Sat, INDEX, MDS-1 (480 kg) • LDREX (182 kg)

Table 1. The use of small satellites in various future space infrastructures

Spin stabilization (nominal) and three-axis stabilization (for experiments) are used in μ-Lab Sat. A fiber optical gyro (FOG) is used because of compact size and no mechanical contacts though its temperature drift is large. The star tracker available is large in size, weight and consumption of power, but a compact star tracker is expected.

μ-Lab Sat is controlled by an integrated controller unit (ICU) with a "three voting system" which has countermeasures for single event upsets in orbit.

3.2 *Communication*

In order to realize constellation and formation "flying", proper radio frequency plan should be established. Many up and down frequencies and many cross frequencies among satellites are needed in most cases. Not only S-band but also lower frequencies (such as UHF and VHF) are desirable. Even in the case of a small satellite, it is necessary to submit advance publication and coordinate with other administrations which suffer from interference from the small satellites. This is a very time-consuming process. The allocated services are not sufficient to use for many small satellites. WRC-2000 adopted agenda item 1.12 for WRC-2003 which includes the allocation up to 3 MHz to space research and space operation services on the primary basis in the range 100 MHz to 1 GHz for the implementation of telecommand links as shown in resolve 1 of Res.723 (Rev.WRC-2000). In order to allocate some bandwidth in 100MHz to 1GHz in WRC-2003, it is necessary to conduct sharing studies and to prove that sharing with other services is feasible. The community of small satellite users may need to promote this issue. Sufficient bandwidth is not available below 1 GHz, and it is very difficult to use a different service band, such as amateur services and mobile satellite services. Unwanted emissions should be restricted based upon radio regulations Appendix S3 and ITU-R recommendations. In the case of harmful interference to other services, it is necessary to switch off the transmitter by command. Examples of some frequency plans of small satellites are shown in Table 2.

3.3 *Countermeasures for Debris*

With satellite constellations and formation "flying", debris may increase. Even small debris is very dangerous for the manned International Space Station (nominal altitude 400 km, inclination 57 degrees). Based upon criteria of design to prevent space debris, it is necessary to change to proper orbit after the end of a satellite's mission. In most cases, not much fuel is available on small satellites and it is very difficult to change orbit. In the case of ETS-VII, finally chaser and target are connected. In the same way, it is desirable to have docking among satellites in formation flight after the end of mission, but this also needs much fuel.

4. Conclusion

Small satellite activities in NASDA and other research organizations in Japan have been introduced. Technological progress in various small satellites will enable to realize efficient and important space infrastructure using constellations and formation "flying" to be created for the benefit of humankind.

Name of satellite	Earth-to-space (up link)	space-to-Earth (down link)	space-to-space (Inter-satellite link)
μ-Lab Sat (NASDA) (50 kg)	2081.083 MHz (50 kHz)	2260 MHz (300 kHz)	N/A
μ-Lab Sat-2 Mothership satellite (30 kg) 2 Daughtership satellites(10 kg x 2) (NASDA)	2081.083 MHz (tentative)	2260 MHz (tentative)	TBD
WEOSS (50 kg, CIT)	401-403 MHz	2285.0 MHz	N/A
FedSat (58 kg, CRCSS)	313.55MHz(30kHz) 2030.4MHz(132kHz) 29.93GHz(6MHz)	400.4MHz(30kHz) 2205.0MHz(5MHz) 20.13GHz(6MHz)	N/A
ST-5 (NASA)	2079.8556MHz or 7209.125MHz	8470.0MHz	2090.3MHz (Forward) 2270.0MHz (Return)
GRACE (NASA/Germany)	2051.0MHz, 2073.5MHz	2211.0MHz 2260.9MHz	24.53GHz 32.7GHz

Table 2. Examples of frequency plan for small satellites; the bandwidth is shown in brackets

Acknowledgements
The author would like to thank the staff of the Satellite Program and Planning Department and Micro Space Systems Laboratory of NASDA.

References

1. Maeda, K., Ohishi, K., and Kawada,Y.: *Small experimental satellite with super conductor submillimeter wave limb sounder,* IAA Symposium on small satellites for Earth observation, Berlin, 1996
2. Maeda, K. and Noda, A.: Disaster monitoring using multi-satellite system, 19th ISTS, May, 1994
3. Maeda, K., and Moriuma, J.: Constellation of communication and positioning satellite system, AIAA/ESA workshop on international cooperation in satellite communications, Noordwijk, The Netherlands, 27-29 March, 1995
4. Maeda, K.,et al.: On-orbit servicing systems and robots, 6th ISCOPS, December 6-8, 1995
5. Noda, A.: Future vision of micro/nano satellite in NASDA μ–Lab Sat and TANPOPO (Dandelion), ISTS 2000, pp.2302-2305, 2000
6. Mizuno, T., et al: Piggy-Back satellite INDEX for advanced technology demonstration, ISTS 2000, pp.2310-2316, 2000

Master of Space Studies 2001 Views and Visions

T. Radcliffe, L. Summerer, International Space University, Strasbourg Central Campus, Parc d'Innovation, Boulevard Gonthier d'Andernach, 67400 Illkirch-Graffenstaden, France

e-mail: radcliffe@mss.isunet.edu, summerer@mss.isunet.edu

Abstract

This paper presents the work of ISU's current class of 38 Master of Space Studies students which comprises the two-part ISU Team Project. Part I of this Team Project, concluded in December 2000, involved a comprehensive review of the literature concerning small satellites. Part II begins with this Symposium and will consume the class for the following 8 weeks. This paper presents a summary of Part I and visions for the future of small satellites by describing the path for Part II.

"The International Space University's Small Satellite Interdisciplinary Survey" (ISIS) presents an overview of small satellites, and addresses their role in today's world. Its intercultural, international and interdisciplinary focus covers commercial and political trends, and addresses five categories of applications with their necessary enabling engineering challenges. "The Proposal for International Spacecraft Constellation for Education and Science" (PISCES) was chosen by the MSS6 class as the Part II project. It involves the design of a Mars-bound small spacecraft and the construction of an international competition to engage space enthusiasts worldwide in a joint space science mission.

1. Introduction to MSS6 Team Project - Smaller Satellites: Bigger Business?

On the first day of MSS6, the class was presented with the seemingly straightforward question: *Smaller Satellites: Bigger Business?* Disillusioned, the first three months of the program were spent constructing a knowledge base to complete the activities of the last two months successfully. Thus Part I of the Team Project, concluded in December 2000, involved a comprehensive review of literature concerning small satellites. Part II begins with this Symposium "Smaller Satellites: Bigger Business?". This paper presents a summary of the scope and conclusions from Part I, and it outlines the chosen path for Part II.

2. ISU Small Satellite Interdisciplinary Survey (ISIS)

In an attempt to construct a useful and comprehensive foundation on which to build Part II, the ISIS literature review asked three questions: *Is there still a role for small satellites today? What scientific, technological, or commercial returns may be expected from small satellites? How will small satellites affect the global space market?*

M. Rycroft and N. Crosby (eds.), Smaller Satellites: Bigger Business?, 127–132.

2.1 *Role of Small Satellites*

ISIS's research and analysis overwhelmingly led to the answer *Yes, there is quite obviously a clear role for small satellites today.* After investigating various other classification methods (program cost, program development, use of state-of-the-art technology), the term *small satellite* was defined for the purpose of this project, and in accordance with the most common categorizations, as a spacecraft having a wet mass of less than 500 kg [Reference 1]. Regardless of its specific prefix or classification typology, and as a result of ongoing miniaturization, the new generation of small spacecraft has proven itself entirely capable of providing valuable service and performing exceptional science. Both potential and realized, the inherent benefits leading to these returns result in lowering the barriers to orbit. The once seemingly insurmountable orbital-access obstacles are now somewhat assuaged by increasingly favorable conditions relating to program cost, capital financing, project management tools, enabling industrial/competition policy, launch opportunity, certainty of market demand, applications awareness among the extended user community, and proprietary flight technology.

This encouraging environment reduces the delays between launch and social/scientific return, and has in turn (re)ignited the participation of diverse actors. Through increased and more affordable access to data, the science community has been revitalized. Similarly, students have been progressively engaged by hands-on experience; their attention being held by the promise of taking an idea from conception to realization through to analysis and publication. Furthermore, even though the development objective sets of countries differ, small satellites play a primary role in enhancing sustainable growth opportunities for developing countries and facilitating international cooperative relationships between universities and small and medium enterprises. [Reference 2] As a result of the cooperative arrangements between manufacturers and development institutions, these growth ambitions are addressed through both the transfer of technology and the on-orbit provision of small satellites designed to meet specific development needs. Some successful examples are analyzed in ISIS [Reference 3].

2.2 *Return on Investment (ROI)*

Whether the investment is technological, short-term, or long-term, small satellites provide numerous social, scientific, technical, and commercial returns to be further utilized and enjoyed. To arrive at this conclusion, the Part I interdisciplinary survey analyzed business models, stakeholders' priorities, policy frameworks, legal regimes, five categories of applications, and trends in engineering subsystems. In a strictly commercial sense, ROI was analyzed from each perspective along the value chain through the eyes of the procurers, manufacturers, launchers, operators and end-users. To answer the original question set, the commercialization of telecommunications and remote sensing activities provided the models from which to draw numerous conclusions.

Critical success factors for commercialization were identified to be the reliability of inexpensive launchers, and the success of current strategies for lowering mission cost (organizational culture, risk appetites, management tools, and procurement strategies). Trend analysis was performed with cost models built factoring payload complexity, bus capability, and technological heritage. To complete the multi-perspective examination of the politics and economics of small spacecraft, the additional social interests of governments were investigated. Complicating this assessment of the industry's drivers is the fact that most satellite programs are still largely national efforts in the majority of space faring countries, and thus subject to issues of national pride and prestige. Furthermore, almost all are civil programs focusing on educational interests, technology transfer and specific national applications. (i.e. in the field of disaster relief to mention just one example) [Reference 3]

This discussion of Part I's analysis and conclusions must not be taken out of context. The scope of ISIS embraced five areas of applications, selected on current and proposed activities:

- Telecommunications
- Earth science
- Space science
- Technology demonstration
- Engineering training.

The areas of space science, Earth science, and telecommunications show the most advancement and the best prospects for the future for small satellites. The following Earth-focused applications seem most promising:

- Low data rate messaging systems
- High data rate, broadband and Internet services
- Store-and-forward data relay
- Earth and near-Earth environmental observations.

With constellation architecture and formation flying to simulate extremely large apertures, future small spacecraft will be indispensable for interplanetary exploration, and for interferometric astrophysics missions [Reference 4].

With respect to technology demonstration and teaching state-of-the-art engineering, the diversity of applications with their corresponding operational environments and requirements has resulted in specifically focused small satellites systems and sub-systems engineering practices. Specific structural changes are most evident in the solid-state devices in on-board computers, data handling devices and in Attitude Determination and Control Systems (ADCS) and Guidance Navigation and Control Systems. These developments have resulted in increased autonomy and integrated design solutions with many redundant and interacting systems. Large improvements and innovations in the area of propulsion subsystems, lightweight structures and advanced thermal control methods are also vital small spacecraft developments.

2.3 *Impact*

Without question, small satellites will enable new players to enter the global space market across the phased spectrum of activities. Existing needs will be satisfied and nascent applications will precipitate industrial growth in unforeseen directions. Furthermore, the activity of these new entrants will create new opportunities for international cooperation and place increasing demands on the relationships that exist today.

The aforementioned advantageous characteristics and potential contributions were set against the limitations of small satellites, especially their inherent debris potential and associated environmental cost, their limitations in capacity and control, and the uncertainty of both their reliability and profitability. From this, it was concluded that compromise is not always best. Earth's diversity of needs requires tools at both extremes of the scale, not simply settling on the middle ground. Given that their risk profile is extremely attractive for some communities, small satellites have a specific role to play alongside their larger more expensive and more capable cousins. As a result, when questioning issues of investment and return throughout ISIS, the focus remained fixed on the potential complementary role of small spacecraft, not slipping into a *big versus small* mentality.

3. Proposal for International Spacecraft Constellation for Education and Science (PISCES)

Based on Part I conclusions and following the lengthy consideration of seven equally innovative, focused, and workable proposals, the MSS6 class agreed to pursue the *Proposal for International Spacecraft Constellation for Education and Science* (PISCES). After numerous iterations, Part II of the Team Project will take the form of a design competition for an international space science mission to Mars. A small spacecraft bus will be designed by MSS6 to the subsystem level and will carry 20 nanoprobes (including landers, penetrators, or aerobots) to Mars. After deploying the payload in Mars orbit, the mothership will serve as a data relay satellite, limiting the necessary capability and autonomy of the probes. These 20 probes will be the subject of the design competition. Requirements and constraints will be defined within which the competitors must design their proposals.

The science objective sets pursued by the probes will be pre-defined by the requirements, and may take the form of Martian atmospheric (profiling, meteorology, or radiation) or lithospheric (mineralogy, morphology, or microscopy) studies. More focused scientific aims follow on from existing private and agency-sponsored studies in order to prepare for a manned mission to Mars to search for life there. Specifically, with onboard instrument packages providing observational, spectroscopic and analytical tools and techniques to reduce the potential for ambiguous interpretations of results, the probes will provide crucial data for future manned missions. They might scout for landing sites with exobiological potential (building on Surveyor information) or perform in-situ sample gathering and analysis.

Beyond the aforementioned size and science objectives, the requirements and constraints may involve competitor nationality, funding opportunities, or professional assistance. Despite the recognized programmatic disadvantages (mothership risk of single point failures, nanoprobe size limiting options, and stringent Martian planetary protection requirements), early specific advantages of PISCES include:

- multiple scientific objectives for one mission
- nanoprobe architecture, for creativity and ingenuity in design
- nanoprobe size, cost, and complexity, being manageable by universities and developing countries
- mothership design, saving on redundant subsystems (interplanetary propulsion, GNC, transit power, communications)
- drawing upon proven technology from a host of other missions.

Beyond the engineering challenges and scientific aims, issues under study will include: the competition's organizational structure, role of the launching and possible sponsoring agency, liability uncertainty, dispute resolution, intellectual property protection, ethical assessments (planetary protection, competitor sponsorship), and commercialisation aspects.

PISCES will ambitiously encourage the participation of cultures long-excluded from space exploration, stimulate new approaches, and open the way to democratize humanity's further steps into outer space. The intended worldwide participation in the contest will make PISCES a truly international space venture, a scientific precursor mission before humankind visits Mars. PISCES will foster interdisciplinary and intercultural cooperation, re-igniting latent enthusiasm, and stimulate modern societies' shared dreams for peaceful and cooperative space exploration.

4. Conclusions

Employing the ISIS analysis of the current technical, political and economic environment for small spacecraft and balancing their numerous advantages against their inherent limitations, a new project, PISCES, has been developed by the MSS6 class. The time has come for the first truly international effort beyond low Earth orbit, and the most suitable way to achieve this goal is a scientific mission to Mars based on a mixture of cooperation and competition. A global competition will select nanoprobes to be launched together on a small spacecraft to explore and support humanity's next steps toward the Red Planet.

References

1. da Silva Curiel, R.A. "SSHP - Satellite Classification" Small Satellites Home Page. Nov. 2000. Surrey Space Centre, November 2000 <http://www.ee.surrey.ac.uk/SSC/SSHP/sshp_classify.html>
2. International Academy of Astronautics (IAA): *Subcommittee on Small Satellites for Developing Nations* (September 2000), <http://www.iaanet.org/commite/9b.htm>. IAA, November 2000
3. Bovet N., Hair J., Kennedy G., Milani P., Pavek M., Schingler R.: *The International Space University's Small Satellite Interdisciplinary Survey*, MSS Team Project Part I Literature Review. International Space University, 2000, and Annex published here, 2001
4. "The Planet Imager Home Page." Planet Imager. 2000. Jet Propulsion Laboratory. 5 December 2000, <http://origins.jpl.nasa.gov/missions/planetimg.html>

Report on Panel Discussion 2:

Micro/Nanosatellites - Benefits, Services and Applications

A. Ismail, M. Rycroft, International Space University, Strasbourg Central Campus, Parc d'Innovation, Boulevard Gonthier d'Andernach, 67400 Illkirch-Graffenstaden, France

e-mail: ismail@mss.isunet.edu, rycroft@isu.isunet.edu

Panel Chair: R. Blott, DERA, Farnborough, UK

Panel Members:

C. Bouzat, CNES, France
J. Burke, NASA JPL (retired), California, USA
T. Iida, Communications Research Laboratory, Japan
M. Mokuno, NASDA, Japan
J. Oberright, NASA GSFC, Maryland, USA
T. Ratcliffe, MSS student, ISU, France
M. Sandford, Rutherford Appleton Laboratory, UK

The Panel Chair suggested that the discussion might try to answer two main questions. Where are the benefits that will turn small satellites into big business? And what will be the "main driver" for this?

The panel considered that both scientific (research) applications and telecommunications (business) applications could flourish. Perhaps Orbcomm would not have failed had they understood better what their customers needed. Where is the commercial market for small satellite communications? **T. Iida** noted that small satellites operating at low data rates are economical, but that high data rate satellites need much more power and hence they are large satellites. An array of scientific satellites or spacecraft needs appropriate intersatellite communications links, but only one communications channel to ground. A significant benefit of small satellites is the international cooperation engendered.

J. Oberright considered the elasticity of the market for smaller satellites. Would the market increase if prices were further reduced? For the science community having well defined objectives, he felt that the answer is no. However, for communications, the answer is yes, if first the technology is right to achieve the right performance and then the price is right.

M. Rycroft and N. Crosby (eds.), Smaller Satellites: Bigger Business?, 133–134.

J. Bouzat opined that a launch cost which was one tenth of the total mission cost was acceptable to all concerned. **M. Sandford** pointed out that access to the geostationary transfer orbit was now relatively cheap, but that scientists often needed access to interplanetary space. Special conditions of celestial dynamics sometimes permitted this from GTO, **J. Burke** said, if time constraints were not severe, and delays of several months could be accommodated.

An MSS student asked whether there were any constraints on technology transfer to developing countries. From the floor, **M. Sweeting** offered the view that the transfer of "know-how" was acceptable provided that the stability of, or balance of power in, the region was not adversely affected. The work of Surrey Space Technology Ltd. is discussed with the British government, through the British National Space Centre, before a possible "dual use" project is embarked upon.

Also from the floor, **C. Belleval** considered that the case for small satellites was based upon their being the test bed for new technologies, rather than platforms for operational telecommunications. **T. Ratcliffe** showed his colours by stating that science would never be the main driver for small satellites.

Keynote Address

A Trajectory to Commercial Microspace

R. Fleeter, AeroAstro Inc., 520 Huntmar Park Drive, Herndon, Virginia, 22066, USA

e-mail: rick@aeroastro.com

Abstract

Norm Augustine when he was CEO of Lockheed pointed out that the history of aerospace companies in pursuing commercial business is one "unblemished by success". A factor contributing to this record of perfect failure is our pursuit of space as an arena for technology development. Little in space addresses the customer's desires, maybe because they seem impossible to satisfy. For example, we have so far been incapable of delivering a product affordable by individuals and on a time scale of hours or at most weeks. What is needed is less a technological miracle, than a shift in focus away from the next technological frontier, and toward finding within our technological repertoire applications which generate more revenue, more consumer value, than they cost to produce. Examples are given in space communications, transportation and entertainment.

1. What is Microspace ?

We tend to define objects according to their attributes - appearance, functionality or maybe price. But things have an evolutionary path which also describes them. Car models are dynamic - they evolve into longer, heavier, more powerful and in some ways less charming versions, mimicking the evolution of their market demographic. A plant evolves from seed to sprout, to flower to fruit and eventually to seed.

Maybe more relevant to commercial microspace, most electronic products evolve ever smaller, more efficient of electricity, more capable, less expensive and more reliable. The color television of 1963 cost, in present US $, about 25 times what a battery powered LCD TV costs today. It came in a huge box filled with vacuum tubes and wire harnesses, consumed hundreds of Watts of power, and needed servicing for adjustments at installation and regularly thereafter. The same could be said of the computing power now available in a diminutive, AAA battery powered Palm Pilot. Early computing power, and our ideas for its applications, were limited by the MTBF (mean time between failures), and the current consumption, of vacuum triodes.

The logic of electronics miniaturization is that as we achieve ever-higher levels of integration and hence simplification and miniaturization, every product becomes cheaper, more reliable and more useful. The smaller color TV freed of the AC umbilical cord can accompany you to a baseball game or onto the bus to work. Portable color monitors are ubiquitous - in automated teller

M. Rycroft and N. Crosby (eds.), Smaller Satellites: Bigger Business?, 135–141.

machines, aircraft cockpits, laptop and palmtop computers. A sort of color TV is becoming common in cell phones which have grown beyond simple wireless phones into highly capable pagers, point of sale terminals and the emerging medium of choice for remote text chatting among kids. Billions of dollars of business have been created by miniaturized products doing the work of their older, bulkier and more expensive counterparts. Many more billions have resulted from new applications made possible because of greater portability, reliability, autonomy and lower price.

Like the color television, microspace's early evolution has been focused on doing the missions which large satellites have done, but shrinking the hardware, mostly through use of more modern components. When AeroAstro proposed the ALEXIS spacecraft in 1988, with a bus mass of 45 kg, orbit average power of 70 Watts and 0.2° stabilization, at a cost including the ground station of a couple million US $, we were ten to twenty times smaller in cost and mass than conventional astronomical platforms. Our biggest obstacle to making that first sale was skepticism. In fact, on the HETE program 3 years later, NASA engineers estimated that all the components necessary for the mission, if poured individually, without integration, structure or wiring, into the launch envelope, would not fit. We had to build a full-scale model showing the components, boxes, packaging and mechanisms to eventually win the chance to build HETE. Both spacecraft fit in their launch envelopes and are now working on orbit, ALEXIS more than 8 years after launch, and HETE (version one lost in a launch vehicle failure), rebuilt and now operating perfectly in its first year in equatorial LEO.

The initial reluctance of the space community to embrace microspace is understandable given our necessarily low tolerance for risk. But space devices have additional and stronger motivations to be small, low power and lightweight. Consumer electronics do not face US $ 10,000 per kg transportation costs, a reliable physical constant for spacecraft engineers since the beginning of the space age. Power is arguably the largest single driver for spacecraft cost, whereas we do not even think about the cost of plugging in a large TV. Reliability is paramount to space systems - they must operate for years without physical human contact. Highly integrated systems do the same tasks as discrete systems with many fewer components, wires, connectors and housings. They gained acceptance in the space world by combining small size, low mass, design simplicity and reliability.

2. A Personal View

Nobody can say when the microsatellite movement got started. But here is a quote I found illuminating from Dr. Benjamin Spock, the famous author of Baby and Child Care (in print continuously since 1946), "All the time a person is a child he is both a child and learning to be a parent. After he becomes a parent he becomes predominantly a parent reliving childhood". Microsatellites were our first satellites by necessity – the first launch vehicles had payloads under 10 kg. Some organizations, like Amsat, remained absorbed in microspace, as mainstream space grew up, due to financial constraints and the hands on experience which draws people to the hobby. Now mainstream space is reliving its childhood and, while enjoying its virtues, having to live with its issues. While NASA pursues lower cost missions with faster schedules, it has not been willing to sacrifice mission complexity proportionately.

While I think it's great to remember youth oversprayed with a patina of nostalgic affection, at least as a counterbalance to teenage nihilism, I don't live on a diet of 1960's rock music, meatloaf, or Tang for breakfast. The television would still be just a television, albeit cheaper, more reliable and smaller, if not for forward thinking engineers who projected it into our telephones, banking, automobiles, advertising, communications and computing, aircraft, even our musical instruments and kitchen appliances.

Tomorrow I have the opportunity to take the side of the debate on commercial microspace in which I personally believe - that it is still a virtual business which does not yet exist. There are plenty of us out there saying that, while Iridium, Globalstar, ICO and their less well known brethren, and numerous remote sensing businesses have all been financial failures, by shrinking the spacecraft, and hence the cost, we will turn that sow's ear into a silk purse. I'm pleased at their enthusiasm, but I do not believe that doing a money-losing mission cheaper will make it into a great commercial success. As Russia and France are proving in remote sensing, even a supply of "free" space assets (spacecraft built for and financed by government missions) might not make imagery sales a strongly profitable business.

The experience of electronics manufacturers in driving down the costs of, say, PCs, of highly efficient farming, and of NASA offering lower priced missions to Congress, is that market price levels will simply erode to erase any benefit of miniaturization and attendant lower prices. There is scant profit in manufacturing color TVs, even though they are 25 times cheaper than they were in 1963. There is zero and sometimes-negative profit margin in carrying long distance phone calls, about 20 times cheaper than they were just 20 years ago, or in selling satellite pictures of the ground. Nor for that matter is there much profit in selling corn, despite that it is more than an order of magnitude cheaper and the product itself is bigger, healthier, more nutritious and tastes better, than it did a generation ago.

These are all examples of cheap commodities. Making them more cheaply is a treadmill, which many in our industry find technically stimulating. But to a businessperson, and a person who believes that the way to unlock the benefits of space is to unleash the power of commercial markets upon it, which in turn requires attention-getting profits, reducing commodity product costs is anathema.

The four decades of microspace which precede the present are the:

- 1960s, "doing it because that's all there is"
- 1970s, "doing it because it's a fun thing to do"
- 1980s, user communities starting to notice it might be useful
- 1990s, recognition of microspace as one valid way of addressing space.

The challenge facing microspace in this, its fifth decade, is not just to keep doing more with less. "Doing more with less" is "faster, better cheaper", and is what we have been doing since long before the aqueduct - even since before agrarianism. It is a synonym for engineering, which needs no catch phrase for definition or popularization. The challenge facing us is to create the applications that make microspace an industry, and not just a curiosity or a means to carry on despite ever-deeper budget cuts.

Sometimes, maybe most of the time, new applications are accidental. The oil companies faced virtual dissolution when the Edison lamp appeared, eliminating the demand for oil for lighting. Luckily, the automobile, dominated by Diesel and other internal combustion engines, came along to save the day. The development of plastics, another innovation completely external to the petroleum field, further boosted the importance of - and demand for - refined products. In the process the oil companies grew from small niche suppliers of lubricants and waxes to some of the world's largest, least popular, and most profitable industrial giants.

PCs have sold a lot of TV monitors, and driven the technology of flat panels. Apple did not invent Visicalc, the first micro spreadsheet software, but it's a good thing that someone did. How many geosynchronous communications satellites owe their existence to global TV stations such as

MTV, CNN and ESPN? Yet Hughes has never created a TV show, let alone a transponder-filling network.

Each of these applications – the automobile, the PC display, the spreadsheet and the array of entertainment programming - drove its enabling industry – oil companies, display manufacturers, PC makers and large satellite builders. What will be the applications that will drive our industry? If history is any guide, maybe none of us has any idea.

3. The Logic of Microspace

The logic of microspace is seeing spacecraft not in terms of program requirements and specifications, but via the trajectory along which they evolve and via the enhanced and new applications which, one day, they will enable. Every technical product and service considered indispensable today started out as an expensive, unreliable, complex and largely useless curiosity, exciting mainly for its oddity. This description would apply to the early automobile, aircraft, telephone, electric light bulb, bicycle, computer, radio, cardiac artery bypass operation, air conditioner (whose inventor thought he was curing malaria), microwave oven, organ transplant, and ballpoint pen. The trajectories of these every day miracles of modern life into the societal mainstream should define the logic of microspace. Its low cost, simplicity, short development times and new technologies positions for a transition which has not yet happened. Most likely, we lack not more technology and still lower prices, but the creative applications that ultimately make a commercial product a commercial success.

The applications niches which microspace serves today reflect the evolution of the technology more than a commercial demand segment. A lot of microsatellites are used to teach people about microsatellites. One used to need to be mechanically inclined to drive a car too, and we can only hope this activity soon becomes vestigial.

Microsatellites are used as miniature versions of communications satellites and remote sensing platforms, of Earth science and space science missions. This trend will continue, just as PCs have supplanted mainframes for most, but not all, computing applications. But this doesn't create a product which enters mainstream society – it replaces the tools in a technicians hands with newer tools.

Historically, engineers imbedded in a certain technology have proven uniquely unqualified to see where it will lead. The Wright brothers never conceived of military aircraft, leading to the US airpower weakness of World

War I. They didn't understand airfreight and airline applications either. The creators of television thought that it would be used to bring live performances of theater and classical music into the home, elevating the cultural sophistication of the world. They were exactly wrong. Refrigeration was developed as a failed attempt to cure malaria by a physician who noticed that people with malaria were almost always living in warm places. After refrigeration failed to save lives, the concept was abandoned for almost 50 years as being useless. In the US railways went westward to carry gold back to the eastern markets. Very little metallic gold now leaves California, but California is the world's seventh largest economic engine.

What we should do is make the microspace technology cheap, simple, accessible and scalable enough to get a large number of people, (not just aerospace engineers, a few astronomers and some communications engineers) to start using it, as happened with the PC and the Internet. But until that happens, here are thoughts on what might become real space businesses.

Entertainment. Engaging the general public in the thrill of space has been a part of the activities of the space industry since Sputnik, enhanced by John Glenn, and consummately leveraged in Apollo. The thrill lived on in the hundreds of millions of web site hits to see the Mars rover. But today's consumers are rather unimpressed that astronauts float in zero-g, having seen those images for 40 years. Space entertainment needs a facelift.

The Encounter mission, which AeroAstro is developing with Celestis and Encounter, will fly millions of messages out of the solar system and onto a 10 million year coast through our Galaxy. Encounter is exciting technologically – using a solar sail for the final push out of the sun's gravity well, and sending back imagery of the sail deployment and acceleration. But the fantasy that the millions of passenger DNA samples might signal to someone our existence, might result in a cloning of humans in a distant place and time, while perhaps being unrealistic, is viscerally appealing. This is probably the mission's strongest drawing card. Combined with a mission that is ground breaking, quick and fun, Encounter is a more inclusive, and if nothing else at least a new, form of space entertainment.

Communications. The Newton and Windows CE together spent five years and a US $ billion investment in palmtop computing, and only had losses to show for it. It took Palm Computing just over a year and a US $ 3 million investment to return a profit on the palmtop platform that has created (revived) that industry. Palm realized that people don't do the same things with a handheld computer that they do with a laptop or desktop computer. And they focused on utility - date book, address book, notes, to-do lists - in a platform priced at the level the consumer wanted to pay - much lower than cost of a PC. AeroAstro's SENS program is trying to be the Palm of microsatellite communications. We don't believe that all clients for handheld terminals for satellite communications want e-mail, faxes, web surfing and the other fancy applications. They have very basic, unglamorous needs, such as keeping track of the household pet, that can be addressed with a 2 x 2 cm transponder that has no user-controllable parts - not even an on-off switch. When purchased, it, like a digital watch, is already running and will remain so for months or years.

Transportation. Small rockets, (like small jet aircraft), will always cost much more per kg launched than big ones. Our clients want to know why rockets can't be cheap like Chevys, but cars are expensive and inefficient compared with trucks and buses. Despite a century of huge technology investment, an old-fashioned yellow school bus gets twice the seat-miles-per-gallon (or seat-km-per-litre) of a brand new hybrid propulsion car. AeroAstro's SPORT will launch satellites as secondary payloads on big rockets, take advantage of weekly flights to GTO and the scale economies of large launch vehicles, and still get to their desired orbits. A secondary slot plus a SPORT module costs less than half the cost of a dedicated launch to the same orbit.

These applications are both old and new. There has always been entertainment in space, and there has always been LEO messaging. But none of the applications on the Palm Pilot are unique nor even original. But they are tailored to the consumer to be useful solutions to real problems, not as technical features. They must be packaged at a price / size / mass, that fits their customers' requirements, and they must be marketed appropriately for their target audience.

In the future, microspace should be overrun by consumers wanting to do consummately simple things. We can only hope that one day, rather than doing ever more in space, that we are asked to do less, but to do it more cheaply, more reliably and more transparently. Space will then become as controversial as a drawer of socks and as complex to use as a telephone. When that happens, we won't have transformed ourselves into a commercial industry. Rather, our customers will have finally, kicking and screaming, dragged us there.

Session 3

Market Potential – Big Business versus the Entrepreneur

Session Chair:

J. van der Ha, Cost-Effective Space Mission Design and Operations, USA

Measuring Complexity and Risk of New Small Satellite Endeavors

D. A. Bearden, The Aerospace Corporation, PO Box 92957, Los Angeles, CA 90009-2957, USA.

e-mail: david.a.bearden@aero.org

Abstract

The space industry has greatly increased the use of small satellites to conduct civilian and commercial space missions. As recently as ten years ago, small satellites were not applicable to operational endeavors consisting primarily of simple unstabilized, battery-powered, single frequency store-and-forward spacecraft used for technology demonstrations. The irony is that small spacecraft designers, in their quest to reduce costs through streamlined development and use of "off-the-shelf" technology, have in many cases pioneered the use of advanced devices in space. The functionality achievable in small packages has taken a leap forward due largely to the availability of increased space-compatible computational power and memory. While it is clear that the current spectrum of small satellites offers a rich suite of capabilities, it is debatable whether the associated complexity of building these systems has outstripped the budget and time that industry and its sponsors are willing, or able, to commit.

Recent failed or impaired small satellite missions have amplified concerns about the applicability of small satellites to operational and commercial applications. Experience shows time and again that risks often do not manifest themselves ahead of time or in obvious ways. However, when examined after the fact, loss or impaired performance is often found to be the result of mismanagement or miscommunication in deadly combination with a series of "low probability" events. Such occurrences, often when the program is operating near the budget ceiling and/or under tremendous schedule pressure, result in failures due to a lack of sufficient resources to test, simulate or review work and processes in a thorough manner.

While small satellites arguably result in lower absolute costs and shorter development times, these benefits may have been achieved at the expense of increasing performance risk. One key question to be answered is when does a mission become resource constrained such that it is prone to failure? This paper takes an objective look at small satellites using a complexity index to normalize development time and costs. An examination of the failure rates of small-satellite missions and conclusions regarding dependence on system complexity are presented. Finally, an example of the methodology to a future small-satellite endeavor is presented with a commentary on the applicability of the approach to risk planning.

1. Background

In the late 1980s, a new paradigm, the advent of modern small satellites, arose and opened up a new class of space applications. Low-profile, low-cost satellites, funded primarily by BMDO, DARPA, STP and university laboratories, were built with maximal use of existing components and off-the-shelf technology and minimal non-recurring developmental effort. Flight hardware and the ground segment of small spacecraft were typically based on the most modern hardware and software available that required a minimum of technology development and flight certification. Manufacturers relied on the emerging advances in microelectronics, software and lightweight subsystem

M. Rycroft and N. Crosby (eds.), Smaller Satellites: Bigger Business?, 145–156.

components to provide system-level downsizing. The relatively small physical scale of small satellites obviated the need for large facilities and equipment such as high bays, hoists, large clean room areas, test facilities, and large special handling equipment and containers. Wide design margins were used to avoid costly analysis, and system-level tests were emphasized since the components were readily accessible and the design lifetimes were short (typically less than five years). Engineering development units were typically not built; instead a "protoflight" approach was favored where a single unit served as both the engineering model and the actual flight item. Quality parts were used where possible but strict adherence to space-rated parts and pedigree/testing was avoided. Redundancy was also minimized in favor of single-string designs with multi-functional subsystems and software intended to provide operational modes of degraded performance [Reference 1].

As a result of these philosophical changes, small-spacecraft programs became perceived as fast-paced, streamlined operations that were intolerant of non-essential bureaucracy. Dedicated subsystem leaders with small teams were given full technical and budgtary responsibility, with goals tailored around what could be done inexpensively in a short time. Fixed price contracts became the norm. Changes to requirements were held to a minimum to avoid analysis and a re-assessment of multiple design options [Reference 2]. These trends continued to gain momentum in the early 1990s, with an increased interest in small satellites from traditional procurers of space systems such as NASA and DoD. The shift was evident in programs like Clementine, SMEX, SSTI, and STEP, among others. Budgetary constraints mandated small satellites as the primary vehicles for science and technology demonstration, without incurring large development and launch costs, and NASA's Administrator Dan Goldin extolled the virtues of a "faster-better-cheaper" approach to scientific satellite missions. Many of the programs implemented by NASA, which in addition to SMEX, included Discovery, MIDEX, and ESSP, dictated "faster" and "cheaper" by specifying launch year and imposing a firm funding cap.

2. The Current Climate

During recent years the traditional approaches to spacecraft design, driven almost solely by performance characteristics and high reliability to meet mission objectives, with cost and schedule as secondary concerns, have been completely reversed. Designers have been asked to meet mission objectives within rigid cost limits and tight schedule constraints, in an environment characterized by rapid technological improvements and unprecedented budgetary constraints. While the original tenets of the small satellite paradigm were characterized by low cost, maximal use of existing components, off-the-shelf technology, and reduced program-management oversight and developmental effort, small satellites have become the platform of choice for more ambitious endeavors with increasingly demanding requirements.

A number of failed small spacecraft such as Lewis and Clark, WIRE, and STEX have fueled an ongoing debate on whether the US government's experiment with small satellite missions is working. The loss of MCO and MPL within a few months of each other sent waves of anxiety throughout government and industry that the recipe for success had been lost. Impaired missions or "near-miss" successes such as MGS and NEAR have been the cause of concern as well. Recent evaluations of NASA's faster-better-cheaper approach to science missions have been critical regarding a lack of *a priori* understanding and assessment of the risks involved [Reference 3]. Some are now wondering whether programs currently on the books, or late in development, are too ambitious for the time and money allocated.

3. Measuring Complexity

To understand how technical and programmatic complexity relates to cost and schedule, a complexity index may be derived based on performance, mass, power and technology choices to arrive at a top-level representation of the system [Reference 4]. A significant data set exists with which a thorough examination of the relationship among risk, cost and schedule can be conducted for small-satellite missions. We might expect that a correlation between complexity and cost or development time, based on actual program experience, would be apparent. Furthermore, a method to assess complexity at the system-level should allow informed overall decisions to be made for new systems being conceived.

Data were assembled for a number of small-satellite (less than 1500 kg) missions launched over the past decade (1990 to 2000). The basis for the relationships suggested is a database of technical specifications and operational status for the small spacecraft shown in Tables 1, 2 and 3.

Failures are categorized as catastrophic (completely lost), impaired (failed to operate as designed but met at least 50% of mission objectives), launch related (lost or impaired due to launch vehicle failure), and programmatic (never realized due to cost or schedule overruns). In some cases impaired missions ultimately overcame early on-orbit problems through operational workarounds to be deemed successful. Future missions where success has yet to be determined are also shown.

Program	Sponsor	Spacecraft Contractor	Launch Mass (kg)	Launch Date	Launch Vehicle	Mission	Failure
Clementine	BMDO/NASA	NRL	494	Jan-94	Titan II	Lunar mapping	I
Wind	NASA	NASA GSFC	1195	Nov-94	Delta II	Solar wind ionospheric	
NEAR	NASA	JHU/APL	805	Feb-96	Delta II	Asteroid mapping	I*
MGS	NASA	Lockheed Martin	651	Nov-96	Delta II	Mars mapping	I*
Mars Pathfinder	NASA	JPL	890	Dec-96	Delta II	Mars lander and rover	
ACE	NASA	JHU/APL	785	Aug-97	Delta II	Low energy particle	
Lunar Prospector	NASA	Lockheed Martin	296	Jan-98	Athena II	Lunar science	
DS1	NASA	JPL/Spectrum Astro	486	Oct-98	Delta II	Technology Demonstration	
MCO	NASA	JPL/Lockheed Martir	629	Dec-98	Delta II	Mars remote sensing	C
MPL	NASA	JPL/Lockheed Martir	576	Jan-99	Delta II	Mars science	C
Stardust	NASA	JPL/Lockheed Martir	385	Feb-99	Delta II	Comet sample return	-
Mars01 Odyssey	NASA	JPL/Lockheed Martir	758	Apr-01	Delta II	Mars orbiter	-
GENESIS	NASA	JPL/Lockheed Martir	380	-	Delta II	Solar wind collection	-
MAP	NASA	NASA GSFC	800	-	Delta II	Cosmic wave background	-
CONTOUR	NASA	JHU/APL	906	-	Delta II	Comet tour	-

Notes: C = catastrophic P = programmatic * = successful through operational work-around
I = impaired L = launch - = yet to be determined

Table 1. NASA Small Planetary Missions

Program	Sponsor	Spacecraft Contractor	Launch Mass (kg)	Launch Date	Launch Vehicle	Mission	Failure
SAMPEX	NASA	NASA GSFC	161	Jul-92	Scout	Science experiments	
MICROLAB	NASA, Orbital	Orbital	75	Apr-95	Pegasus	Lightning experiment	
METEOR	NASA	CTA (Orbital)	364	Oct-95	Conestoga	Microgravity experiments	L
Polar	NASA	NASA GSFC	1300	Feb-96	Delta II	Particle physics	
TOMS-EP	NASA	TRW	295	Jul-96	Pegasus XL	Ozone mapping	
FAST	NASA	NASA GSFC	191	Aug-96	Pegasus XL	Auroral measurements	
HETE	NASA	MIT/AeroAstro	128	Nov-96	Pegasus XL	High energy experiments	L
SAC-B	CONAE, NASA	CONAE	191	Nov-96	Pegasus XL	Science experiments	L
Seastar	NASA	Orbital	372	Aug-97	Pegasus XL	Ocean color	
SNOE	NASA GSFC	LASP	132	Feb-98	Pegasus XL	Space physics	
TRACE	NASA	NASA GSFC	250	Apr-98	Pegasus XL	Solar coronal science	
Clark	NASA	CTA (Orbital)	266	Feb-98	Athena I	Science experiments	P,C
Lewis	NASA	TRW	385	Jul-98	Athena I	Hyperspectral imaging	C
SWAS	NASA	NASA GSFC	288	Dec-98	Pegasus XL	Astronomy	
WIRE	NASA, JPL	NASA GSFC	255	Mar-99	Pegasus XL	Astronomical telescope	C
TERRIERS	NASA, BU	AeroAstro, LLC	288	May-99	Pegasus XL	Space physics	I
FUSE	NASA, APL	Orbital	1360	Jun-99	Delta II	Space science	
QuikSCAT	NASA, NOAA	JPL/Ball Aerospace	870	Jun-99	Titan II	Ocean wind measure	
ACRIMSat	NASA, JPL	Orbital	115	Dec-99	Taurus	Sun-Earth atmosphere	
IMAGE	NASA GSFC	Lockheed Martin	494	Mar-00	Delta II	Neutral atom/UV experiment	
HETE-2	NASA	MIT/AeroAstro	124	Oct-00	Pegasus XL	High energy experiments	
EO-1	NASA GSFC	Swales	588	Nov-00	Delta II	Earth imaging technology	
QuikTOMS	NASA	Orbital	166	-	Taurus	Ozone mapping	-
TIMED	NASA GSFC	JHU/APL	600	-	Delta II	Upper	-
ICESat	NASA GSFC	Ball Aerospace	970	-	Delta II	Ice and sea-level lidar	-
HESSI	NASA GSFC	Spectrum Astro	296	-	Delta II	High energy experiments	-
GRACE	NASA	JPL	432	-	Cosmos	Gravity measurements	-

Notes: C = catastrophic P = programmatic * = successful through operational work-
I = impaired L = launch - = yet to be determined

Table 2. NASA Small Earth Orbiting Missions

Program	Sponsor	Spacecraft Contractor	Launch Mass (kg)	Launch Date	Launch Vehicle	Mission	Failure
PEGSAT	DARPA	DSI (Orbital)	68	Apr-90	Pegasus	Message relay	
POGS/SSR	STP, ONR	DSI (Orbital)	68	Apr-90	Atlas I	Geomagnetic survey	
SCE	ONR	DSI (Orbital)	56	Apr-90	Atlas I	Communications	
TEX	STP, ONR	DSI (Orbital)	67	Apr-90	Atlas I	Communications	
MACSAT	DARPA	DSI (Orbital)	61	May-90	Scout	Communications	
ORBCOMM-X	Orcomm Inc.	DSI (Orbital)	17	Jul-90	Ariane ASAP	Communications	
REX	STP	DSI (Orbital)	77	Jun-91	Scout	Radiation	
LOSAT-X	SDIO	Ball Aerospace	76	Jul-91	Delta II	Sensor experiments	
MICROSAT	DARPA	DSI (Orbital)	26	Nov-91	Pegasus	Communications	I,L
MSTI-1	BMDO	JPL/Spectrum Astro	144	Nov-92	Scout	Sensor experiments	I
ALEXIS	DOE	AeroAstro	113	Apr-93	Pegasus	X-Ray mapping	I,L*
RADCAL	STP, NRL	DSI (Orbital)	91	Jun-93	Scout	Radar calibration tests	
DARPASAT	DARPA/AF	Ball Aerospace	161	Mar-94	Taurus	Classified	
STEP 0	STP	TRW	489	Mar-94	Taurus	Autonomy experiments	
MSTI-2	SDIO	Spectrum Astro	170	May-94	Scout	Sensor experiments	I
STEP 2	STP	TRW	180	May-94	Pegasus	Signal detect/modulation	I,L*
STEP 1	STP	TRW	352	Jun-94	Pegasus XL	Atmospheric physics	L
APEX	STP	Orbital	209	Aug-94	Pegasus	Power experiments	
ORBCOMM FM-1&2	Orbital	Orbital	48	Apr-95	Pegasus	Commercial Messaging	
STEP 3	STP	TRW	295	Jul-95	Pegasus XL	Science/communications	L
REX II	STP	DSI (Orbital)	110	Mar-96	Pegasus XL	Radiation	
MSTI-3	SDIO	Spectrum Astro	212	May-96	Pegasus	Hyperspectral imaging	
FORTE	DOE	LANL/SNL	210	Aug-97	Pegasus XL	Science	
STEP 4	STP	TRW	386	Oct-97	Pegasus XL	Science/communications	C
GFO	U.S. Navy	Ball Aerospace	357	Feb-98	Taurus	Radar altimetry	
MIGHTYSAT	STP	CTA (Orbital)	69	Jul-98	STS/GAS	Science	
STEX	NRO	Lockheed Martin	592	Sep-98	Taurus	Tether experiment	C
TSX-5	BMDO/STP	Orbital	67	Jun-00	Pegasus XL	Remote sensing	I

Notes: C = catastrophic; I = impaired; P = programmatic; L = launch; * = successful through operational workaround; - = yet to be determined

Table 3. Other US-built Small Satellite Missions

Launch-related failures are included in the complexity calculations, but excluded from the risk analysis. Missions with non-US sponsorship that relied on unknown foreign-nation contributions are not considered. Also, Space Shuttle science experiments and university-developed spacecraft are not considered since these approaches are typically not relevant to operational and commercial applications.

To estimate spacecraft complexity, we

- identify the parameters that drive or otherwise contribute to the system design
- quantify the identified parameters
- combine the parameters into an aggregate complexity index [Reference 5].

The complexity index uses a matrix of technical factors to place in rank order a new system relative to a baseline data set. The parameters that contribute to the calculated spacecraft complexity index, statistics of interest, and an example calculation for satellite are summarized in Table 4. Drivers include both general programmatic factors (e.g., heritage, foreign involvement, and amount of redundancy), and demonstrable technical parameters (e.g., mass, power, pointing knowledge, data-storage size, and technology choices). The mean complexity index is determined by averaging the rank order of the individual factors. Equal weighting is used so that the analysis does not differentiate between parameters or allow a particular parameter to dominate. The index is then normalized relative to the maximum and minimum to arrive at a percentage between 0 and 100%. The total flight system (payload instruments and bus) costs, spacecraft bus costs and development times (period of time from contract start until ready for launch) are the independent variables against which the complexity is compared.

Factor	Unit	Min	Mean	Max	Example	
Total Flight System Cost	(FY02$M)	0.3	71	599		
Carrier/Spacecraft Bus Cost	(FY02$M)	0.2	32	282		
Development Time (actual)	(mos)	11	39	96		
Payload Mass	(kg)	0	67	780	75	71%
Payload Average Power	(W)	0	83	436	40	45%
Payload Data Rate	(Kbps)	0	7361	90500	6000	77%
Number of Instruments	-	1	3	11	7	88%
Foreign Partnership		None, GS, LV	SC Bus	PL, mult PL	0	0%
Mission Design Life	(mos)	0	27	96	36	71%
Satellite Launch Mass (Wet)	(kg)	5	375	1360	750	87%
Spacecraft Bus Dry Mass	(kg)	26	184	889	500	95%
Heritage	(%)	0%	48%	100%	50%	56%
Level of Redundancy	(%)	0%	22%	100%	40%	83%
Orbit Regime		GEO	LEO/MEO, Lib/Lunar	Interplan (au)	1.0	2%
BOL Power	(W)	12	377	2500	300	61%
Orbit Average Power	(W)	3	161	1140	200	70%
EOL Power	(W)	3	320	2370	250	67%
Solar Array Area	(m^2)	0.2	3.8	12.8	4.5	68%
Solar Cell Type		Si	GaAs, GaAs-mult	GaAs-conc	GaAs	50%
Solar Array Configuration		body-fixed	deployed	articulated	D	50%
# Articulated Structures		0	1	9	1	73%
# Deployed Structures		0	2	15	2	67%
Battery Type		Lead-acid	NiCd, SNiCd	NiH2, Li-Ion	SNiCd	50%
Battery Capacity	(A-hr)	1	23	175	20	68%
Structures Material		Aluminum	Al w/Comp-face, Exotic	Composite	Al	0%
ADCS Type		None, Magnetic	Grav-Grad, 3-axis-(ST)	Dual-mode	spin	40%
Pointing Accuracy	(deg)	0	3.1	0.0001	0.10	81%
Pointing Knowledge	(deg)	0	1.5	0.00003	0.05	69%
Number of Thrusters+Tanks	(#)	0	5	25	9	74%
Propulsion Type		None	Cold-Gas, Mono	Biprop-(blow,pres), Ion	Cold-gas	20%
Total Impulse (delta-V)	(m/sec)	0	285	5845	100	72%
Downlink Comm Band		UHF/VHF/SHF	S-band, L-band	X-band, Ka/Ku-Band	S	25%
Max Downlink Data Rate	(Kbps)	1	1714	80000	1000	63%
Max Uplink Data Rate	(Kbps)	0.1	16.0	500	5.0	80%
Transmitter Power (peak)	(W)	1	10	60	10	64%
Central Processor Power	(Mips)	0	19.4	100	10	33%
Onboard Software Code	(KSLOC)	2	67	650	70	77%
Flight Software Reuse	(%)	0%	30%	75%	50%	25%
Data Storage Capacity	(Mbytes)	0	2008	72000	128.0	53%
Thermal Type		passive	heaters, semi-active	active	passive	0%
Entry/Landed Elements?		N	-	Y	Y	100%
Mean Complexity Index		4%	39%	72%		57%
Normalized Complexity Index		0%	49%	100%		78%

Table 4. Factors Contributing to Complexity Index Calculation

4. Examination of Trends

Using the process above, complexity indices have been calculated for all programs in the baseline data set. Launch-related failures are included in the complexity calculation, but excluded from the comparative risk analysis since system performance is undemonstrated (i.e. it is unknown whether these systems would have failed or succeeded once on-orbit). The complexity plotted against total flight system cost, spacecraft bus cost and development time are shown in Fig. 1, Fig. 2 and Fig. 3.

There are several conclusions that can be drawn from these figures. First, a correlation between complexity and costs and schedule is evident. Second, the average complexity of missions that failed or were impaired is significantly higher than that of successful missions. And, third, a threshold is apparent beyond which allocated project resources (time, funds) may be insufficient; in some cases it appears that a failed or impaired project is constrained on schedule but not on cost. While the complexity index does not predict the manner in which a failure will occur, it does define a framework within which a new mission being considered may be compared with past missions. It is not known whether the allocation of additional resources (cost, schedule) will increase the probability that a given mission will succeed. However, it is clear that when small satellites fail they are beyond a discernable threshold where the resources allocated are insufficient.

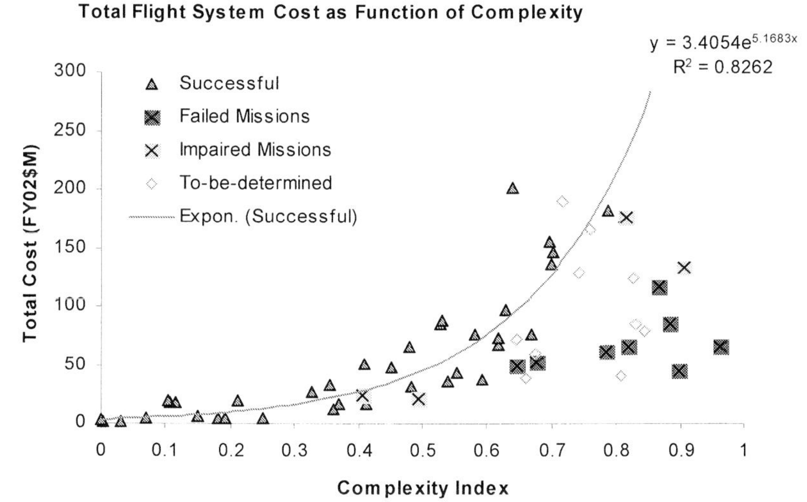

Figure 1. Total Flight System Cost versus Complexity

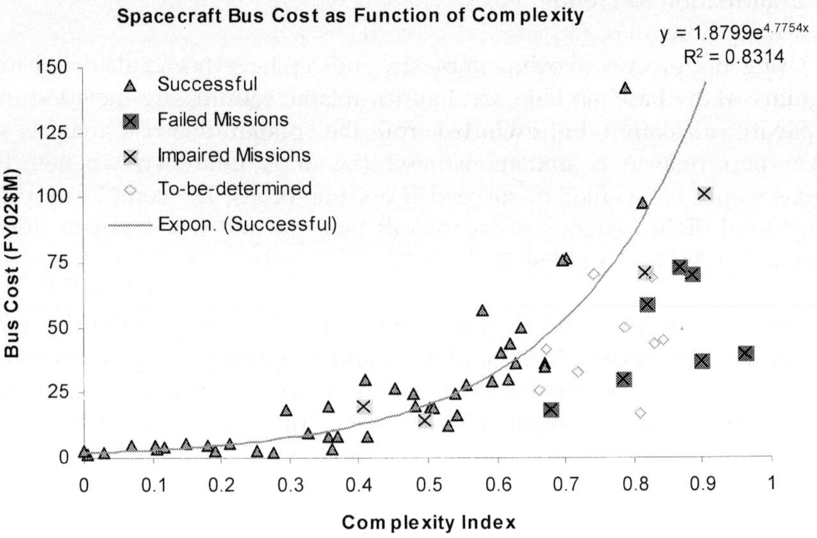

Figure 2. Spacecraft Bus Cost versus Complexity

5. Application to Future Small-Satellite Endeavors

The advantage of the complexity-based risk assessment lies in its "top-down" approach. System requirements and top-level design specifications are all that are required to complete a risk assessment. The method is therefore well suited for use in effectiveness studies to estimate how cost or schedule vary as a function of system requirements, to support developmental planning studies and trade-off studies, and to perform technology-impact assessments. Furthermore, the approach may play a role in specific program studies assessing cost versus capability, measuring individual parameter sensitivity, and cross checking proposals.

At the heart of estimating risk for new small satellite endeavors is allocation of funds and the schedule. The critical question to be asked, and answered, is the following: When do performance and technology requirements reach a threshold that, while ostensibly being achievable within the allocated budget or schedule, leads to failures due to unexpected circumstances? The process for assessing a future small-satellite endeavor is straightforward:

- calculate the complexity index based on available technical specifications
- overlay the proposed costs and schedule onto the figures
- input budget or schedule data into regression equations to calculate allowable complexity for a set budget or development time (or visa versa)
- identify descope options which simplify the design

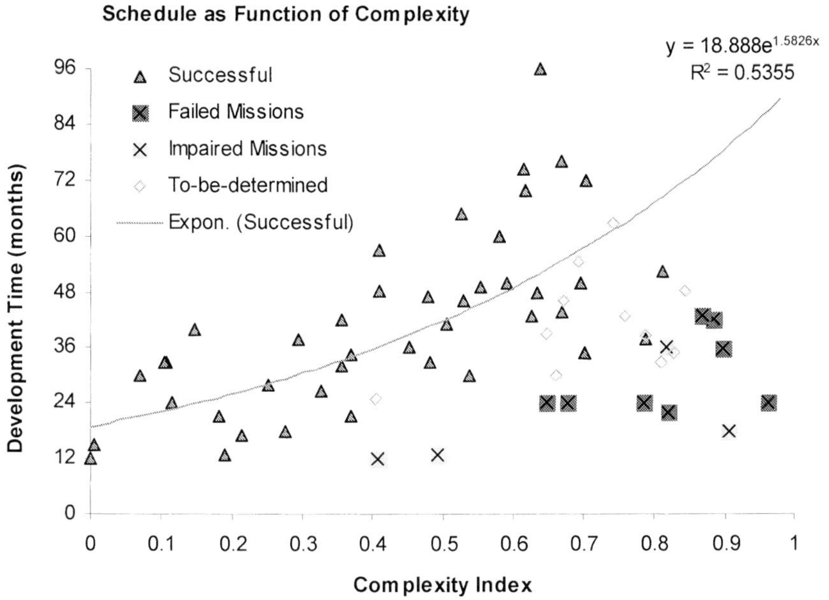

Figure 3. Development Time versus Complexity

6. Summary

Under constrained budgets and rigid schedules, the US government has embraced small satellites for conducting space-based investigations. Recent failed or impaired small satellites have brought into question this approach. While small-satellite missions have resulted in lower costs and shorter development times these benefits have, in some cases, been achieved at the expense of increasing performance risk considerably. To address the issue of when a new small-satellite endeavor is too ambitious relative to the resources which it has been granted such that it will be prone to failure, a new measure of complexity is presented here to estimate the development time and costs required. A clear dependence of success rate on system complexity is also identified.

References

1. Bearden, D.A.: Small Satellite Costs, *Crosslink Magazine*, The Aerospace Corporation, Winter 2000-2001
2. Bearden, D.A. and Abramson, R.L. *Small Satellite Cost Study - Risk and Quality Assessment*, 2nd International Symposium on Small Satellites Systems and Services, Biarritz, France, June 29, 1994
3. Dornheim, M.: Aerospace Corp. Study Shows Limits of Faster-Better-Cheaper, *Aviation Week and Space Technology*,. pp. 47-49 June 12, 2000
4. Sarsfield, L.: *Federal Investments in Small Spacecraft*, Report for Office of Science and Technology Policy, RAND Critical Technologies Institute, DRU-1494-OSTP, September 1996
5. Bearden, D.A.: *A Methodology for Technology Insertion Assessment Balancing Performance, Cost and Risk*, Ph.D. Dissertation, University of Southern California, May, 1999

Acronyms

ACE	Advanced Composition Experiment
ACRIMSat	Active Cavity Radiometer Irradiance Monitor Satellite
Al	Aluminum
ALEXIS	Array of Low-Energy X-ray Imaging Spectrometers
APEX	Advanced Photovoltaic and Electronics Experiment
APL	Applied Physics Laboratory
ATP	authority to proceed
au	astronomical units
A-hr	Ampere hours
Biprop	bipropellant
BMDO	Ballistic Missile Defense Organization
BOL	Beginning of Life
BU	Boston University
CONAE	Argentine Space Commission
CONTOUR	Comet Nucleus Tour
CTA	Computer Technology Associates
DARPA	Defense Advanced Research Project Administration
deg	degrees
DOD	Department of Defense
DOE	Department of Energy
DSI	Defense Systems, Inc.
DS1	Deep Space 1
EO-1	Earth Observing 1 Mission
EOL	End of Life
ESSP	Earth System Science Pathfinder
EUV	Extreme Ultra-Violet
FAST	Fast Auroral Snapshot Explorer
FUSE	Far Ultraviolet Spectroscopic Explorer
FORTE	Fast On-orbit Recording of Transient Events
FY	fiscal year
GaAs	gallium arsenide
GEO	Geosynchronous Earth Orbit
GFO	Geosat Follow-on
GRACE	Gravity Recovery and Climate Experiment
GSFC	Goddard Space Flight Center
HESSI	High Energy Solar Spectroscopic Imager
HETE	High Energy Transient Experiment
IMAGE	Imager for Magnetopause-to-Aurora Global Exploration
JHU	Johns Hopkins University
JPL	Jet Propulsion Laboratory
kbps	kilobits per second
kg	kilogram
KSLOC	thousand software lines of code
LANL	Los Alamos National Laboratory
LaRC	Langley Research Center
LASP	Laboratory for Atmospheric and Space Physics
LEO	low Earth orbit
Lib	Libration Point
Li-ion	Lithium-ion
LLC	Limited Liability Corporation
LOSAT-X	Low-Orbit Satellite Experiment
LV	Launch Vehicle
M	meter
MACSAT	Multiple-Access Communications Satellite
MAP	Microwave Anisotropy Probe
Mbytes	megabytes
MCO	Mars Climate Orbiter

MEO	Medium Earth Orbit
MIT	Massachusetts Institute of Technology
MGS	Mars Global Surveyor
MIDEX	Medium-Class Explorers
MIPS	Milliens Instructions Per Second
m	metre
MPL	Mars Polar Lander
MSTI	Miniature Sensor Technology Integration
N	No
NASA	National Aeronautics and Space Administration
NEAR	Near-Earth Asteroid Rendezvous Mission
NiCd	Nickel Cadmium
NiH_2	Nickel Hydrogen
NOAA	National Oceanic and Atmospheric Agency
NRL	Naval Research Laboratory
NRO	National Reconnaissance Office
ONR	Office of Naval Research
PEGSAT	Pegasus Satellite
POGS/SSR	Polar-Orbit Geomagnetic Survey Solid-State Recorder
PL	payload
QuickTOMS	Quick Total Ozone Mapping Spectrometer
QuickSCAT	Quick Scatterometer
R	Correlation Coefficient
RADCAL	radiation calibration
REX	Radiation Experiment
SAC-B	Satelite de Aplicaciones Cientificas B mission
SAMPEX	Solar Anomalous and Magnetospheric Particle Explorer
SC	Spacecraft
SCE	Selective Communications Experiment
SDIO	Strategic Defense Initiative Organization
sec	second
SHF	Super High Frequency
SMEX	Small Explorer Program
SNiCd	Super Nickel Cadmium
SNL	Sandia National Laboratory
SNOE	Student Nitric Oxide Explorer
SSTI	Small Spacecraft Technology Initiative
ST	Star Tracker
STEP	Space Test Experiment Platform
STEX	Space Technology Experiment
STP	Space Test Program
SWAS	Submillimeter Wave Astronomy Satellite
TERRIERS	Tomographic Experiment using Radiative Recombinative Ionospheric EUV and Radio Sources
TEX	Transceiver Experiment
TIMED	Thermosphere Ionosphere Mesosphere Energetics and Dynamics
TOMS-EP	Total Ozone Mapping Spectrometer - Earth Probe
TRACE	Transition Region and Coronal Explorer
TRW	Thompson Ramo Woodridge
TSX	Tri-Service Experiment
UHF	Ultra-High Frequency
VHF	Very-High Frequency
W	Watt
WIRE	Wide-field Infrared Experiment
Y	Yes

The Future Satellite Business Is Small Satellite Business

B. Doll, Astrium GmbH, Friedrichshafen, Germany

e-mail: Bernhard.Doll@astrium-space.com

Abstract

The small satellite initiative created by NASA about 10 years ago was to capitalize on space experience in terms of cost reduction. In the meantime, many missions sharing the "faster, better, cheaper" slogan have evolved - with mixed results. In this paper, Astrium's interpretation of the small satellite slogan is explained, e.g., that "small" means small in cost but not necessarily small in size or mass. Consequently, it will be shown how Astrium - as a leading space company with a reputation of nearly 100% in-orbit success - achieves low costs while maintaining quality, reliability and performance.

An overview of small satellite platforms developed for several market segments is given together with an insight view of the *Flex*Bus - a concept based on standardized modular construction aiming for mission tailored low cost satellites. Finally, a classification of conventional "large" satellites, "small" (low cost) satellites, "micro-" and "nano-" satellites is given. Based on examples of current and future programs a clear market trend towards smaller satellites can be demonstrated, while a more reserved assessment of the market potential for micro- and nano-satellites is shown. Micro-satellites will stay focused on universities or pave the way on export markets, while nano-satellites will remain a technological target, with no significant business in the near future.

1. Introduction

As the demand for low cost missions increases due to the tight budgets of government customers and the growing competitiveness of commercial customers, the terms "faster, better, cheaper" and, especially, "small satellites" have become a motto within the space community. Nowadays, it is generally acknowledged that for a huge variety of applications small satellites are the most cost efficient means to bring science and applications into orbit.

Astrium, the leading European space company, merged from the former DaimlerChrysler Aerospace and Matra-Marconi Space, has successfully responded to the challenge of the complex and changing markets with different small satellite concepts, reflecting the heritage and special market focus of its formerly independent companies.

Rather than using physical parameters as mass, volume or power consumption of a spacecraft, Astrium defines the term "small satellite" in terms of low cost, and, to be precise, low cost of the overall mission, not only the cost of the satellite itself. As such, the emphasis is to provide the customer with a

M. Rycroft and N. Crosby (eds.), Smaller Satellites: Bigger Business?, 157–164.

satellite bus that is not only low cost in development and production, but also gives a low cost of the overall mission.

2. "Faster, Better, Cheaper"

In the following paragraphs Astrium's understanding of the term "faster, better, cheaper" is given. Apparently, this definition is in line with the general understanding; the deductions derived might be unique, though.

2.1 *Faster*

It has become a kind of trademark for the space industry that the time span between a scientific or business idea and the actual sending into space of the spacecraft covers many years, if not decades, as in the case of the International Space Station. Astonishingly, it should have been a fair assumption that, with increasing experience in satellite development, manufacturing and test, the time to completion should become shorter. Actually, this effect has not been evident; rather the opposite effect occurred, admittedly due to the fact that satellite missions were becoming more and more complex.

However, for both scientific and commercial missions, the time to "pay-off" has become a more and more important issue for both kinds of customers. Today, mission realization times far below 5 years are requested. Nevertheless, there is a limit to which the overall development time can be compressed without significantly increasing the development risk and cost. This limit, in Astrium's experience, is about 3 years.

Within these 3 years the "faster" aspect can be realised in the efficient execution of work packages for design, analysis, development and testing. Due to the experience gained over decades of successful space venturing in a company like Astrium, necessary and unnecessary activities can be clearly distinguished and concentration on the essentials is possible. Assuming a realistic team, i.e. not everybody can do everything, a contradiction between a minimum development time on the one hand and the faster execution of work packages on the other will appear. Gaps in the work flow of each member of the team will show up as a natural consequence. In order not to allow people to fill those gaps with non-essential activities, a "two-projects-for-one-team" approach, i.e. each engineer is working on two projects in parallel, has to be implemented.

2.2 *Better*

There used to be an undeniable tendency to think that complex spacecraft were better spacecraft. In our understanding, better satellites, as well as other classifications of small satellites, should be stated in terms of cost only: a better satellite provides a better performance-to-cost ratio. In our experience, the best performance-to-cost ratios are definitely not be found in super-complex multi-payload satellites such as were developed by all the major space agencies (NASA, ESA, NASDA) in the last few years. It much to often turned out that the relationships and interference between the different payloads require extremely high handling efforts, exceeding by far the savings from having them all flown on one mission, i.e. with one launch.

It is our experience that focussing on one single mission objective or on a few objectives with non-interfering accommodation and operational requirements will allow the development complexity and cost of the spacecraft to be kept within commercially sound limits. Moreover, the resulting effect of physically compact satellites can be the utilization of small launchers which are now available at attractive prices.

Another essential characterization for a better spacecraft is the maintenance of high quality and reliability. With its successful heritage, a company like Astrium cannot afford to lose its reputation of nearly 100 % in-orbit success by delivering cheap satellites with compromised quality/reliability. It has become obvious that the art of being successful in the small satellite business is dependent on the way in which a company can make use of its experience, and the transformation of this experience into streamlined development, manufacturing and test processes. Many examples show that starting, without experience, with low prices or with the unprofessional use of existing experience bear a huge potential for mission losses, potentially jeopardizing the overall idea of "faster, better, cheaper".

2.3 *Cheaper*

The present market requires satellite costs reduced by a factor of 3 compared to previous developments. We think that such cost reductions are possible without impacting the spacecraft reliability, as already successfully demonstrated in the German CHAMP [Reference 1] and US GRACE [Reference 2] mission.

The majority of space companies are presently probing into the best methods for reducing the prices of their space products and some have utterly failed. The key to technical success is the avoidance, identification and mitigation of risk areas early enough in the project to avoid bad surprises during the assembly, integration and testing phase. Since mistakes are major cost drivers, manufacturing of the spacecraft or its components should never be started with unproven technologies or inexperienced suppliers. Design risks should be avoided by having safe margins and flexible performance requirements, or should be identified early and mitigated by peer reviews.

3. The Astrium Approach

The main characteristics of Astrium low cost small satellites are a full failure tolerant design, the concentration on extensive and demanding system environment testing as well as the utilization of proven technology, wherever possible, from experienced and high-quality suppliers.

The envisaged cost savings are achieved by following standardized and refined processes for design, reviews, analysis and documentation. With confidence in the selected suppliers at hand and an experienced team, final verification will be concentrated during the system environment testing; the effort spent on unit level testing and controlling will be minimized. Efficient use of the system team on more than one project not only avoids distraction to non-essential activities, but also allows seamless utilization of manpower, thus keeping the team and its experience together. This system team typically is not only responsible for overall system work but also includes subsystem activities such that contracts to suppliers will be issued for unit level rather than subsystem level. Last but not least, the selection of a low EEE parts quality standard, with an adequate tolerance to the radiation dose expected during the mission, as well as extensive use of "burn in", i.e. accumulation of operating time during unit and system assembly, integration and testing, are valuable ways to limit the overall cost of a satellite program.

Applying the above major principles, Astrium has developed reliable, but low cost, small satellite platforms aiming for different market segments as summarized in Table 1.

4. The *Flex*Bus Concept

The basic concept of *Flex*Bus is to provide tailor made solutions for scientific and Earth observation space missions at competitive prices under firm fixed contract conditions. It has been demonstrated that this concept allows Astrium to prepare binding C/D proposals within 2 months after key requirements definition.

Platform Type	Target Mission	Characteristics
Eurostar	GEO Telecom	Low cost due to serial production with minor modifications for different customers (typically 3 t, 5 kW)
Globalstar	LEO Telecom Constellation	Low cost due to serial production of about 60 identical spacecraft for a single customer (0.5 t, 1.5 kW)
Leostar	LEO Earth Observation (High Pointing Accuracy)	Low cost due to standard platform, serving individual customers with minor modifications (typically 0.5 t, 0.5 kW)
***Flex*bus**	Science and Earth Observation Missions, with Unique Bus Requirements	Low cost due to standardised modular construction kit on the unit level. Mission tailored structure / thermal subsystems to serve individual missions / payload needs (Typically 0.5 t, 0.5 kW)

Table 1. Astrium Small Satellite Summary

The main idea of *Flex*Bus is to maximize the recurring effort in engineering and hardware, and consequently to use the available resources to focus on the particular mission requirements and the uncompromised implementation of the mission payload. This has been achieved by having a basic electrical system architecture capable of easily accommodating a set of:

- **Core elements,** used for all *Flex*Bus applications, e.g., S-Band RF equipment, data handling unit, heaters and thermistors, frame software
- **Selected elements** from a pool of hardware in accordance with the specific system performance requirements, e.g., power control/distribution units, batteries, altitude and orbit control sensors/actuators, high rate transmitters, software library
- **Mission tailored elements** to be newly developed for each particular mission, e.g., mechanical/thermal configuration and structure, solar array, application control software

With the mechanical/thermal configuration unique for each mission, a *Flex*Bus spacecraft can hardly be recognised by its outward appearance as demonstrated in the variety of mission applications depicted in Fig. 1, comprising scientific missions as well as operational applications in high resolution optical and SAR imaging. This wide range of applications shown here clearly demonstrates the versatility of the *Flex*Bus concept.

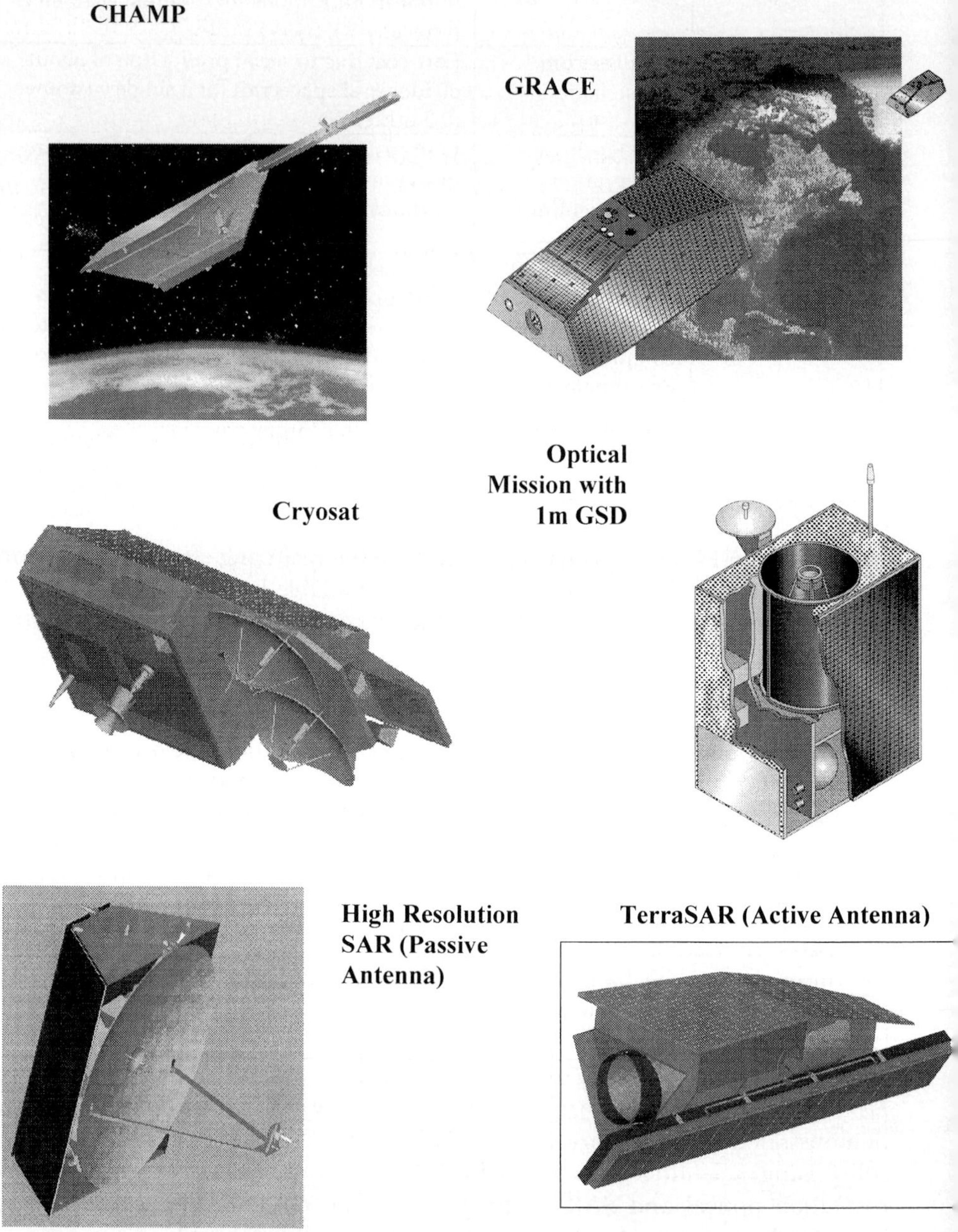

Figure 1. *Flex*Bus Mission Examples

5. The Current Space Market and its Future Potential

The satellite market as it appears today can be classified as presented in Table 2. There are 4 major classes of spacecraft; large satellites, small satellites, micro-satellites and nano-satellites.

As for the large satellites, i.e. "classical" satellites like ESA's ENVISAT and METOP, they still form the major portion of the space market in terms of turn-over and business. However the share of small satellites is already significant, with a clear trend to further growth. More and more, classical complex spacecraft will be substituted by a number of dedicated small satellites, each with less complexity and proven technology, with shorter development times at lower cost.

	Large Satellites	**Small Satellites**	**Micro-Satellites**	**Nano-Satellites**
Typical Launcher	Ariane	Rockot/Soyuz	Piggy Back	Piggy Back
Typical Cost	800 Million €Euro	100 Million €Euro	15 Million €Euro	15+ Million €Euro
Typical Technology	Complex, including new technologies	State-of-the-art, no technology risks	Simple mission, cheap technology	Advanced technology
Typical Customer	Institutional	Institutional, Military, Export and Commercial	Universities (and Export)	Institutional
Typical Missions	Large scale operation, leading edge technology, for science and Earth observation	Operational or scientific low cost missions with state-of-the-art technology	Educational and technology demonstrator, with low performance payload	Technology demonstrator
Present Business	+ + + +	+ + +	+	o
Future Business	+	+ + + + +	+	(+)
Contractor	Industry		Universities (Industry)	Universities (Industry)

Table 2. Present/Future Satellite Market Classification

The business with micro-satellites will stay limited and mainly focused on universities for educational purposes. However, micro-satellites do serve as attractive and low cost means of in-orbit technology demonstration and missions with no stringent demand on full performance/operations. As such, this kind of spacecraft may ease the access to export markets, especially for countries with smaller purses that want to join the space faring nations.

Nano-satellites can be considered an interesting target from the technological point of view. However, there is no clear trend yet visible for a significant operational performance and as a consequence, no substantial business can be expected in the near future.

References

1. <http://op.gfz-potsdam.de/champ/index_CHAMP.html>
2. <http://essp.gsfc.nasa.gov/grace/index.html>

Market Growth Potential for Small Satellites: An Industry View

A. Lewin, Orbital Sciences Corporation, 21700 Atlantic Blvd., Dulles, VA, 20166, USA

e-mail: lewin.andy@orbital.com

Abstract

The potential of small satellites has frequently been likened to the advent of the personal computer. The drive towards smaller space systems has unquestionably opened the doors to space for dozens of companies and countries. Indeed, the "faster, better, cheaper" mentality embraced by small satellite producers became an industry mantra. However, nano-, micro-, and mini-spacecraft remain a very small part of the overall space industry landscape. This paper examines an industry view of the potential of small spacecraft to create a PC-style revolution in the satellite industry.

Increases in the size of the small satellite market can come in two ways. Either small spacecraft will supplant larger ones in existing markets or they will create large new markets. To evaluate current and future small satellite penetration into the existing satellite market, the market is segmented into a matrix of three basic satellite types: LEO, GEO, and interplanetary, and four applications: communications, remote sensing, science, and weather. Although small satellites have made important strides, they do not dominate any of the identified market segments. Furthermore, small satellites suffer from fundamental weaknesses that are unlikely to be overcome by medium-term advances in small satellite technology.

If existing markets will not provide sizeable growth then the only remaining possibility is the creation of new markets. Historically, the market potential of most revolutionary advances does not come from the replacement of old technology but from the generation of entirely new products and services enabled by this new technology. The catalyst for this transformation is typically one new capability that itself warrants adoption of the new technology.

Advances in small satellite technologies will support modest advancements of the small satellite industry, but it is considered unlikely that they will cause any major shifts in the space industry in the near- to medium-term.

1. Introduction

The last 20 years have witnessed an explosive growth in the capability and number of small satellites. Just 40 years ago, space was accessible only to the world's superpowers. Advances in small satellite technology have enabled dozens of nations to join the space age. Even individual universities can now afford to launch small spacecraft.

Despite the proliferation of micro- and nano-satellites, the business impact of these systems is minimal. In the commercial arena, these spacecraft are generally built by relatively small companies. The world's aerospace giants have shown only limited interest in small spacecraft.

M. Rycroft and N. Crosby (eds.), Smaller Satellites: Bigger Business?, 165–171.

2. Small Satellite Market Penetration

The expansion of commercial small satellite providers must be evaluated in the context of the overall satellite industry. For this purpose, the market is first broken down into two major elements, manned and unmanned. The unmanned spacecraft market is split into three types of spacecraft: interplanetary, low Earth orbit (LEO), and geosynchronous (GEO). Each of these spacecraft can carry out one of four mission types: communications, science, remote sensing, and weather. For each combination of satellite type and mission, the market size is rated qualitatively using one of the following five categories: very small, small, medium, large, or very large. Similarly, the small satellite (nano- and micro-satellite) market penetration is estimated using: none, very little, some, moderate, significant, dominant. The results of this analysis are shown in Table 1.

	Market Size	Small Satellite Market Penetration
MANNED SPACE	Very large	Very little
UNMANNED SPACE		
Interplanetary		
Communications	Very small	Very little
Remote Sensing	Very small	None
Science	Medium	Very little
Weather	N/a	N/a
GEO		
Communications	Large	None
Remote Sensing	Small	None
Science	Small	None
Weather	Large	None
LEO		
Communications	Large	Some
Remote Sensing	Large	Some
Science	Large	Some
Weather	Small	Some

Table 1. Qualitative satellite market summary

The market segmentation approach described above yields seven primary space markets: manned space, interplanetary science, GEO communications, GEO weather, LEO communications, LEO remote sensing, and LEO science. In financial terms, small satellites have made only limited inroads into these marketplaces, with nearly all activity focused on LEO applications.

2.1 *Manned Space*

Manned space represents an enormous, but highly concentrated market. Vast resources are channeled into relatively few projects. Nations have spent US $ tens of billions building the International Space Station (ISS). The ISS remote manipulator arm alone cost US $ 1 billion, and each Space Shuttle launch costs more than US $ 0.5 billion. Due to the safety concerns and the inherent scale of manned space activities, micro-spacecraft have little presence in this market. One potential application is small robot spacecraft for ISS maintenance, although there are no current plans to build such a system.

2.2 *Interplanetary Science*

To date, no micro-spacecraft have performed interplanetary science missions. The primary problem is that the propulsion requirements generally drive the mass outside the scope of micro-spacecraft. In addition, the long mission durations and complicated mission profiles typically lead to a complexity level beyond the capacity of small satellites.

However, small spacecraft are entering the interplanetary arena. NASA recently completed a competition for the Mars Micro-Satellite Bus (MMSB). This concept uses two slots on the Ariane Structure for Auxiliary Payloads (ASAP). The competition required a common bus concept for performing a variety of communications and science missions. The Mars micro-missions represent an exciting new realm for small satellites. Unfortunately, NASA cancelled the MMSB program after completing Phase B and is again focusing on larger Mars missions.

2.3 *GEO Communications*

Micro-spacecraft are not used in geosynchronous communication missions. In this market, a "small" satellite weighs more than 1000 kg. The large size of GEO spacecraft is driven by several factors. First, the large distance from GEO to Earth requires high transmit power and large antennas to focus signals at their designated locations. Second, they require long lifetimes - generally 15 years - which forces a redundant architecture. Finally, the spacecraft require sizable propulsion systems to boost from geosynchronous transfer orbit to the final GEO orbit as well as to maintain the orbital position over the satellite life.

2.4 *GEO Weather*

The challenges facing micro-satellites in the GEO weather satellite market are almost identical to those in the GEO communications market. The need for propulsion and long design life are again issues. In addition, many of the instruments needed for gathering weather data are too large for a micro-satellite. Furthermore, several instruments need to be co-located to take simultaneous measurements. Although this is possible with small satellites, it is generally less efficient than placing all instruments on a common large platform.

2.5 *LEO Communications*

LEO communications has been one of the most successful markets for small spacecraft. The most prominent example in this category is the ORBCOMM constellation of 35 micro-satellites which provides global near real-time two-way messaging. However, small satellites command a relatively small market share. Deployment of the ORBCOMM system cost less than US $ 200 million, compared to the US $ 1 billion spent fielding the Iridium constellation.

Small satellites will continue to play a role in LEO communications applications. Improvements in small satellite technology will permit more powerful small spacecraft to perform increasingly complex missions. It is conceivable that these improvements will allow modest growth in the micro-satellite market share, but larger satellites will still dominate this segment.

2.6 *LEO Remote Sensing*

Advances in small satellite and instrument technology have been crucial to recent developments in this area. Small spacecraft have now demonstrated imaging capabilities with less than 10m resolution. Despite this progress, small spacecraft still comprise a relatively small portion of this market. Continued progress is largely dependent upon further miniaturization of the remote sensing payloads. Small spacecraft can be expected to make additional inroads in this segment over the medium term.

2.7 *LEO Science*

Micro-spacecraft have been especially successful in LEO science missions. The use of smaller (although not necessarily micro- or nano-) satellites has been at the core of NASA's "faster, better, cheaper" procurement approach. In the last decade, dozens of nations and universities have launched their first spacecraft, usually performing science missions. Small spacecraft are an ideal platform for small science instruments and for pathfinder missions. The lower cost of space access enabled by micro- and nano-spacecraft has dramatically increased the number and variety of space missions performed by the scientific community. While the number of small satellite science missions can be expected to increase, the overall market share is likely to remain relatively flat, particularly in the wake of NASA's re-evaluation of its "faster, better, cheaper" philosophy.

Looking across all market segments, two broad themes emerge. First, in many cases, mission requirements naturally drive towards larger satellites as the optimum solution. Second, even where small satellites have become more common, the low costs of these missions relegates them to a small percentage of the total market share.

3. Small Satellite Market Growth Potential

An examination of the current space market indicates only limited growth potential for small satellites. Therefore, substantial increases in the small satellite market size must be fueled by the development of new markets.

The promise of small satellites has often been likened to that of the personal computer (PC) in the 1970s. The analogy is very compelling. Prior to the advent of the PC, relatively few large businesses could afford expensive mainframe computers. The low cost of the PC enabled even small companies to put a PC on everyone's desk, sparking a dramatic increase in business productivity.

In the satellite industry, the high cost of large satellites limits their availability to all but a few large companies and governments. Micro-spacecraft provide space access at a fraction of the cost and schedule of their larger counterparts. It has been suggested that inexpensive micro- and nano-spacecraft can create a space revolution comparable to the PC revolution of the 1980s. Despite the seemingly strong promise of small spacecraft, a PC-style transformation of the satellite business has not yet occurred. Therefore, it is valuable to consider what fueled the explosive growth of personal computers and evaluate its applicability to the current position of small satellites.

In most cases, new technology growth is driven by a new capability that itself warrants adoption of the new technology. This catalyst is sometimes called the "killer application", or "killer app". For the PC, the initial killer app was the spreadsheet and, to a lesser extent, word processing and desktop publishing. For the first time, managers, accountants, and engineers could easily perform complex series of calculations. Then, they could change one number and have the entire result automatically re-computed. Along the way, many other indispensable computer applications have emerged: e-mail and the Internet are two of the more significant examples.

If the growth in the small satellite market is to parallel that of the PC, it too requires a killer app. This is where the analogy begins to break down. At one time, the small satellite killer app appeared to be LEO communications constellations. In the early 1990s, three companies had been approved to build constellations of low-Earth orbiting micro-satellites, and several others were seeking spectrum to build additional systems. Teledesic proposed an "Internet in the sky" comprised of more than 600 satellites. It appeared that small spacecraft were about to come of age. However, the terrestrial build-out of wireless communications capability exceeded everyone's expectations. In the end, only one micro-satellite constellation was ever fielded - ORBCOMM. Despite the technical success of the constellation, ORBCOMM is now attempting to emerge from Chapter 11 bankruptcy. After attracting financing in excess of US $ 800 million, ORBCOMM was sold for less than US $ 20 million [Reference 1].

If LEO communications constellations are not the industry's killer app, another must be found. The market survey shows that small satellites have successfully performed a wide variety of missions. The problem is one of volume. We assume, for the sake of argument, that a typical micro-satellite is sold for US $ 3 to US $ 15 million. At this price, it takes 67 to 333 spacecraft to earn US $ 1 billion in revenue—the same as one ISS remote manipulator arm. Therefore, if small satellites are to achieve significant success, the industry must find an application with the broad, mass-market appeal necessary to drive dramatic growth.

4. Conclusion

To date, advances in small satellite technology have substantially increased the number of organizations with access to space. Micro- and nano-spacecraft have demonstrated their viability for a wide variety of applications. However, limitations imposed by their small physical size prevent small satellites from being used in some of the industry's largest market segments. In addition, the large number of small spacecraft launched is offset by their low cost, relegating micro-spacecraft to a minor market share in the segments in which it competes.

Despite the promise of small satellites, they have failed to generate the PC-style revolution of the satellite industry. The central problem is the lack of a "killer app" that can spur the sales of the thousands of micro-satellites required to make an impression on the overall spacecraft market.

Small satellite technology will continue to improve, permitting modest growth in the number of applications and the overall number of spacecraft.

However, the improvements will not be sufficient to overcome the challenges which small satellites face in traditional large spacecraft markets, nor are they likely to enable the killer app required to spawn major growth in the small satellite market.

References

1. Foley, T.: Bidders Plot Comeback for a Satellite Network, *New York Times, April 2,* 2001

The COCONUDS Concept: a Low-cost, Distributed Network Approach to Satellite Earth Observation

F. F. Verduijn, T. Algra, W. A. Brokx, National Aerospace Laboratory NLR, Anthony Fokkerweg 2, 1059CM Amsterdam, Netherlands

e-mail: algra@nlr.nl; brokx@nlr.nl

G. J. Close, C. Lee, Science Systems (Space) Ltd, Chippenham, UK

B. J. Denore, GEOSYS, Madrid, Spain

J. B. Williams, Natural Resources Institute, Chatham, UK

Abstract

The overall objective of the COCONUDS (Co-ordinated Constellation of User Defined Satellites) initiative is to ascertain the practicality of a radically different, low-cost, distributed network approach to satellite Earth observations. Thinking from a user's perspective, and bringing together European expertise in low-cost [small, mini and micro] satellite and ground station developments in a co-operative venture, it should be possible to finance and develop a cost-effective, sustainable, and user-friendly constellation of Earth observation satellites, to meet many of the as yet unsatisfied needs of local decision makers for recent local environmental information.

The COCONUDS constellation of up to 9 satellites observes the Earth from LEO with a resolution of 30 metres, four spectral bands, and a swath width of 350km; This ensures global one-day coverage. High spatial and spectral resolution should not be the primary drivers. Using advanced on-board data compression techniques, the EO data can be directly received by commonly available low-cost commercial-off-the-shelf receiving stations. This allows the local decision-maker to collect data independently, without interference by the satellite operator, large central data processing and distribution entities or other administrative bodies.

This paper assesses the user needs and requirements, both in developed and developing countries. The conceptual design of the space and web-based ground segment of the COCONUDS constellation is described, including the direct-reception front-end and key sub-systems.

It concludes that COCONUDS must be realised in one form or another, if the needs of a large group of disenfranchised users are ever going to be met, and if multiple local environmental management processes around the world are to be based on proper information.

1. Introduction to the COCONUDS Concept

To test the COCONUDS (Co-ordinated Constellation of User Defined Satellites) hypothesis, three primary objectives were to establish the user's needs before trying to deflect mainstream EO (Earth Observation) development onto a new path, to assess the technical feasibility of meeting those needs through a suitable constellation of micro-satellites and to assess its economic viability through an exploration of pertinent financial, political, social and institutional issues. The technical feasibility of the COCONUDS space and ground segments

M. Rycroft and N. Crosby (eds.), Smaller Satellites: Bigger Business?, 173–177.

was considered in a new fashion, with the user's needs forming the starting point. Options for the end-user station are evaluated and their implications for satellite communications and imaging functions considered.

This paper summarises the results achieved in the European Commission funded COCONUDS project in the two years since it started. It illustrates the users' needs, how these can be met, and some of the problems facing the physical implementation of the COCONUDS vision. The user needs obtained during the first workshop of the COCONUDS Concerted Action [Reference 1] are aimed at local decision-makers in developing and developed countries who require information to improve local environmental decision making. The user needs for operational Earth Observation applications are:

- Appropriate, reliable and accurate information
- Fast and easy access to data (i.e., real-time and web-based)
- Timely data (i.e., daily global coverage)
- Affordable data and information products
- Availability of services

Commercial Earth Observation systems can only be made successful with all of the needs listed above addressed simultaneously. The information needs can be translated into a set of high-level system requirements:

- 30 m resolution plus a possible complementary 100 m wide-swath sensor
- Four spectral bands for vegetation monitoring, similar to SPOT XS/XI
- Effective revisit time of five days, taking into account cloud coverage
- 400 km swath (not necessarily all at once)
- Broadcast capability using low cost local S- or X-band antennas
- Geometric precision and radiometric precision of secondary importance
- Web-based dissemination of data and information products

The second workshop of the COCONUDS Concerted Action dealt with the feasibility of meeting the user requirements with a small satellite constellation. It started with an inventory of the performance of current and planned remote sensing systems. Fig. 1 shows that COCONUDS fills the gap of high revisit time requirements without restrictions in resolution or coverage.

The COCONUDS space segment must consist of about eight satellites for daily global coverage. The sensor has a focal length of 136 mm and an aperture of 15 mm, when a resolution of 33 m and a swath width of 400 km are chosen. The used CCD-array consists of 12000 pixels with a pixel size of 6.5 μm.

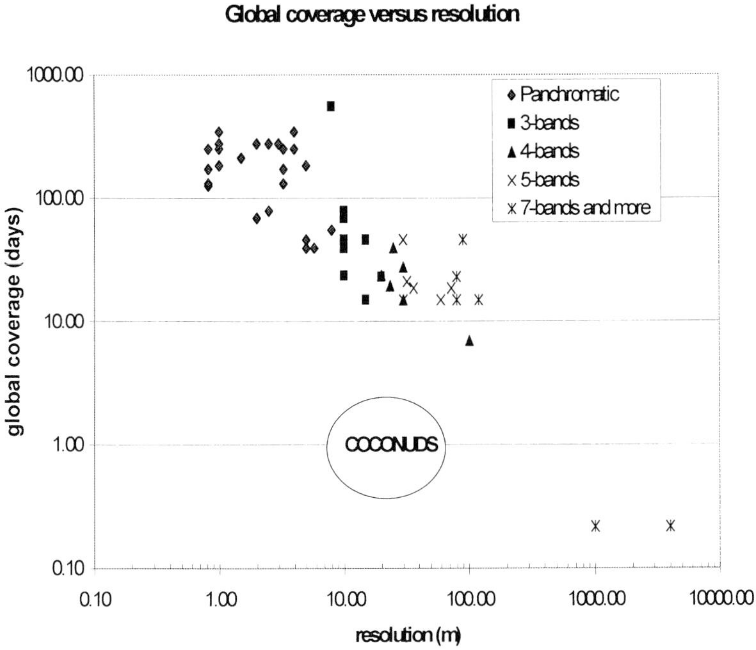

Figure 1. Global coverage (revisit time) versus spatial resolution of current and planned remote sensing systems

The attitude control system must have a pointing accuracy of 0.2 degrees (with the satellites at an altitude of 700 km) and an attitude determination accuracy of 0.02 degrees. This can be met using a 3-axis control involving momentum wheels and magneto-torquers, and star sensors and/or gyros for attitude determination.

Low-cost standardised end-user ground stations should be used. A promising candidate is the upcoming METOP Advanced High Rate Picture Transmission station (AHRPT) able to receive up to 3.5 Mbps using a tracking dish antenna. For a 2.5 m dish antenna and a minimal elevation angle of 45 degrees, the required transmission power is only 2 Watts, which is indeed suitable for micro-satellite implementation. This type of station is the follow-up of the currently wide-used NOAA HRPT ground-station, with a down-link rate of 0.66 Kbps. Other possibilities would be the 8 Mbps S-band station that will be used by Surrey Space for their Disaster Monitoring Constellation and the 120 Mbps mobile X-band RAPIDS station (Real-time Acquisition and Processing - Integrated Data System) of the NLR/BURS/NRI consortium.

The resulting down-link data rate is a key parameter for the overall system performance. Direct data readout means that all data are immediately transmitted without storage on board. Therefore, a high-performance on-board data reduction system will directly influence the size of the complete constellation. We assume 8 bits/sample and automatic gain control (line by line). Without data reduction, a four-band optical imager and a target rate of 3.5 Mbps would allow a resolution no better than ~140 m, assuming a satellite velocity of 7 km/s. However, sophisticated data reduction should be applied, consisting of cloud editing, spectral data reduction, and nearly loss-less data compression. Cloud editing means that data of cloud-contaminated areas are not transmitted, resulting in an additional average compression ratio of about 2. In this way the overall data rate can be reduced resulting in a down-link data-stream of 1.75 bits/pixel and a resolution of 33 m. An image strip with an effective swath width of 320 km can be down-linked in this way. This calls for a constellation from 7 to 9 satellites, for daily coverage at European and equatorial latitudes, respectively.

Alternatively, a second ground segment concept was studied based on a web-integrated approach. In this case a few RAPIDS stations (120 Mbps) are used to read out images which are requested by users via the Internet. Only requested data is stored on-board and down-linked subsequently. This concept allows a user to acquire images of any part of the globe within an average of 21 hours. Advantages are that fewer satellites are needed (4 or 5) and that the user only needs a web connection. Disadvantages are a larger acquisition delay and possibly network congestion.

2. Economic Viability of the COCONUDS Concept

The first two COCONUDS analyses have shown that the capital costs for the micro-satellites, instruments and launch are relatively modest (approximately US $ 5 million for each). However, the uptake and use of data by target groups are likely to remain very modest unless operational activities are guaranteed for 5 to 10 years with easily available data (and preferably free), using the NOAA meteorological model [Reference 2].

Different operational models have been examined. One approach has been to consider a sustainable local ground segment through making the satellite data essentially free for service providers and recouping costs through other incentives (e.g., value-added services). Indeed, this was found to be entirely feasible for adequate quantities of such data being directly broadcast to low cost reception systems, if this feature were to be "designed in" at the outset.

Another issue that arose in the discussions was the value of the benefits to different stakeholders, and who should pay for them. For example, if all countries in Europe were to use the same methodology (including satellite imagery involving additional expense) for assessing their coastal zone environmental change, it would be much easier and less expensive to pull all the coherent information together for development/evaluation of appropriate European-wide policies. The individual countries themselves do not benefit directly and would probably not use the more expensive methodology for themselves. Concerning environmental change, such as deforestation and global warming, regional and global (rather than local) government attention is needed. Thus, it is appropriate for Europe to support public information tools such as the Global Monitoring for Environment and Security initiative (GMES) or COCONUDS for the benefits accruing from common standards. We must learn how to manage Earth Observations better if we are ever to cope successfully with global environment change.

3. How to Proceed

In the future we must continue to feed ideas into the GMES process, to track developments and to offer comments at appropriate points in favour of inclusion of the interests of civil society, the developing countries and use of the Internet. Or we could focus on future political needs of Kyoto protocol compliance procedures to a European micro-COCONUDS system, which would also aid developing countries in their Kyoto related actions. A more pro-active option would be to drive the complete multi-layered COCONUDS concept forward with actions promoting the development of a user web, a service web and an environmental sensor web as a human rights or democratic issue, with a clear focus on user needs rather than a technology driven approach.

References

1. Denore, B. J.: *On how Future Low-cost Remote Sensing Satellites could meet the Information Needs of Environmental Managers*, Proceedings of the XIX ISPRS Congress, Amsterdam, Holland, July 16-23, 2000
2. Friday, E. W., Gross, E. M. and Yerg, M.: The Public-Private Partnership in the USA for the Delivery of Weather Services and its Relationship to the International Exchange of Meteorological and Related Data and Products, *WMO Bulletin*, Vol. 45-1, January 1996

A Microsatellite-based Demonstrator Mission for Agriculture and the Environment

B. Coquil, Ph. Campenon, Astrium SAS, 31 avenue des cosmonauts 31402 Toulouse CEDEX 4, France

e-mail : bernard.coquil@astrium-space.com; philippe.campenon@astrium-space.com

Abstract
This paper introduces a microsatellite project which Astrium intends to launch in oreder to trigger the development of agriculture and environment applications using remote sensing data. The concept for a multispectral instrument aboard a Myriade microsatellite is described in outline.

1. Introduction: Opportunities in Agribusiness

Among the many different application domains that Earth observations can cover, agriculture is potentially one of the largest. This is because of the temporally evolving nature of crop field (spatial) information which is required for the huge acreage devoted to agriculture around the world. Remote sensing observations have been used for some time for crop inventories based on classification methods and for local assessments of current biophysical conditions using empirical spectral indices.

Convinced that the agricultural market was the most promising sector to emerge, Astrium decided four years ago to launch an initiative (the XSTAR project) in order to study the capabilities required by a new satellite system. The system has to provide the right information for commercial end-user services, to be supplied to farmers, co-operatives or retailers.

2. From Remote Sensing Raw Data to Useful Products

An innovative concept of advanced information products based on the use of powerful reflectance models has been developed in co-operation with world-leading agribusiness companies and agronomic research institutes. The goal is to provide growers with information enabling them to monitor plant growth, identify stress, and predict yield. This technique has been validated through extensive airborne campaigns (see Fig. 1 and Fig. 2) simulating the acquisition conditions of future satellites.

M. Rycroft and N. Crosby (eds.), Smaller Satellites: Bigger Business?, 179–184.

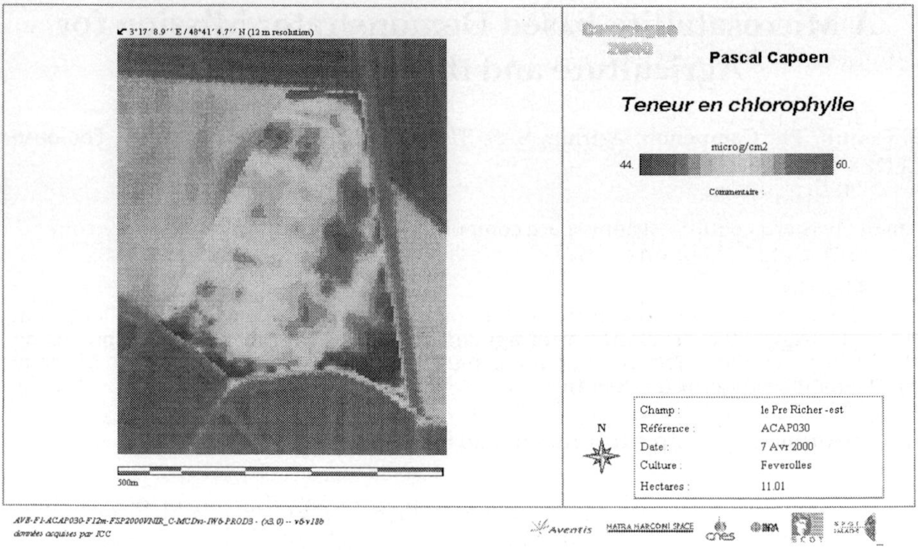

Figure 1. Example of the Chlorophyll content as a result of image processing

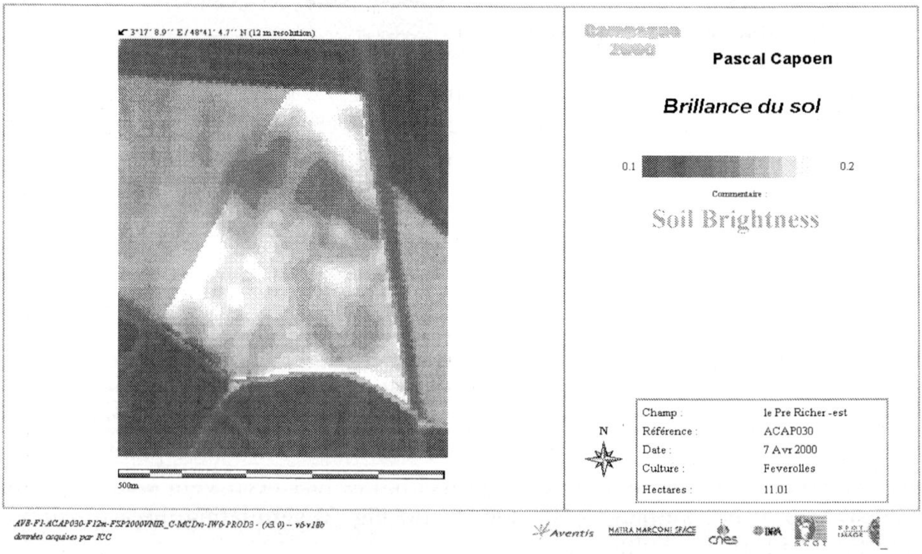

Figure 2. Example of soil brightness derived by image processing, for the same field

Such products are based on the use of generic products (biophysical parameters) which characterise the vegetation and soil state:

- Chlorophyll content
- Leaf Area Index
- Soil Brightness.

3. End to End Demonstration

Astrium is continuing research to develop new products, but this promising market still has to be triggered. In particular, the end-users are waiting for the demonstration not only of the product, which has to be valid throughout a crop season, but also of the reliability of the service. It is expected that the information will be updated on a weekly basis.

The performance of existing remote sensing satellites is generally limited by their low revisit frequency, or by their limited spectral resolution (SPOT, LANDSAT, IRS, IKONOS), or by their poor spatial resolution (AVHRR, MODIS sensors). Hence, a consortium of companies, including CNES, Spot Image, Scot, INRA and Astrium has been created with the objective of continuing research and development work and validating new products. Another objective is to launch dedicated new satellites within a few years.

Thus the consortium is today studying the possibility of mounting a "super-spectral" instrument on a microsatellite platform, and testing this spacecraft under nearly operational conditions. This microsatellite would then be integrated within a fully operational system of two mini-satellites to be launched less than two years after the micro-satellite.

Most applications in the fields of agriculture and the environment require :

- high revisit frequency (every 2 to 4 days)
- high thematic content
- rapid product delivery
- high spectral resolution.

On the other hand, a demonstrator mission has to cope with several constraints, including compatibility with a microsatellite, short development time (and thus the re-use of existing designs as far as possible), and a limited budget. A tentative set of specifications is as follows:

- 8 spectral bands (so-called "superspectral")
- high revisit frequency, thanks to pseudo-resonant orbits (see Fig. 3)
- 20 m spatial resolution
- 120 km swath width
- one (or only a few) off-nadir platform tilts per orbit, to acquire the selected image strips
- Imaging capacity per day of 140,000 km^2
- solar array generating 100 W
- satellite mass of 120 kg.

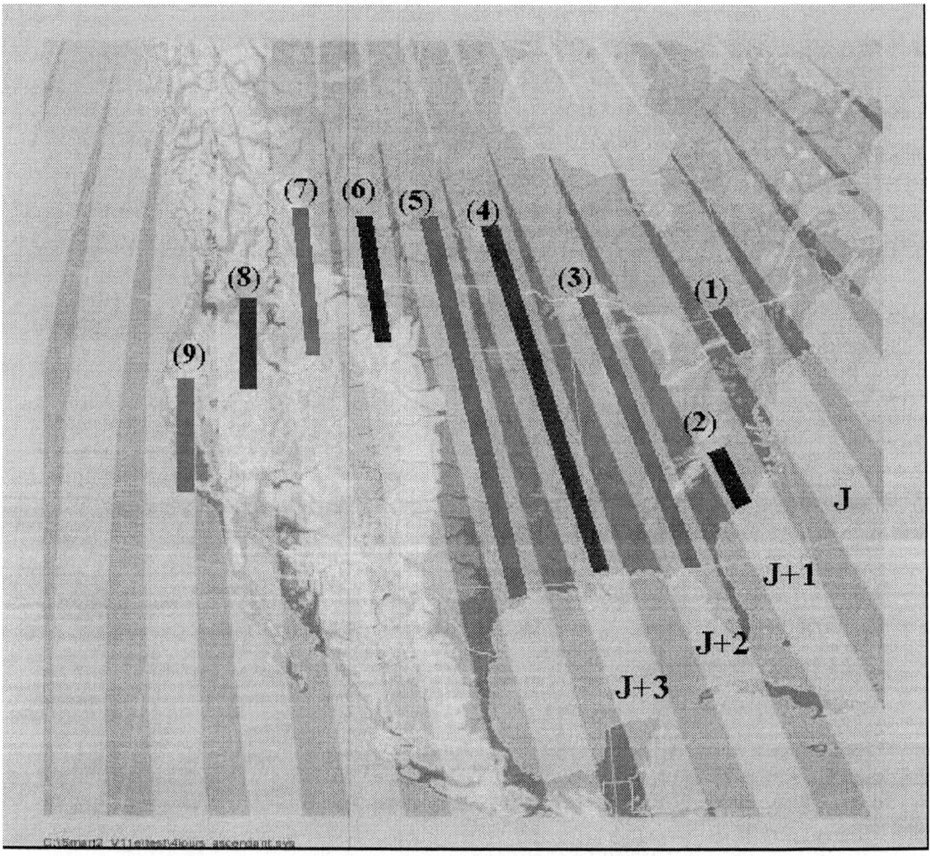

Figure 3. Example showing selected image strips (numbered) for consecutive satellite orbits (J, J+1, ...) over North America

The satellite itself (see Fig. 4 and Fig. 5) is based on the Myriade microsatellite product line currently developed by CNES together with Astrium. CNES is using this for several scientific missions, the first of which is Demeter, and Astrium is using a set of four such platforms in a constellation for the French Defence Ministry.

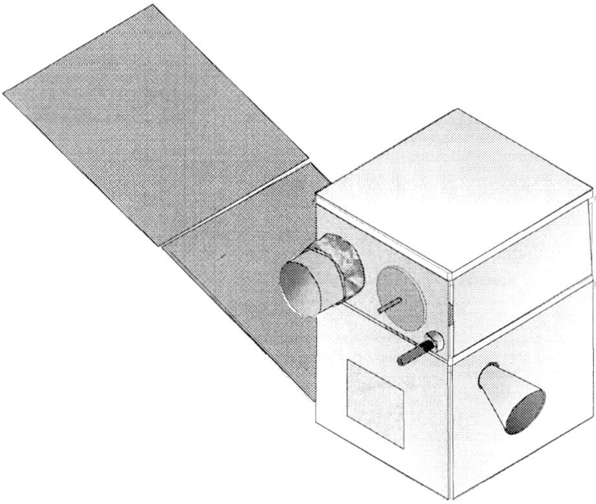

Figure 4. The microsatellite in its orbital configuration

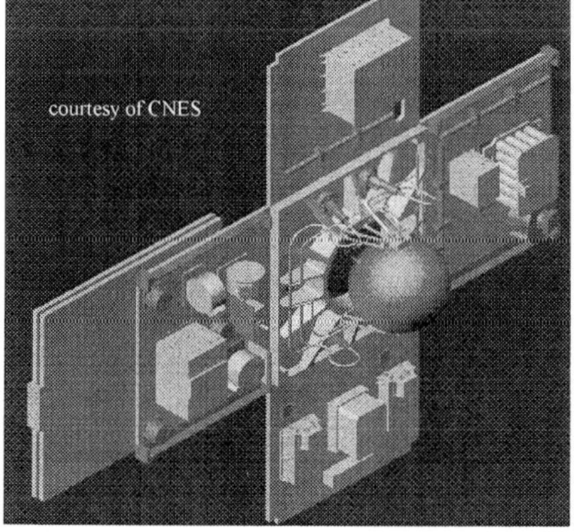

Figure 5. Exploded view of the platform

The super-spectral instrument (see Fig. 6) uses the heritage of the "High Resolution Stereoscopic" (HRS) instrument to be aboard SPOT 5 from next year. This imager, with a set of two 4-lines CCD detectors and strip colour filters, will provide a high quality, 8-band image sampled every 20 m over a 120 km swath, as prescribed.

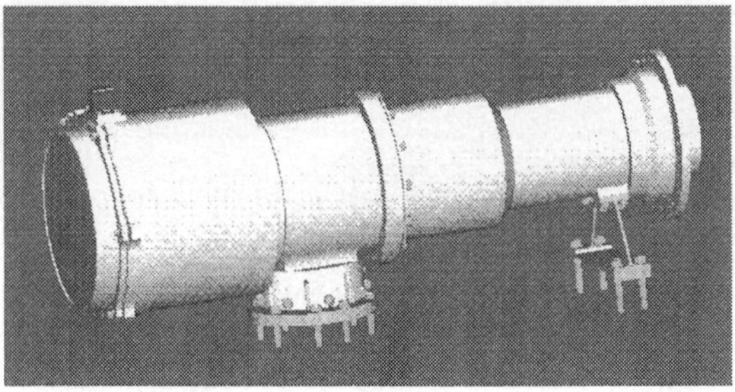

Figure 6. The Superspectral Instrument

4. Conclusion

The microsatellite demonstrator discussed here is a first step. It is essential to trigger the emerging agriculture market, by showing the full adaptation of the space infrastructure to the market needs. As the first satellite of a constellation of six microsatellites, it will already demonstrate the high revisit capabilities on targeted zones. The complete system will be deployed later. Last but not least, such a system should also prove to be well fitted to other environmental applications.

Some Problems Specific to Small Satellite Project Financing

J. Kreisel, GENES GmbH Venture Services, Koelner Strasse 27, 50226 Frechen, Germany

e-mail: joerg.kreisel@t-onlinede

Abstract

The Financing of space-related business ventures is different from that of most non-space businesses, particularly when involving the space segment (e.g., satellite business such as telecommunications, navigation or remote sensing). It therefore requires dedicated know-how and structuring. The aim of this paper is to discuss the basics of financing in general and the problems associated with the financing of small satellite projects in particular.

The paper gives an introduction to the business and finance-related characteristics of small satellite projects and the major finance drivers. Financial implications are discussed considering both financial feasibility and an optimum finance mix. The concluding remarks focus on those routes, which best overcome present hurdles for small satellite project finance by taking finance into account as a design parameter, and by considering appropriate countermeasures early in the new business creation process.

1. "Finance", not "Budget"

In the past, space projects were mostly government funded, as the whole space sector was dealing primarily with public funds, and they are still to a large extent today. This has led to a budget-driven culture, which is inappropriate for the financing of new commercial space ventures, be they satellite services, space infrastructure, or even spin-offs and technology transfer. Finance for most commercial space ventures is the key to success and - besides technology and market issues - remains one of the major challenges in space commercialization [References 1]. When talking finance, much more than figures and funds is involved, and this has only barely been addressed in the space community to date. However, there are standard rules in the financial community, which have to be known and to be applied for space businesses as well as in other domains. Particularly in times of booming financial markets, good concepts will in general find the funds which they need so long as they meet the requirements of the financiers. Finance, therefore, is a crucial consideration early in the process of creating a commercial satellite business, since to quite a large extent it determines the roadmap to business by transforming projects into new business ventures. The greatest obstacle in space finance is the fact that there are almost no success stories yet, and the space industry's learning curve with respect to the setting up of commercial ventures, involving private sector financing, is still underdeveloped.

M. Rycroft and N. Crosby (eds.), Smaller Satellites: Bigger Business?, 185–195.

1.1 Terminology and Definitions

One has to distinguish between the generic kinds of capital, namely equity and debt (see Fig. 1). These two sorts of money play different roles in the financing of a business; which is used, can have severe impacts on the commercial viability of a planned business.

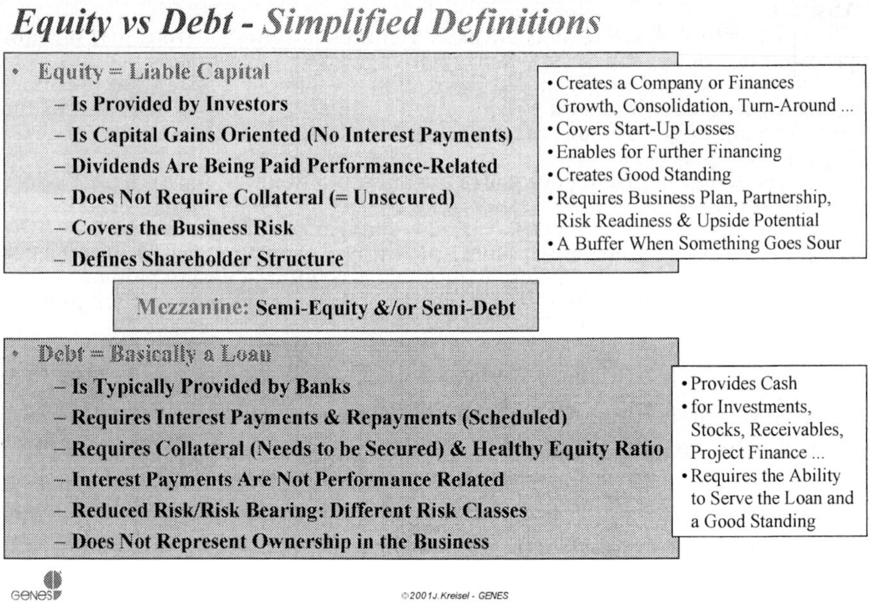

Figure 1. The basic features of equity and debt

The term "project finance" may be somewhat misleading, since in the financial world it typically describes big volume debt financing, say greater than approximately Euros 20 million. A better term would be "corporate finance". Equity and debt are usually provided by different types of financiers; they are of different relevance regarding the particular stage of development of a new business venture [Reference 3]. In terms of costs, one has to differentiate between capital investments (e.g., the space segment, product development, etc.) and operating expenses or costs (e.g., for personnel, administration, travel, marketing, etc.), because the nature of the spending directly interferes with the type of financial needs (equity vs debt). Cost estimation models used in space do not yet address this issue appropriately and, besides have not yet been proven to be appropriate in aerospace (e.g., for hypersonic transport).

The "financials" of a business are a set of three tables of figures, the profit and loss account, the balance sheet and the cash flow statement providing an overview of the development of the business within a planning period (several years). These documents can be supported by a liquidity plan and a financial plan (financing needs vs financial sources). The financing strategy is the professional structure, which defines the infusion of equity and debt to the business over time ("rounds of finance") based on certain achievements (milestones).

1.2 *Linking Finance to the Stages of Development: Optimum Finance Mix*

In a technology-based startup, the critical phase is the early stage of development. Regardless of whether such a business is space-related or not, the prime focus should be on equity financing to ensure, that early losses can be covered and that the company's balance sheet remains healthy (see Fig. 2). Following this philosophy of linking finance to the stages of development of a new business venture, debt financing will be applied the more that the business matures. By applying financial engineering one can reach an optimum finance mix, which optimizes the cost of capital and "dilution effects" (decrease in ownership of the founding partners by giving up shares to new investors). In many cases, professional financial structuring makes a business case viable.

1.3 *New Business Creation and Finance*

The new business creation process, in which an idea matures into a business plan (see Fig 3), is a time-consuming and high-effort approach requiring dedicated expertise and practical experience, as well as the early consideration of financial issues.

Reasons for developing a business plan are:

- to structure the business envisaged
- to obtain financing or partnerships
- to run, and monitor, the business.

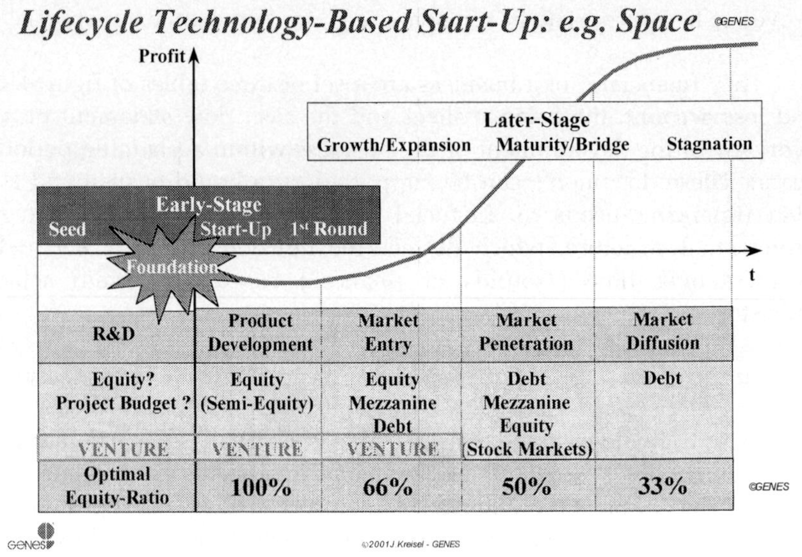

Figure 2. Finance according to the stages of development

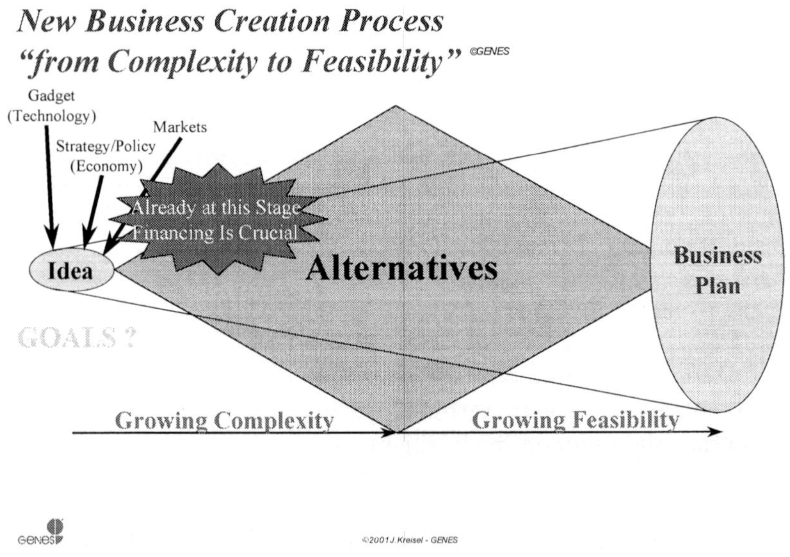

Figure 3. Transforming an idea into a business plan

2. Satellite Finance

From a business and financial perspective, are there really small satellite and big satellite projects? The latest evidence is, that small satellites are becoming more and more capable, while the formerly much more capable big satellites shrink in size and mass. Thus, whether a satellite project - besides constellation: orbit, size, launch, etc. – is classified as small or big may be in terms of the services envisaged, in the overall financing needs, or in the size of the revenues expected. In the case of commercial satellite services in telecommunications (the only real commercial space business so far), somewhat standardized financing schemes evolved during the 1990s, when numerous constellations went out fund raising. Such financing strategies (see Fig. 4) became possible due to "flagship" and "deep-pocket" sponsors and customers supporting those promising ventures (high growth markets and revenue potential). In addition, the specialized debt markets for satellite and launch deals (including high yield bonds), which evolved simultaneously, provided substantial financing. In the light of this, traditional bank financing for the commercial satellite business is applied only when the business matures.

These financing strategies are mostly applied to either big satellites in GEO or big LEO constellations of small satellites. However, the climate for satellite financing has changed significantly due to the Iridium failure of 1999 and the more recent downturn in the technology markets. These events have caused financiers to look even more carefully at anything related to space, and it is now much more difficult to raise money. The current financial sources for satellite ventures are few and far below market demand, since only a few financiers focus on satellite deals and have the necessary relevant expertise. When it comes to equity there are even less potential investors, particularly in the early-stage; here there are almost no funds available, except for strategic sponsors and a few venture capitalists (SpaceVest in the USA, GENES of Germany, SpaceInvest of Switzerland, etc.). Thus, it is only new satellite ventures based on a professional business plan, and selling a "sexy" story to investors, which have any chance to get finance and progress through their embryonic stage. The space sector should therefore foster the development of dedicated, early-stage financing and coaching activities for commercial space ventures to fill this gap.

2.1 *Major Satellite Finance Drivers*

The financing of satellite businesses (see Fig. 5) is characterized by massive up-front investments, extensive time-to-market, and by long-lead payback periods. In return enormous revenues (e.g., in telecommunications) can be generated, which for numerous satellite businesses represent a sound business case. Nevertheless, the risks involved (technology, market, regulatory, legal, business, financial, etc.) are high, and professional structuring and sound management is essential for successful financing.

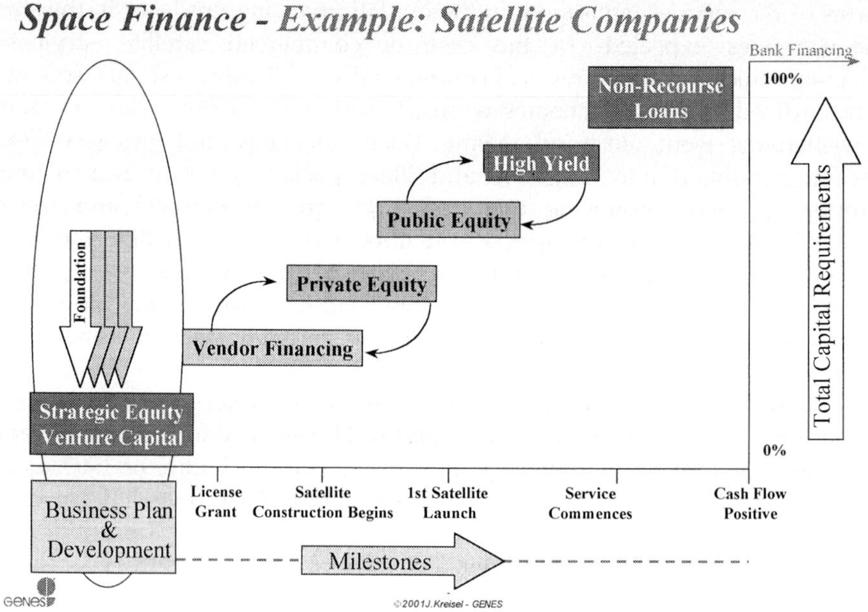

Figure 4. Example of the financing of a communications satellite system (financing corresponding to milestones reached)

Although the financing of a satellite business is basically the same as that for any other business, special considerations apply. The financing is primarily driven by the investments (satellite bus, payload, integration, launch, insurance), by the revenues and by the costs (operating expenses, depreciation and interest, etc.). This results in the financing needs (equity and debt) and the financial world as the source of finance, as Fig. 5 shows. Satellite ventures in general suffer from the burden of substantial depreciation due to the expensive space segment, which leads to a late break-even. From a financial standpoint this can only be counterbalanced by extraordinary revenue expectations and an appropriate profit potential once the business takes off. Another difficulty is to provide debt financiers with appropriate collateral, since most of the assets are the space segment and are therefore not fungible. Given a market, the revenue potential of most satellite ventures is driven by the capability of the space segment (number of satellites and launches, power, capacity, etc.). Therefore, higher revenues can be achieved, only with higher investments in the space segment, resulting in higher depreciation, higher costs and lower, and later, profitability. Likewise higher revenues require more efforts in marketing and sales (personnel), resulting in additional costs. Nevertheless, even higher spending may make sense, if the overall business case becomes more viable in that way. Cost-effectiveness is crucial, but the common discussions on cost reduction in space projects are missing the point, because it is a good business case that counts; the actual costs are only of secondary interest. Depending on the particular satellite business to be financed, the actual situation and perception in the financial community (e.g., investment climate for space, risk profile, professionalism of analysts and investors, success stories in the field, etc.) will determine when and how to approach which investors. The financiers themselves will base their decisions on the business plan, taking into account factors such as management, market, product/service (the "unique selling proposition" USP) and return potential. Finally, only convincing stories offering appropriate risk-rewards ratios will be financed. This means, that a particular satellite venture can be financially feasible, or not, depending on its business plan and on the financing strategy; this situation calls for the best structuring and professional support linking the space project with the finance.

3. Finance Issues Related to Small Satellite Business

For small satellite businesses the case is even more complex. In general, the boundary conditions in terms of finance are as described above. However, there are a number of difficulties particularly related to small satellite projects; fortunately there is business potential for the genre of small satellite business. Since the addressable market is directly proportional to the capability of the satellites (payload), which is driven by complexity and size (e.g., power), there is only a limited market, and hence limited revenue potential or growth in revenues, for small satellites. However, with the number of satellites, revenues can be increased, but that requires higher investments. Most small satellite applications use low Earth orbits (LEO) and are based on a satellite lifetime of from 5 to 7 years, whereas big satellites in geostationary orbit (GEO) function for up to15 years. Therefore, the relative depreciation of the space segment (LEO small satellites) is about twice as high as for big satellites in GEO, resulting in a less favourable depreciation to revenues ratio.

In terms of technology, small satellites are either use a COTS approach or introduce a range of new technologies; they can be simple, but very complex, systems. The technologies used and the complexity of the system also have an impact on the financial feasibility, which can only be verified on a case-by-case basis. To date, there is no evidence that small satellites are generally cheaper than big satellites, when taking into account the revenues to be generated by each satellite. Even in terms of size and mass, they range from small to rather big, and big satellites these days are becoming smaller and cheaper. Thus, there is probably no such thing as the "small satellite business", by definition. Small satellites are launched either piggy-back (low cost) or on a dedicated launcher (higher cost). The relative costs or investments associated with ground stations for small satellite businesses are definitely higher than those for big satellites.

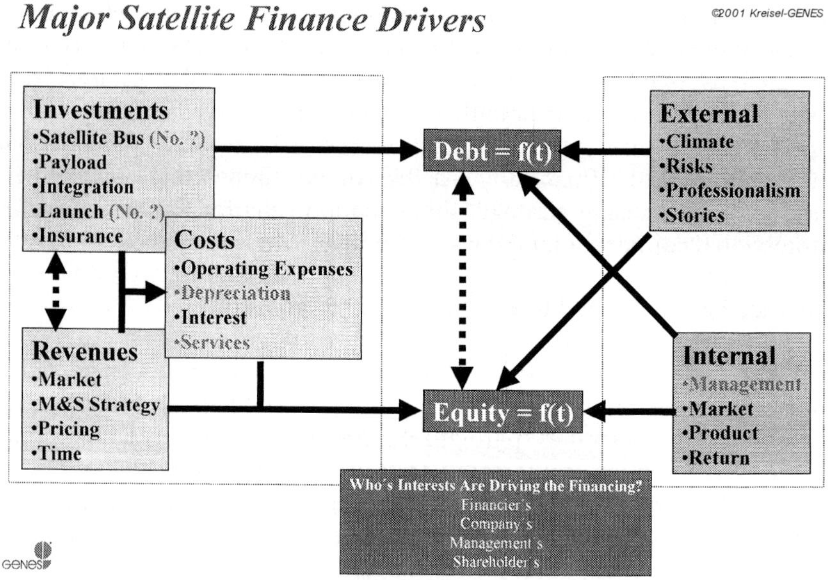

Figure 5. Interactions between various factors driving the satellite financing

Fund raising efforts for small satellites are basically the same as for big satellites, and should not be underestimated. Financial services can easily cost in the range of 3% of the financing. They are partly due as up-front retainers and it usually takes 6-12 month to close a single round of finance. Financial services for small satellite projects are therefore relatively more expensive in terms of time and money.

Another problem is that the volume of financing sought by small satellite businesses is mostly under the critical mass for most satellite financiers - the current players prefer to get mandates above Euros 200 million. On the other hand, small satellite businesses are well positioned for niche markets such as, e.g., the Germany-based RapidEye AG (http://www.rapideye.de). This targets the agricultural insurance sector with a multi-spectral, high-resolution, high-revisit system of four satellites (315 kg each) in a 600 km altitude and 12 hr Sun-synchronous orbit (SSO).

4. Financing Strategies

When looking at potential financing strategies one should select potential "venture backers" (equity investors) according to their value adding capabilities in that stage of development of the business when financing is needed [Reference 2]. Depending on the round of finance, as illustrated in Table 1, there may be different partners to provide an optimum finance mix for the venture. Adequate structuring of equity infusions will also facilitate the debt financing needed and provided by banks. Theoretically, there are various potential investors available, but each individual small satellite business requires a specially tailored approach to best fit the financial partners. Since many of the presently proposed small satellite projects lack the major ingredients for a "sexy" business, offering a higher percentage of shares (=> "dilution") in return for the associated risks could help to bring in those elements (partners), which are required to get the business off the ground. Public private partnerships are a serious alternative for small satellite businesses, at least during their early stages when the risks involved are too big to be borne by the company alone, and when the project is in the public interest. However, some small satellite projects are not commercially viable at all. It is then better that these remain as pure projects within an existing company, for strategic reasons, or are run for the government.

5. Conclusion and Recommendations

There is no evidence, that small satellites are either cheaper or smaller than big satellites given the current convergence of technical capabilities, size and cost of those two segments. However, big satellites typically generate more revenues per Euro spent for the space segment; they live longer, which is an advantage in terms of financing (depreciation, profitability, etc.). Nevertheless, at least theoretically, there are promising niche markets and good business opportunities for small satellites. Currently, a prime finance problem associated with small satellite projects is, that there are only few proper business cases around, and we find almost no success stories yet. This may change.

The development of professional and commercially viable business plans and management issues, in parallel with optimum financing strategies based on entrepreneurial approaches, are the real challenges. This may be possible when addressing high-profit niche applications (e.g., B2B) and particularly by increasing revenues per satellite (advanced technology or business case). To lift the financial feasibility of small satellite projects, a sound understanding of the structures and procedures in the financial community is mandatory. Use must also be made of contractual know-how being directly linked to finance and the commercial success of such ventures.

The actual climate for financing small satellite projects is now rather unfriendly, since specialized finance (particularly early-stage equity) is almost unavailable. However, depending on the individual business case, financing will be possible, when sound "homework" has been conducted, the right financing strategy chosen and an innovative means of finance (such as, e.g., venture lease, or vendor/customer financing, etc.) employed. Therefore, finance should be considered as some sort of a design parameter early on during the process of creating a commercial small satellite venture, combined with the awareness that this will require different resources.

For "small satellite projects of the hybrid type" (with commercial potential and public interest), innovative PPP (public-private-partnership) schemes will be an option. The training of key personnel is mandatory here to provide necessary basics of business and finance and to build business acumen. Thus, the famous paradigm of "faster, better, cheaper" should be changed to "faster, better, value-balanced"! Once businesses, such as, e.g., RapidEye or Sirius get off the ground, many more small satellite businesses will evolve and financing will be facilitated. There probably is a future in the small satellite business.

Contributions of Venture-Backers

Small Sats	VC	Mgmt	Strat. Inv.	PE	Space Ind.	Space Ag.
Capital (Equity)	++	+	++	+++	+	+
Vision	0	+++	0	-	-	-
Business Culture	+	++	0	0	-	-
Business Skills	++	+++	0	+	0	-
Experiences (Early-St.)	+++	?	?	-	-	-
Connectivity						
•Finance	++	?	0	+++	-	-
•Industry	++	?	+++	?	?	?
•Markets	++	?	+++	?	?	?
Distribution Channels	-	?	+++	?	?	-
Technology/Expertise	+	+	?	0	+++	+++
Production Capabilities	-	-	++	-	++	0
Operator Capabilities	-	-	++	-	+	+

GENES

Table 1. Plus signs indicate the value adding capability of the various potential backers of a satellite venture (PE=private Equity; VC=venture capital)

References

1. Kreisel, J.: *Finance and its Impact on New Commercial Space Ventures,* STAIF2001, Session D02 Financing and Funding the Enterprise, Albuquerque, N.M., USA 2001
2. Kreisel, J.: *Optional Approaches for Venture-Backing of New Space Ventures,* IAA-00-IAA-1.4.06, 51st IAF Congress, Rio de Janeiro, Brazil, 2000
3. Kreisel, J.: *Space Finance Issues Driving Future Space Industries,* Space Industry Symposium, Tokyo, Japan, 2000

The CNES Micro-satellite MYRIADE Program as a Laboratory for Innovative Methods of Space Project Management

C. Belleval, Université Louis-Pasteur, BETA, 61 avenue de la Forêt Noire, F-67085 Strasbourg, France

e-mail: cbe@bigfoot.com

Abstract

In this paper, we discuss the management of large hi-tech projects (LHTP) and, more particularly, space programs. The purpose of a LHTP is to give birth to a complex product or system (CoPS) with unique properties, through an uncertain technological path. Such characteristics preclude mass production methods. Instead, we argue that the management of a LHTP requires a better understanding through modeling of what complexity and uncertainty of an open system are, as opposed to the analysis of complication and risk related to a closed system. We consider that cost reduction and optimization are not valid processes for LHTP. Instead, our main proposal is that the management of a LHTP and more specifically a space project requires a strategy focused on the acquisition and development of core competence. We illustrate how such a resource-based strategy can influence space projects with the case study of the CNES MYRIADE Program.

1. An Overview of the MYRIADE Program

The Micro-satellite MYRIADE Program was launched by CNES to fulfill three goals: to provide the scientific community with a new type of access to space in LEO, to provide a test-bed for innovative technologies, to enable CNES to bring a new contribution to the French and European space policy. Another main objective of the MYRIADE Program is to foster a renewed innovation process in a space agency where the "Mission Success First" spirit was so strong that it led to an excessive technological caution.

We consider MYRIADE to be an ideal laboratory to experiment new methods in space project management. Not only has the product line all the characteristics of complexity and technological intensity of its "big brothers", but each mission project life cycle is short (a goal of two years between design and launch), thus allowing an increased pace of learning by experience. The idea of validating experimental processes aimed at producing an actual product (as opposed to a prototype) is a proven way of selling know-how to the whole organization. Because MYRIADE has the potential of impacting the performance of the whole space agency, it can be considered as a strategic process.

M. Rycroft and N. Crosby (eds.), Smaller Satellites: Bigger Business?, 197–204.

2. The Nature of Space Projects

Space projects are particular members of a family called large hi-tech projects (LHTPs). The purpose of a LHTP is to give birth to a complex product or system (CoPS) starting from a concept. CoPS are "high cost, technology-intensive, customised, capital goods, systems, networks, control units, software packages, construct and services" [Reference 1]. CoPS can be distinguished from mass products by:

- their unique configuration tailored for a specific customer or a specific market
- their emerging properties during the processes of conception and production
- the direct involvement of the client, since the supplier does not sell a product, but its ability to meet the customer needs.

Two essential properties are associated with a LHTP: the complexity of the system architecture and the uncertainty of the process emergence.

3. Complexity Modeling and Uncertainty Management vs Complication and Risk Analysis

The notions of complication and complexity are completely different. A complicated system can be analyzed, segmented and simplified to provide an understanding whereas a complex system has to be modeled. Therefore the right question in relation to a complex system is not "What is it made of?" but "What is its purpose?". The development of a complex system cannot be managed through an analysis, for the emergence of the technological path, the combination of components and subsystems, and the arbitration during the project life-cycle are unpredictable: a complex system is an open system. Modeling is a means to understand the complex system and to anticipate the consequence of possible actions. On the contrary, risk is a notion linked to a closed system, with a limited set of events, each one of them linked to a probability (and possibly a cost). Risk management means risk analysis. Risk is therefore not a variable adapted to the management of a LHTP, such as a space project, though it is possible to conduct local risk analyses. For a complex system, risk is typically associated with certain technologically mature components whereas uncertainty lies in the combination of components, subsystems and the architecture.

4. Strategies for Dealing with Complexity and Uncertainty

The failure rate of a LHTP typically increases with the size and complexity of the products. We noticed similar patterns in space projects. While studying failures of certain NASA "Faster, Better, Cheaper" scientific missions through reports [References 2, 3, 4, 5] we noted that except for one document [Reference 6], the documents only discuss direct management problems which in our view are a consequence of a misunderstanding, namely what a complex emerging system is and how to manage it.

We have identified three main issues to be addressed. First, we challenge the strategy of simplifying a complex system. As an example, simplifying the Mars Polar Lander meant reducing telemetry, a choice resulting in an absence of feedback during the supposedly failed atmospheric entry. Moreover, designing smaller missions is not a valid strategy for certain goals such as outer planetary exploration, landing or sample-return. Second, there is a strong belief that the process of building a CoPs can be optimized. Such a view inherited from the classical schools of economics and management has the potential to generate destructive project management mistakes: how can someone optimize the production of an object only existing as a concept when the project is about to start? The project manager's latitude of freedom decreases as the knowledge of the project increases; i.e.: when crucial decisions are to be taken (in the conception phase), there is nothing to optimize. Third, the idea of subcontracting the conception and production of a CoPS as a means of getting better quality, increased efficiency and decreased costs is in our view nothing else than a belief. One of the reasons is the high risk of opportunism from the supplier since CoPS require specific investments and a limited number of transactions between the client and its suppliers. Such dependency requires some mechanisms of coordination and control to prevent opportunism: they may generate a significant level of transaction costs. Moreover there is a well known captive relationship between space agencies and the limited number of main contractors: there is no real competition between suppliers. Indeed any attempt to use the market as an optimizer is meant to fail. We believe the decision for a space agency to manage internally or to externalize the main contracting should be decided on different grounds. Since CoPS emerge from a coordinated process of conception, what is important is the management of innovation.

5. Good Strategy and Good Practices in Space Project Management

Because system simplification, resource optimization and the systematic externalization of main contracting are not valid answers to the management of space projects, we have attempted to propose new ways of dealing with the two characteristics of CoPS: the complexity of the system architecture, and the system emergence through an uncertain technological path.

In an economy driven by innovation as well as in a project where the innovation process is crucial, the corporate strategy must primarily focus on knowledge management, not on cost reduction. In fact an innovative strategy should focus on the process of organizational learning: how individuals coordinate their knowledge to fulfill a common goal. Individuals acquire competence through the learning process using factual knowledge, know-how (tacit knowledge acquired in action) and relational (cooperative) knowledge. The process (a project being a particular case) is the realization of the competence as a potential. The investment in knowledge of an individual or an organization allows uncertainty management, since it delays irreversible choices later in the project cycle when the system knowledge is improved. Because the acquisition of competence is a long and cumulative process, individuals thus organizations must identify which competence will bring a competitive advantage, then give them a strategic priority. PRAHALAD and HAMEL [Reference 7] call the result of this choice "core competence". More generally, the resource-based theory, as opposed to the classical management school of optimization, identifies the source of competitive advantage in the possession of unique assets (difficult to imitate and hardly substituted) that are managed through knowledge, coordination (between individuals and the organization) and cooperation (arbitration between the individuals' interests and their participation to common organizational goals). Hence our main proposal:

The management of LHTP and more specifically space projects requires a strategy focused on the acquisition and development of core competence.

For the space agency, the choice of project organization is a matter of deciding who should master the technological trajectory. An external main contractor means that the agency loses control of the system architecture. However, since the agency retains the prime management, it must maintain some architecture-related competence: using an external contractor is certainly a source of duplication, hence increased costs. Moreover, surrendering the knowledge embedded in the architecture means losing control of the innovation process. In the case externalizing would become a systematic policy, we identify two major risks: one is the increase of space project costs due to competence asymmetry (leading to bad coordination and increased supplier opportunism); second, innovation would tend to dry up because of gradually increasing budgetary pressures on space agencies; aerospace companies are unable to cover R&D expenditures without state support. Therefore, our view is that the organization of prime management, main contracting and general subcontracting should be decided as a result of a knowledge sharing strategy that distinguishes different levels: architecture subsystems and components; the contractual arrangements as well as the nature of partnerships should be arranged according to the distance from the core competence involved: full integration, strategic alliances, supplier networks or subcontracting.

6. Management of the MYRIADE Program

Since CNES uses MYRIADE as a strategic process aimed at fostering innovation, it was logical that the agency decided to maintain the product line main contracting inside its organization. However, it established a strategic partnership with ASTRIUM and ASPI for the co-development of certain product line equipment. Moreover, CNES selected an aeronautics (not a space) subcontractor joint venture to manage the product line integration (LATECOERE - SOTEREM). Because the selected companies had no competence in satellite technology but some in other hardware integration, the contract included a technology transfer protocol between CNES and its supplier. It was understood such a process would include concurrent engineering.

The MYRIADE management has set up a concurrent engineering workshop named "Atelier μCE" targeting the conception phases of each mission. One of the current issues is to extend the co-exploration and concurrent engineering to the study, pre-selection and designs of payloads.

Due to a large success, the scope of the Program has broadened to a number of heterogeneous missions differing by the degree of innovation as well as by the nature of the initiative (science, technology demonstrator, international cooperation, military). One of the critical questions in the near future is to adapt the Program organization to the variety of project configuration. We proposed to solve the problem using the spectrum of committed core competence. Since the CNES strategy is to focus on highly innovative processes and to leave recurrent operations to industry, we studied four different scenarios according to the amount of innovation: the more a mission requires innovative inputs, the more CNES is involved in its management (see Table 1). A mission designed for a main contractor client would see the space agency playing a relatively secondary role. In addition we studied and discussed the opportunity to create a subsidiary housing the product line in partnership with aerospace companies. This issue has both pros and cons. On the one hand, externalizing favors adaptability, performance and knowledge exchange between partners; furthermore, it brings a unique structure whatever the mission type (agency or external customer). On the other hand, CNES risks externalizing the innovative learning, thus diminishing the impact of the Program on agency methods. So far the CNES management has decided to keep the product line management inside the Centre Spatial de Toulouse, though it has experienced some difficulties reconciling the product line versatility requirements and some delays in human resource allocation and management.

7. Conclusion – Future Developments

In this paper, we have summarized some (but not all) of the main conclusions that were derived from our immersion in the MYRIADE Program which formed the basis of our thesis [Reference 8]. The confrontation between our professional experience in large industrial projects, the academic background of a PhD in management, and the context of the case study led us to focused as well as to general interpretations of LHTPs, and more specifically, space programs. The most important conclusion is an urgent need to reconsider the characteristics of such projects: emergence, complexity and uncertainty.

In our view the crucial point is not budget constraints but knowledge management through a set of core competence and processes. Using the now familiar vocabulary, our results can be put like this: while using "cheaper" as a way to manage space projects will probably lead to failures, giving priority to "better" respects the logic of a technology intensive emerging system, increases process quality by reducing redesign loops, therefore leading to a faster project life-cycle and lower expenditures.

Acknowledgements

This paper is dedicated to the memory of Charles BOUZAT, who left us on 1 January 2002, too early to see the accomplishment of the Myriade Program he was heading. May he get all my gratitude for the special support he gave to my work.The author also wishes to thank CNES, particularly J. BLAMONT, S. JANICHEWSKI, A. OUSTRY and the whole Program team, for having given him this outstanding opportunity to be involved in the MYRIADE adventure.

Organization 1 : Product line main contracting by CNES; training of the product line integration subcontractor by CNES (this company being the future functional chain assembler).
Organization 2 : Satellite main contracting by CNES, functional chain integrator company operational.
Organization 3 : Payload support and satellite conception (phases A-B) under CNES authority; in phases C-D, new equipment supply and assembler managed by a space main contractor company.
Organization 4 : Main contracting by a private company.

Mission		Organization 1	Organization 2	Organization 3	Organization 4
Product line development (DEMETER satellite)					
Science					
Technology	Equipment main contracting				
	Functional chains main contracting				
	Payload main contracting				
	Technological mission : satellite main contracting				
	Experimental mission (e.g. Interferometric Wheel)				
Cooperation	Joint main contracting (e.g. FBM)				
	Prime management / joint mission (e.g. Canada)				
Training universities and schools					
Others (mission prime management outside CNES)					

Table 1. MYRIADE project organization according to mission types

References

1. Hobday M., Rush H. and Tidd J.: Innovation in Complex Products and Systems, *Research Policy, vol. 29,* July 2000
2. NASA, *LEWIS Spacecraft Mission Failure Investigation Board,* 1998
3. NASA, *Mars Program Independent Assessment Team,* 14 March 2000
4. NASA, *NASA Faster Better Cheaper Task Force,* March 2000
5. NASA, *Report on Project Management in NASA – by the Mars Climate Orbiter Investigation Board,* March 13, 2000
6. NASA, *Faster Better Cheaper : Policy, Strategic Planning, and Human Resource Alignment,* Office of Inspector General, IG—1-009, 13 March, 2001
7. Prahalad C.K. and Hamel G., The Core Competence of the Corporation, *Harvard Business Review,* May-June 1990
8. Belleval C. : *Le Pilotage des Grands Projets de Haute Technologie dans une Organisation Centrée sur ses Compétences de Base – Le Cas des Programmes Spatiaux – Application au Programme MICROSATELLITES "MYRIADE" du CNES,* Ph.D. Thesis, Université Louis-Pasteur, Strasbourg, France, 2001

Report on Panel Discussion 3:

Market Potential - Big Business versus the Entrepreneur

A. Alale, L. Summerer, International Space University, Strasbourg Central Campus, Parc d'Innovation, Boulevard Gonthier d'Andernach, 67400 Illkirch-Graffenstaden, France

e-mail: alale@mss.isunet.edu, summerer@mss.isunet.edu

Panel Chair: J. van der Ha, Cost-Effective Space Mission Design and Operations, USA

Panel Members:

D. Beaden, The Aerospace Corporation, USA
C. Belleval, Université Louis Pasteur, France
W. Brokx, National Aerospace Laboratory, The Netherlands
P. Campenon, Astrium, France
B. Doll, Astrium, Germany
J. Kreisel, GENES Venture Services, Germany

The discussion focused mainly on the interrelation of Public Private Partnerships and small satellites, possibly "killer applications" that could increase the market for small satellites, and the ways in which big companies could accommodate the concepts of small satellites. Furthermore, the roles of small satellites in the fields of Earth observations and telecommunications were debated.

The panel agreed that today's market is missing success stories. Up to now, there is no "big business" with small satellites. **J. Kreisel** noted that a successful business has to be profitable, not necessarily big. The lack of such success stories is the reason why the financial community is still not attracted by this market. The climate is expected to change as soon as the embryonic phase is past. Then big companies, and governments, will need to learn how to work effectively with small, innovative companies.

It was not unanimous that there is any such thing as a "killer application" for small satellites. Nevertheless, the panel agreed that, if there were such an application, it would not become obvious until the very last minute before entering the market; it may appear in a very unexpected way. It was also

M. Rycroft and N. Crosby (eds.), Smaller Satellites: Bigger Business?, 205–206.

pointed out that no big business started as a big business; normally, it began with a small, niche product.

In general, there has to be more focus on the end users' needs and the real market demands. A shift from the technology focus to the business mindset is necessary. The economy is not driven by engineering optimization, but by the constant innovation of services – small satellites have to find their role. Therefore, "new people" are needed in the space business; teams of capable people who are good communicators must be created. Such small, flexible and entrepreneurial teams could then be combined to generate big business activities in space. **J. Kreisel** believed that it was good that space activities were now near the information technology business.

Session 4

Launch and Support Services for Micro/Nanosatellites

Session Chair:

P. Balaam, Arianespace, France

New Propulsive Module for Nanosatellites

R. Fleeter, AeroAstro Inc., 520 Huntmar Park Drive, Herndon, Virginia, 22066, USA

e-mail: rick@aeroastro.com

Abstract:
Microspace transportation suffers from an asymmetry: 90% of potential low cost secondary accommodations go to an orbit only 10% of microsatellites want - GTO. SPORT, the Small Payload Orbit Transfer module is a joint program of AeroAstro Inc. and Astronautics Technologies SB, Malaysia, to make low cost, frequent, plentiful secondary space on large launch vehicles useful to small satellite missions. SPORT uses a combination of aerobraking via an inflatable ballute plus conventional propulsion to carry up to 70 kg satellites aboard Ariane 5 ASAP to equatorial LEO, or somewhat less mass other orbit inclinations and altitudes. As spacecraft become smaller, the practicality and cost effectiveness of dedicated inevitably declines, making utilization of secondary accommodations increasingly important to the small satellite developer and user communities. The evolution of the SPORT concept, its first application scheduled for launch in 2003, its performance and its potential to substantially increase the frequency and quality of microspace transportation are described.

1. See you in Stockholm...

It turns out aerospace engineering can resolve one of the fundamental questions of Physics in the 21st century. The universe is completely symmetric. Ninety percent of microsatellites want to go to LEO (Low Earth Orbit), and 90% of large launch vehicle capacity goes to GTO (Geosynchronous Transfer Orbit). Coincidence?

2. Surviving Droughts and Famines

I'll leave that question to the physicists, but meanwhile, what about all that wasted launch resource? Waiting in line for gasoline back in 1974, shivering in my dormitory room the University was keeping at 15C over a New England winter to save heating oil, and swimming in an icy pool with tepid showers, I remember hearing that there was in fact no fuel shortage, just a distribution problem - the oil coming out of the ground wasn't making it to the gas stations and oil depots in Rhode Island. Same was true when gas prices doubled last winter in Chicago. And more seriously, devastating drought and famine is almost always a local phenomenon - there's plenty of food and water, but political, social and infrastructural barriers prevent its delivery.

As smug high tech engineers working in the developed world, we might feel some pity for the poor victims of these tragic, or at least inconvenient shortages. But we see droughts and famine as relics of low tech, poorly

M. Rycroft and N. Crosby (eds.), Smaller Satellites: Bigger Business?, 209–217.

capitalized societies, and gas price shocks as the machinations of greedy companies and inept bureaucrats. Meanwhile we sit in conferences frustrated that a new world of microspace would be ours if only we could get these spacecraft into orbit as cheaply and plentifully as we can build them. NASA wasted ten years and millions of dollars ultimately attempting to prove the negative assertion that as launch vehicle size scales downward to match the smallest satellites, the cost per kg can't also scale down. This inadvertent bit of research went under the program name Bantam Lifter.

Embarrassingly, it takes no money nor even a deep understanding of rockets to draw the graph below (Fig. 1). As the launch vehicle's payload tends toward zero, it's cost doesn't, thus establishing an asymptote. Bigger vehicles are cheaper, per kg, though the smaller ones are of course cheaper per flight. If you put a person on board, you are on a different, higher cost, curve.

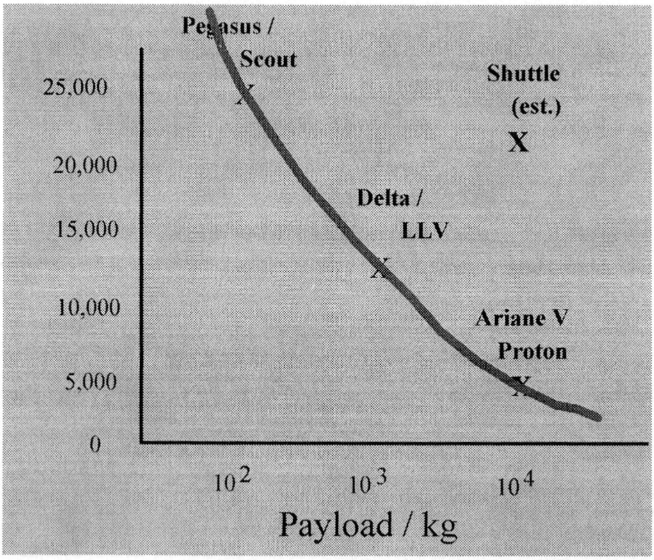

Figure 1. As spacecraft become smaller, dedicated launches become less desirable

Thus the cheapest route to space is on the biggest rockets, and the smaller the spacecraft, the more difficult it is to afford a dedicated launch. Think about air travel. The most efficient jets are the big Airbus 340 and Boeing 777. You can certainly put less people into a smaller jet, like a 737, and because you can more easily and frequently fill it, and since for short flights customers are less sensitive to per-mile cost, offer substantial value for short, less popular routes plus shuttle services where flight frequency is strongly desired. But scaling down further, to flying just one or two people aboard a dedicated jet, the cost can be 10x or more that of the biggest jetliners. When "small" satellites had a mass of 250 kg, Pegasus, rocketry's version of a 737, was attractive. Per kg cost was high, but per flight cost attractive. Now that a small satellite has a mass of 25 or even 2.5 kg, we are entering a new realm – the business jet cost jump for dedicated launches. We could launch 25 kg for maybe US $5 M or US $ 10M, but the potential savings of piggybacking on a US $5000/kg large launch vehicle, for a total launch cost of US $ 250k, is going to attract a lot more passengers.

3. Whereupon The Cavalry Charges Over The Hill

Arianespace alone could potentially launch 60 microsatellites per year via ASAPs on flights to GTO. At their nominal per kg cost, those rides could be as cheap as US $ 250k each – lower than the price of even the cheapest commercially produced micro and nanosatellites. A price so low that transportation ceases to design your mission for you. Unfortunately, transportation would still define your mission because it would have to be carried out at GTO.

At AeroAstro, after trying every possible alternative and failing unequivocally (isn't that the engineering method?) a light bulb went on somewhere. Find a way to exploit this vast resource of untapped cheap transportation.

One reason brilliance didn't flash earlier (let's ignore the implications of that regarding your author's intellect for now) is that size and cost matter very much in this application. Ariane 4 ASAP carried 50 kg, and if the propulsive module necessary to perform the orbit transfer itself weighed 50 kg, the exercise was moot. Similarly if the Ariane slot costs millions of dollars, and an orbit transfer module costs US $ 5M to US $ 10M, then you might as well buy a Pegasus when they were priced in that neighborhood. However Ariane 5 carries over 100 kg per ASAP slot, and the continuous reduction in spacecraft size has enabled building the orbit transfer module within a mass budget of 10 or 20 kg, depending on its payload mass, meaning we can accommodate spacecraft of 50,

60, even 70 kg and carry them to LEO, or some other popular microsatellite destinations. AeroAstro's Bitsy spacecraft program, putting all the functions of a basic spacecraft on a single circuit board and inside a single, simple box, has cut the cost of the propulsive module to well under the cost of the piggyback slot itself.

In its first flight, scheduled for late 2003, we plan to demonstrate the advantages of propulsion assisted piggybacking: low cost; lots of launches to choose from; very high launch vehicle reliability; very benign launch environment; and arrival at the exact orbit desired for the mission.

4. Revenge of the Thermodynamicist

We hate entropy creation. And paying a huge 2nd law cost to drive to GTO, followed only by paying an almost equally huge fare to decelerate down to LEO - my thesis advisor's ghost would have haunted me. At GTO you have about 2500 m/s excess velocity, and the Earth's atmosphere is standing by to dissipate those $E=1/2\ mv^2$ Joules. But most drag-based solutions, perhaps influenced by Mars missions where if you don't slow down in the first pass through the atmosphere, you zoom past the planet, and by missile reentry, use very high drag to reduce Apogee in a single pass. This puts enormous thermal stress on the reentry vehicle, and is hard to stabilize and to execute precisely. Over the years numerous inventors including ourselves (we should more precisely be labeled reinventors) have suggested a ballute - a large drag device deployed behind the spacecraft, to bleed excess kinetic energy.

The ballute has several advantages. It reduces the Δv required to initiate orbit decay since you don't need to dip low into the atmosphere. It reduces heat load on the passenger spacecraft since most of the heat is dissipated on the ballute. It is intrinsically stabile, because the ballute dominates the pressure force, while the relatively massive spacecraft stays out in front - cp behind cg. You can take the time to make very fine adjustments of perigee - ease down into the drag region and experimentally determine the best altitude, rather than rely on sometimes imprecise models of atmospheric density.

Of course the advantage of any drag device is it saves you almost 2500 m/s of Δv - but it costs something to get that savings. Propulsion is still required to lower perigee to the best drag altitude, and then to recircularize when the target apogee is achieved. But instead of 2500 m/s ΔV, we now require a few hundred m/s, a savings of about 80%.

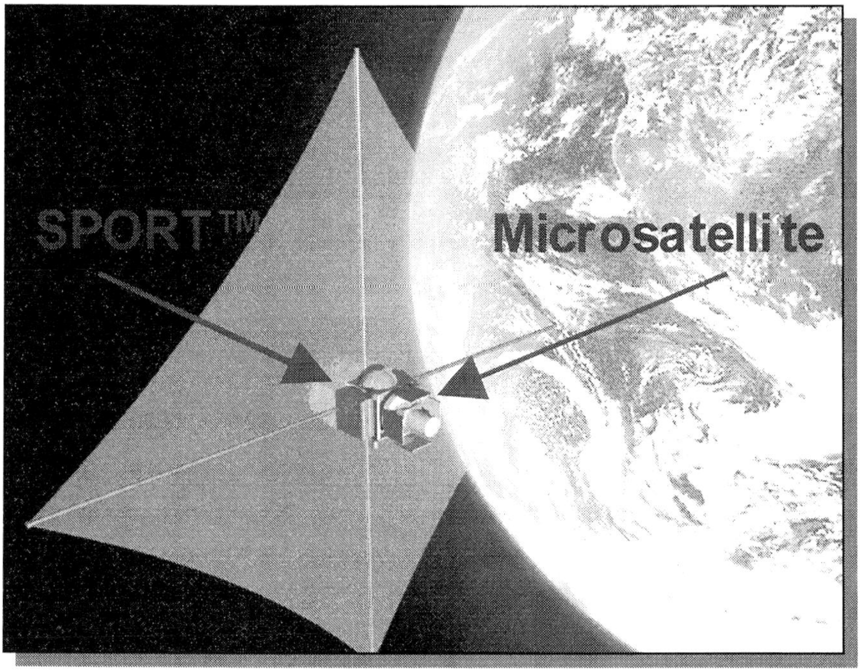

Figure 2. SPORT uses a large drag device deployed aft to dissipate 2500 m/s Δv

5. Putting the Pieces Together

SPORT is in development by AeroAstro (Herndon, VA) and Astronautic Technologies (Kuala Lumpur) in an innovative partnership. Like many countries near the equator, Malaysia's most desired orbit is LEqO - Low Equatorial Orbit, from which their geographically advantaged location provides revisit every 100 minutes. Until SPORT there was no access to that orbit suitable for low cost microspacecraft. Not coincidentally, Malaysia is ideally located to provide communications and control services with SPORT, and perhaps more importantly, to promote SPORT among neighboring countries in the equatorial belt. In our collaboration, ATSB and AeroAstro share in the development of the SPORT, and in the first spacecraft payload, in marketing of SPORT, and in its operations.

SPORT is also supported through launch agreements with Ariane, to ensure our clients an ample, reliable, high quality host to GTO. Fig. 3 shows payloads integrated with Ariane, including a major spacecraft and secondaries with and without SPORT.

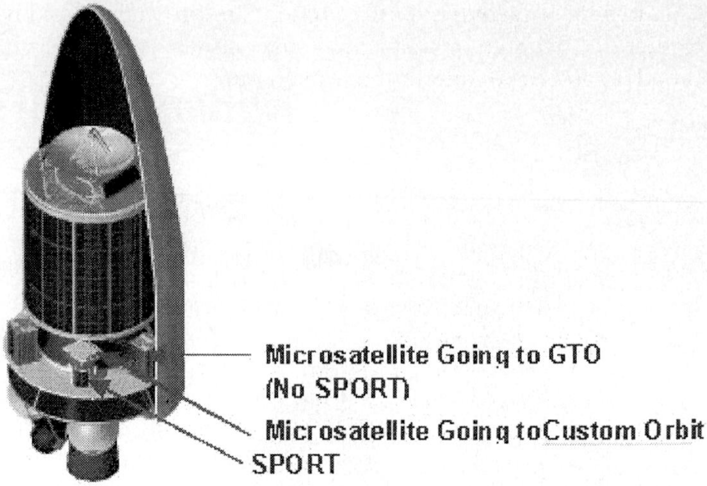

Figure 3. Secondary payloads with and without SPORT

Though most of us live distinctly north of the LEqO accessible region, there are many reasons to fly in it. The MIT HETE satellite, originally developed by AeroAstro, is in that orbit to minimize X-ray background noise. Higher equatorial orbits combine frequent revisit with coverage of a much wider band, potentially extending as far as Northern Europe and South Australia.

Because propulsive ΔV aboard SPORT is so small, we have not needed a costly R&D program to improve on an already mature technology - SPORT uses a welded, off the shelf, highly reliable hydrazine system. The drag device is a simple inflatable - low performance, low complexity. As shown in Fig. 2 (above), SPORT is a wafer sandwiched between its ballute - pictured here as a deployed sail with normal along the orbit velocity vector, but will likely be an inflated balloon - less elegant looking, but also cheaper and more reliable to deploy. SPORT will have very little chance of seeing the sun, but doesn't need to - it is powered by primary Lithium Ion batteries, and has power consumption so low that even a multi-month gradual decent and perigee reboost is readily accomplished without the cost, complexity and attitude constraints inherent in solar arrays.

SPORT is built around the existing AeroAstro Bitsy nanosatellite spacecraft kernel (Fig. 4). Originally developed for the US Air Force and NASA and scheduled for first flight in October, Bitsy contains all the functions of a spacecraft on a single circuit board, in a package the size of a cigar box. Peripheral devices like deployment of the ballute, or on other missions a science payload, momentum storage or additional solar panels, are all accommodated by Bitsy's standard interfaces and on-board software.

Figure 4. Bitsy spacecraft kernel developed at AeroAstro

6. Taking it out for a Spin

You can scrutinize this graph (Fig. 5), or just remember a rule of thumb – for GTO to LEO with minor plane changes, about 50% of the total mass is SPORT. Thus on a 120 kg Ariane V slot, we absorb less than 60 kg, leaving more than 60 kg for the spacecraft itself.

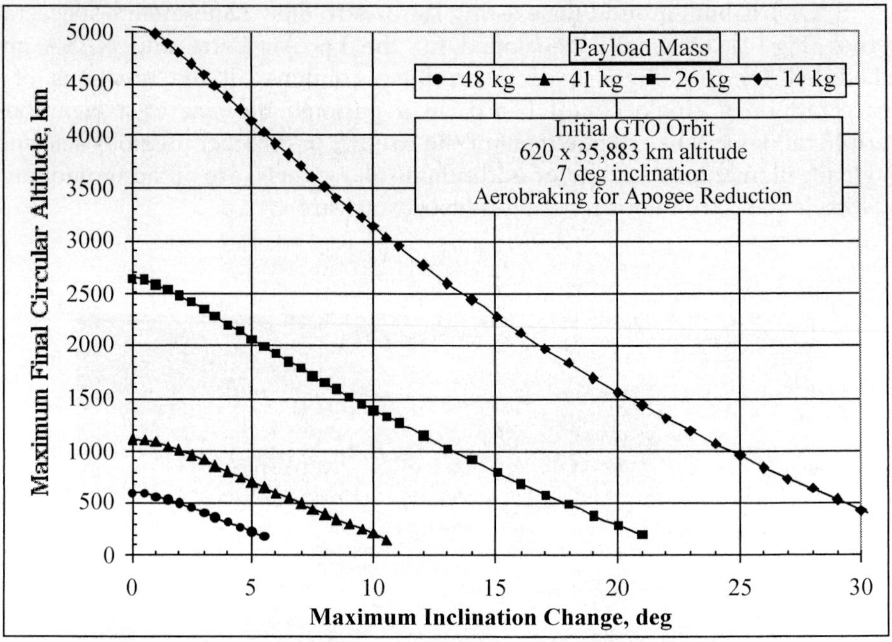

Figure 5. Typical SPORT performance for a 100 kg secondary slot

SPORT can go more places than just LEO. We can take advantage of the low orbital velocity at apogee to do plane changes of 30 degrees or more – easily reaching KSC and WFF orbits from GTO injections. Piggybacked to Molniya SPORT can insert into sun synchronous and other polar inclinations. Without the ballute, SPORT can be used to climb from Shuttle release to a much lower drag, longer lived orbit at 400 km to 1000 km, depending on mission requirements. For geosynch inspection, SPORT can climb from GTO to GEO, but about 80% of mass is propulsion. Fig. 6 illustrates some SPORT orbit transfer capabilities.

To fully leverage the wide range of missions accessible to SPORT, the design is kept highly modular. This enables carrying much more propellant for very high ΔV missions, or just the limited load needed to climb from Shuttle resupply, thus we can optimize for each mission the mass allocation to the passenger spacecraft.

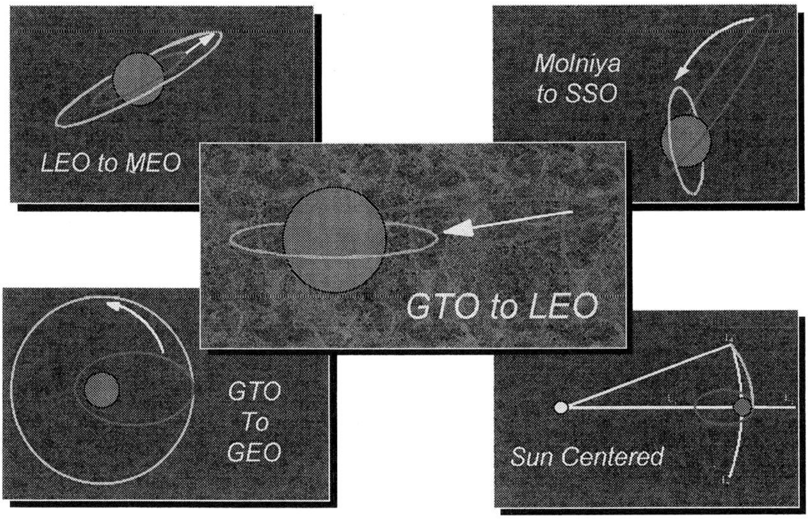

Figure 6. SPORT can provide transportation to a variety of mission orbits

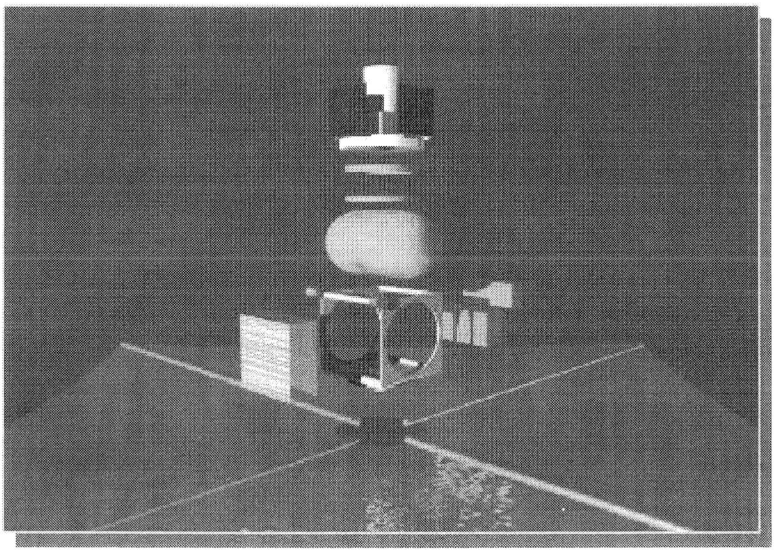

Figure 7. Modular design

SPORT - it's not your grandfather's upper stage.

Military Missions for Small Satellite Constellations

T. J. Lawrence, European Office Aerospace Research and Development (EOARD), Edison House, 223/231 Old Marylebone Road, London, UK

e-mail: tim.lawrence@london.af.mil

M. E. B. France, Force Structure, Resources, and Assessment Directorate (J-8), The Joint Staff, The Pentagon, Washington, DC 20318-8000, USA

e-mail: martin.france@js.pentagon.mil

Disclaimer: The views and opinions expressed in this paper are the authors'; they do not reflect current policy of the US government or any of its departments, agencies, or services.

Abstract

General background information is provided on the various types of US space missions, small satellite applications, potential architectures, and the technologies needed to develop capabilities consistent with current strategy. Examples of current US Air Force-sponsored and international small satellite programs are presented.

1. Introduction

In its report to Congress, released on 11 January 2001, the Commission to Assess United States National Security Space Management and Organization recommended the US Secretary of Defense to "direct the Defense Advanced Research Projects Agency and the Services' laboratories to undertake development and demonstration of innovative space technologies and systems for dedicated military missions."[Reference 1] With the military and commercial value of space assets increasing, as well as our reliance upon their products for routine and crisis operations, the use of small satellites to perform a variety of military missions becomes an attractive alternative to larger, more expensive (or non-existent!) current systems and architectures. Not only would small satellite-based constellations have lower development and launch costs, they would also provide options for spiral development of capabilities (owing to their shorter on-orbit life span) and allow commanders more flexible surveillance and space control options in times of peace and conflict. A review of the vision for how US military space strategy will evolve over the coming decades, along with an examination of possible architectures and current small satellite technology development and demonstration, provides a solid basis for determining the possible military uses for small satellites.

M. Rycroft and N. Crosby (eds.), Smaller Satellites: Bigger Business?, 219–228.

2. Future Military Space Missions

Historically, military forces have evolved to protect national interests and investments, with wars fought on land, at sea, and in the air. Today, due to the increasing value and importance of national interests and investments in the space medium (both military and commercial), space forces are needed to ensure the freedom of operation in space and peaceful use of its products. The United States Space Command's Long Range Plan (LRP) "Implementing USSPACECOM Vision for 2020" [Reference 2] addresses the evolution of military space strategy in the 21st century to address this need by developing four operational concepts from its current assigned missions and the anticipated strategic environment in support of Joint Vision 2010.

Control of space is defined [Reference 2] as the ability to assure freedom of operations within the space medium, and an ability to deny others the use of space, if required. To assure control of space, USSPACECOM identified five interrelated objectives:

- Real time space surveillance precise enough to give complete space object identification, and to maintain full situational awareness
- Timely and responsive spacelift, and secure and effective satellite operations once in orbit
- Enhanced protection from hostile actions for military and commercial systems using active and passive systems, including self-protection
- Prevention of unauthorized access to, and exploitation of, US and allied space systems
- Robust means, both lethal and nonlethal, temporary and permanent, of counteracting hostile space systems that place US and allied interests at risk.

Global engagement [Reference 2] is the combination of worldwide situational awareness and precise application of force. USSPACECOM's tasks in this arena are to:

- Survey high interest areas, providing complete awareness in peace, crisis, or war
- Defend against ballistic and cruise missile attacks
- Hold at risk a finite number of high value Earth targets (e.g., enemy command and control facilities) with near instantaneous application of force from space.

The space forces and space-derived information are integrated with land, sea, and air forces and their information, requiring command protocols, communications standards, and fused databases.

Finally, USSPACECOM would augment military space capabilities [Reference 2] through global partnerships by leveraging civil, commercial, and international space systems—a fundamental change in space operations based on the dramatic growth in international space-based capabilities, constrained military spending, and growth in multi-national operations and alliances. The strategy calls for the sharing of space-based information, for satellite sharing, and for international standardization of satellite protocols.

3. Proposed Mission Architectures

In the near term, deployable, "sortie-type" microsatellites to augment existing Earth surveillance and theater communications and to conduct space control missions seem to be within the realms of possibility considering the current state-of-the-art. Such vehicles, designed for easy integration onto an awaiting booster, would require rapid launch options—reusable or expendable vehicles able to take-off within hours of initial warning—to respond to crises. An architecture incorporating these capabilities would allow National Command Authorities (NCA) to increase US coverage of and presence over crisis areas—possibly deterring the potential aggressor—and "retreat" from these orbits just as quickly as tension eased.

A Space Operations Vehicle (SOV) (Fig. 1) could provide sortie access. The SOV would be about the size of the Space Shuttle, or smaller, and could be launched repeatedly (much as today's high-performance aircraft) for rapid sorties with "turn times" on the order of 48 hours or less. Recent studies have shown that launch frequency is the key parameter in lowering launch costs [Reference 3]. Payload mass could be sacrificed in favor of turnaround time, if necessary. An SOV has the added advantage of being able to operate entirely within the continental US in a sub-orbital mode, ejecting small satellites with small, expendable motors or housed in reusable upper stages to boost them to their final operational orbits. Such an approach would eliminate the need for overflying an unfriendly territory, reduce SOV performance requirements (and therefore cost), and greatly increase the available payload.

Figure 1. Space Operations Vehicle

The Space Maneuver Vehicle (SMV) is an orbital, reusable spaceplane (Fig. 2 and Fig. 3) that is launched by the SOV or by an expendable booster. An SMV would have a small payload bay capable of carrying roughly 500 kg of payload, and on-board propulsion system enabling significant orbit maneuverability. With a number in orbit, the SMVs could deploy or maneuver various small satellite platforms into suitable orbits, or retrieve them if they were damaged or no longer needed, removing the stigma of permanently placing weapons in space—in the case of future space control or force application payloads. A possible predecessor to the SMV is NASA's current X-37 program.

Figure 2. Space Maneuver Vehicle

Enabled by the launch systems mentioned above, entire small satellite constellations could be deployed in a matter of hours or days, depending on the number and location of the vehicles. Based on the USSPACECOM LRP, small satellite constellations would seem to provide excellent utility as Earth surveillance assets—particularly imagery and ground moving target indications (GMTI)—depending on the state of technology. Such a constellation would augment existing assets to provide continuous, near-real-time coverage of crisis regions, as well as improved situational awareness of the space environment. Similarly deployed small communications satellites could improve theater bandwidth, vital to the commander and NCA.

As part of its Project Air Force study, the RAND Corporation investigated the life-cycle costs of a space-based GMTI radar system, comparing fully dedicated systems with those that were deployed on demand, either fully or to augment a more austere on-orbit presence. RAND found that, for a 24 satellite constellation providing full-time coverage equal to airborne assets, life-cycle costs were 36% cheaper for a system where half of the assets were deployed on demand, and then either recalled when tensions eased or left in place to be augmented by newer systems at some point in the future as operational needs warranted [Reference 4]. Presumably, miniaturization of system components to the point where small satellites could accomplish this mission would further reduce launch costs—and therefore life-cycle costs.

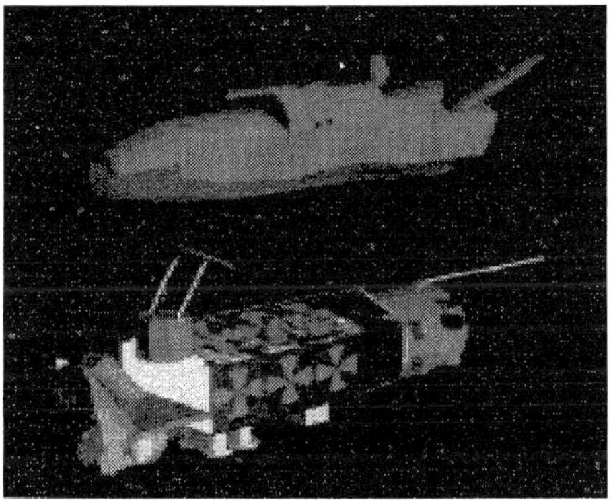

Figure 3. SMV with Defensive Small Satellite

Similarly, constellations of maneuvering satellites could also be deployed to protect high-value low-Earth orbiting (LEO) assets. Near their larger cousins, with an appropriate set of sensors, they could detect enemy jamming, directed energy, or impending kinetic energy attacks. These systems could also inspect for any existing damage, act as defensive shields if their neighbor were to be attacked, or maneuver to neutralize enemy anti-satellite capabilities. Similarly equipped small satellites could inspect enemy satellites for capabilities and signs of hostile intent, equipped with nonlethal means of neutralizing these systems, as policy and NCA direction warrants.

While the development of reusable launch vehicles and spaceplanes is currently the province of NASA, the Department of Defense should continue to develop the small satellites payload portion of the architecture described above in order to provide the NCA with options consistent with the USSPACECOM LRP. Focusing on the technologies and demonstrations that reduce risk and enable the missions described above will enable the Department of Defense to exploit future launch capabilities as they become more widely available.

4. Current USAF Programs

There are several on-going US and international programs that are investigating technologies that would enable some of the capabilities discussed above to become a reality. Some of the technologies needed for these architectures were developed in previous US programs. For example, the Strategic Defense Initiative's (SDI) DC-X program demonstrated the basic technologies needed for a launch-on-demand reusable rocket in 1993. Similarly, SDI's Clementine mission and several other small satellite surveillance missions demonstrated the feasibility of low-cost, comprehensive space control and global surveillance systems.

In collaboration with DARPA and NASA Goddard, the Air Force Office of Scientific Research began sponsorship of a university nanosatellite program in 1996. The goals of this program are to have students build satellites on the order of 1–10 kg to demonstrate the following capabilities:

- Formation Flying
- Tethered operations
- Lightweight sensors
- Advanced communications
- Micropropulsion
- Micro power systems
- Lithium ion batteries

As part of this program, Stanford University has announced a 1-2 kg cubic satellite, selling for US $ 25K - $ 250K.

TechSat-21 is a program run by the US Air Force Research Laboratory's (AFRL) Space Vehicles Directorate (VS). Currently scheduled for launch in late 2004, TechSat-21 will consist of three independent, LEO microsatellites, each less than 170 kg, capable of a one to two year mission life. Its goals are to demonstrate enabling technologies for microsatellite formations and to assess the utility of microsatellite formations for multiple missions. Its key challenges are intersatellite positioning and communication, autonomous formation control, and sparse aperture operation. TechSat-21 will include an early demonstration of space-based radar (SBR) using electronically steerable antenna (ESA) technology developed under Discoverer II. The end goal is an affordable, real-time, on-demand global awareness capability.

The XSS microsatellite series is the Air Force Research Laboratory's (AFRL/VS) sequence of on-orbit experiments to develop a satellite-oriented space logistics and servicing capability. The first mission, dubbed XSS-10, (Fig. 4) expects to demonstrate several capabilities and subsystems, including:

- autonomous and manual control of microsatellite on-orbit
- autonomous navigation and inspection of object of interest (Delta 2nd Stage)
- proof of microsatellite mission control architecture
- lightweight propulsion
- microsatellite guidance, navigation, and control (GN&C)
- communications system (SGLS)
- integrated camera with the star sensor.

Figure 4. The Proposed XSS-10 Microsatellite

The Satellite Test and Research Vehicle (STRV) missions (Fig. 5) are a series of microsatellite platforms designed to test various technologies in the geosynchronous transfer orbit (GTO) environment. The current mission is designed to fast-track the latest space technology into orbit by carrying 25 hardware and 4 software experiments into the harsh space environment of GTO. In this orbit, the satellites receive very high doses of natural radiation which allows the accelerated life testing of new technologies and components. The experiments are sponsored by a wide variety of space organizations in the UK, continental Europe, the US and Canada.

Figure 5. The STRV microsatellite

In addition to the future launch options mentioned earlier, whose predecessors are being developed by NASA, the US Air Force is pursuing other options, albeit much more slowly than their European counterparts. The Evolved Expendable Launch Vehicle (EELV) secondary payload adapter (ESPA) program, managed by the Air Force Space and Missile Center's Space Test Office (SMC/TEL), provides low cost small satellite launch capability for US payloads similar to the capability currently present on the Ariane 4 and 5 launchers.

5. International Programs

Meanwhile, the relatively slow US pace in developing reusable rockets, microsatellite launch concepts, and small satellites is being filled by others. In the late 1980s, several European groups saw the potential for microsatellites coupled with an affordable means of getting them to space. The French-based Ariane 4 and its follow-on Ariane 5 were equipped with the means to launch up to eight 100 kilogram microsatellites as auxiliary payloads during each of their frequent, large commercial satellite launches. These payloads can be launched at minimal cost (several hundred thousand US $ per microsatellite), with no interference with the primary payload(s), and with considerable security. This new capability has ignited a significant European microsatellite industry, perhaps the most advanced of which is the UK-based University of Surrey Space Centre that has successfully built and launched over 20 microsatellites in the past decade and a half. These systems cost a few million US $ apiece.

In June 2000, the University of Surrey launched the next generation of highly capable microsatellites, which they dubbed a "nanosatellite." This device, SNAP-1A (Surrey's Nanosatellite Application Platform, Fig. 6), with a mass of 6.5 kg, was launched as a "double piggy-back" on a Russian booster along with a Russian navigation satellite and a Chinese microsatellite. The SNAP-1A detached from the other payloads and used an on-board propulsion system to maneuver in the vicinity of the other satellites and take high-quality images of them. The SNAP-1A system, costing around US $ 0.5 million, demonstrated three axis control, proximity operations using a cold gas propulsion system, cross linking, and GPS for orbit determination. SNAP modules may now be purchased by US government-sponsored researchers.

Figure 6. SNAP-1A

6. A Way Ahead

In order to field these architectures, the current set of Air Force small satellite programs should be funded to completion. Next, a focused and fast-paced prototype demonstration program is needed to mature launch capabilities that make sortie access to space with small satellites a reality; helping to achieve the promise small satellites hold for accomplishing current and future military missions. Many of the necessary technologies to do this have already been shown feasible through the efforts of NASA and other agencies. However, they have not been integrated into full system demonstrations. Use of commercial-off-the-shelf (COTS) systems is important in attempting to deploy these systems. In many programs, too much effort is devoted to the technology instead of putting existing technology into a working system. Time is lost on trying to get the extra 1%–2% of performance, instead of applying the 80/20 rule—80% of the performance at 20% of the cost. We could demonstrate the necessary capabilities for sortie access to space, deployable intelligence, surveillance, and reconnaissance (ISR), and small satellite-based space control in the next five years.

Global, real-time monitoring of Earth and space, coupled with a commitment to protect and defend our critical space assets, based on full force integration and global partnerships should form the basis of US space strategy during the next two decades. Technologies such as those now being incorporated into small satellites offer a viable means of fielding these capabilities at reasonable costs, and should be a cornerstone of DOD modernization and restructuring.

References

1. Report of the Commission to Assess United Stated National Security Space Management and Organization, Pursuant to Public Law 106-65, January 11, 2001
2. Long Range Plan: Implementing USSPACECOM Vision for 2020, United States Space Command Report, United States Space Command, Colorado Springs, Colorado, 1998
3. Bruno, C. and Czysz, P.A.: An Electro-magnetic-chemical Hypersonic Propulsion System, AIAA 98-1582, AIAA Aerodynamics Conference, Reno, NV,1998
4. Gonzales, D., et. al.: *Implications of Reusable Space Systems for Air Force Space Operations (Draft), Volume I, Executive Summary,* The RAND Corporation, DRR-2015-AF, January 1999

Pegasus - Past, Present and Future

W. E. Frick, Orbital Sciences Corporation, 21700 Atlantic Boulevard, Dulles, VA, 20166, USA

e-mail: frick.warren@orbital.com

Abstract

On the morning of 5 April 1990 the Pegasus launch vehicle embarked upon its maiden voyage to place two small satellites into a 593 x 667 km orbit inclined at 94 degrees. A specially modified NASA B-52 took off from Edwards Air Force Base with the Pegasus rocket attached to its right wing and headed NW over the Pacific Ocean to its pre-determined launch point. After confirmation that all systems were "go" the 14.9 m, 18,800 kg rocket was released from the carrier aircraft and, twelve minutes later, became the first privately developed rocket to place a satellite into Earth orbit successfully. Since that time, the Pegasus program has conducted 30 missions launching over 70 satellites for both government and commercial customers. Many of these missions have included multiple satellite deployments. The vehicle has been upgraded to include an extended length (XL) configuration (16.8 m long, 23,100 kg) for added performance, an optional liquid fuel fourth stage for improved injection accuracy, and a new L-1011 aircraft. In addition a new flight computer, inertial navigation unit, new reaction control thrusters, and subsystem batteries have been incorporated. Many other components have been upgraded from the original configuration.

Currently, the Pegasus program focuses on vehicle enhancements that will continue to make Pegasus the most reliable small satellite launch vehicle in the world. New business opportunities outside the normal satellite launch industry are being pursued; these include quick reaction launch capability and technology demonstrations for the NASA Space Launch Initiative. The paper reviews some of the highlights of the past 11 years and also the current developments and activities for the coming years, especially in relation to launching small satellites.

1. Pegasus History

Pegasus has a very different history than other launch vehicles. It did not start as a government-funded program, or as a spin-off of a weapons system. It started because Orbital Sciences Corporation could not find a small launch vehicle in the late 1980's to fulfill a business plan.

1.1 Concept

In the mid 1980's, the concept of a data-only Low Earth Orbit (LEO) constellation of satellites was pursued by Orbital Sciences Corporation. The satellites needed an inexpensive launch vehicle to make the business plan successful. Such a launch vehicle did not exist. The Scout was too expensive, and was a government, not a commercial, product. The Air Force was not interested in commercializing their F-15 based ASAT program. Orbital would have to develop the rocket.

M. Rycroft and N. Crosby (eds.), Smaller Satellites: Bigger Business?, 229–237.

High level design trade-offs were made. Should the carrier aircraft be a small and fast carrier aircraft or a low and slow aircraft? Should the rocket have a wing or not? Should control be by vectorable nozzles or control surfaces? LEO payload requirements grew from less than 15 kg to 90 kg. Specifications were created for hardware and in September, 1987 rocket motor vendors contacted. One of the potential motor vendors guessed from the expansion ratio of the first stage, indicating a high altitude ignition, that Orbital was developing a rocket to go on top of a Peacekeeper or Minuteman booster. This guess became prophetic as the Pegasus-derivative Taurus and OSP Minotaur vehicles using Peacekeeper and Minuteman first stages were developed later; however, that was not Orbital's intention at the time.

In November, 1987 Hercules agreed to become a joint venture partner with Orbital in Pegasus development. Orbital obtained the motor expertise it needed to make Pegasus feasible and Hercules was the sole provider of a new product line.

In late 1987 the Pegasus program held its Preliminary Design Review. The size of the total package was 59 pages long (exclusive of motors). The payload requirement for the launch vehicle was now 226 kg, and grew to 272 kg after the meeting.

1.2 DARPA Award

In 1988 Pegasus won a Defense Advanced Research Projects Agency (DARPA) award for one launch service with options for five more. NASA would provide the B-52 carrier aircraft and site for an Orbital-constructed Vehicle Assembly Building on Edwards Air Force Base. However, the design, development, and funding of the rocket was on Orbital's budget.

1.3 Development

In February 1988, the detailed work of turning the concept of Pegasus into a working piece of hardware began. Vendors for the subsystems were contacted. The detailed design trade-offs involved in putting the rocket together were started and engineering began on all the interfaces, specifications, budgets, and details *ad infinitum*, that make up a launch vehicle.

The stage 3 motor had a static firing in March 1989, and the final design review was held in April. The Stage 2 motor was static fired in May and, in July, the Stage 1 motor static firing was completed. In August, a complete inert Pegasus was rolled out and in November, this inert rocket was mated to the B-52 and carried to the drop point. Two more successful captive carry flights were carried out and, on 5 April 1990, Pegasus was dropped off the coast of California, successfully making it to orbit. Orbital Sciences Corporation was awarded the United States of America's 1991 National Medal of Technology for the Pegasus program.

1.4 *Flying Pegasus*

Pegasus has now flown 30 times. Over those 30 flights, the program has run into many circumstances that any rocket may encounter, and many that are unique to its air launch heritage. The following is a brief listing of several of the flights that stand out - no attempt is made to summarize each flight, or to list all of the unique events that were encountered.

First Flight – PegSat - 1990

Among its other functions, PegSat was an environmental monitoring satellite that measured the environments of this new rocket. As with any new rocket, there are a lot of unknowns. The initial flight was not flawless, but it did achieve an acceptable orbit, and with the assistance of its telemetry, as well as the environmental information PegSat provided, NASA and Orbital were able to measure payload environments accurately [Reference 1].

Liquid Upper Stage - 1991

For the second launch of Pegasus a liquid fuel fourth stage was added. This stage was inside the cylindrical avionics section and used no payload volume. It was initially called the Precision Injection Kit (PIK) as it was designed to reduce the orbit injection dispersions that the solid fuel third stage rocket had, which limited final orbit accuracy. The acronym for this stage was later changed to Hydrazine Auxiliary Propulsion System (HAPS) as its other uses became evident.

Several serious anomalies were encountered on this flight - the stage 1/ stage 2 separation did not occur nominally, and stage 2 ignition occurred with the rocket in a seriously non-optimal direction. In addition, the fairing did not separate cleanly and, due to the partially attached fairing, control authority was reversed during a portion of flight. Despite these anomalies the payload of seven small satellites achieved a functional but low orbit, and were able to demonstrate the concept of small tactical communications satellites.

Logistical and Operational Challenges - 1993

The third launch of Pegasus demonstrated its transportability and simple design. Fifty seven days after the first hardware arrived at the Vehicle Assembly Building, Pegasus took off on a cross-country journey to a Florida launch site with the Brazilian satellite SCD-1 (Satellite for the Collection of Data, in English) on board. A fin anomaly during captive carry mandated a return to Edwards Air Force Base prior to launch, and the satellite was successfully launched several months later.

This launch was also a challenge to command and control. The launch itself departed from the Shuttle landing strip on NASA's Kennedy Space Center (KSC) in Florida. Range safety was performed by the U.S. Air Force's Eastern Test Range, and the launch control center with all the management and senior personnel was on Wallops Flight Facility (WFF) in Virginia, over 1000 km from KSC.

Rapid Response - 1993

The fourth flight of Pegasus demonstrated one feature designed into Pegasus. The satellite on Pegasus' fourth flight had a cryostat on board that required the satellite be in a vacuum shortly after cryogen loading and payload mate. Five days after the payload was mated to the rocket, it was successfully in orbit.

Guidance and Extended Operations - 1994

On the fifth flight the basic vehicle performed well and the HAPS also performed well. Due to a slight algorithmic error during the HAPS burn the inertial navigation system thought it was somewhere fictional, so the rocket put itself into an orbit that was much lower than intended.

HAPS had been designed to fail on orbit gracefully. Two separate mechanisms were designed in that were predicted to cause the hydrazine to vent safely into space. Due to the guidance anomaly the HAPS ended up with much more fuel on board than expected. After the satellite had separated and after two years on orbit, Orbital was notified that HAPS had become hundreds of trackable pieces. When filament wound tanks fail, they fail spectacularly. The actual cause of the failure has never been completely determined but steps have been taken to make sure it will never happen again.

Stretching a Rocket, Changing the Aircraft - 1994

In the early 1990's Orbital made the decision to stretch the rocket to increase capability and to gain performance. To create the Pegasus XL, the first stage of Pegasus was stretched 140 cm, and nearly 3000 kg of solid propellant added. The second stage was stretched 93 cm and nearly 900 kg of fuel added. The third stage motor had its throat diameter decreased, to increase the effective expansion ratio, increase burn pressure, and increase specific thrust. These changes increased performance from 275 kg to 350 kg into LEO, for only a small increase in cost.

The original Pegasus rocket had the coefficients for its control algorithms determined analytically. Wind tunnel testing was not performed - Computational Fluid Dynamics (CFD) were used, as well as robust control algorithms. Stretching Pegasus changed the coefficients; the rocket became longer, with different thrust and control coupling profiles. On the sixth flight Pegasus failed. The changes proved to be beyond the capability of the CFD analytical techniques for accurate modeling. After the flight wind tunnel testing was used to determine the faults in the analytically derived control laws, showing that CFD does have its limits.

Flight six was also the first mission using the L-1011 as the Pegasus carrier aircraft rather than the B-52. The modified L-1011 had many capabilities that the B-52 was lacking. Orbital modified it with a weight carrying capability well in excess of the B-52 to allow for the Pegasus XL and future Pegasus version's increased mass. The L-1011 also had fully functional flaps, allowing shorter takeoffs and landings - the aircraft would no longer be restricted to long military runways. In addition, air conditioning was now possible for the payload. Pegasus was modified by canting the fins down 23 degrees, allowing the gear doors to close. The aircraft performed well on the sixth launch.

Separation Failure - 1995

The ninth mission was another Pegasus XL. This mission was a failure, due to an interstage separation anomaly. The Pegasus program brought in Aerospace, the US Air Force, NASA, and its commercial customers, and completely changed the way in which it approached operations. This change was later to be very beneficial in achieving the string of successes that it now enjoys, as well as gaining the ISO-9001 certification that is now mandated by NASA for all launch vehicle contractors.

Dual Payloads - 1996

For its fourteenth mission Pegasus needed to put two NASA non-loadbearing satellites into orbit. The program developed a variable-length Dual Payload Attach Fitting (DPAF) to support the satellites separately. The rocket launch was successful but, in the final stages of the mission, one of the launch vehicle batteries suffered a failure, and the satellites could not be separated. The batteries had passed all qualification testing and had significant flight heritage on other launch vehicles. Batteries cannot be fully functionally tested before flight. This was the last time that Pegasus did not successfully place its payload into orbit.

International Operations - 1997

For its fifteenth mission Pegasus was taken to Spain for payload mate, and then ferried to the Canary Islands for a retrograde launch [Reference 2]. In addition to building a launch control center outside Madrid to control the launch, Orbital worked with Wallops Flight Facility to import a NASA mobile range to augment the Spanish assets on Gran Canaria.

External Experiments - 1998

For its twenty fourth launch the primary payload was the Brazilian SCD-2 satellite, a twin satellite to SCD-1. As a secondary payload, NASA put a metallic glove on the wing of the rocket, measuring transonic and hypersonic shock wave transitions. The launch was successful and the program learned quite a bit about the post-separation behavior of the winged first stage as the NASA telemetry system stayed operational much longer than predicted.

Solid Hydrogen - 1999

For its twenty sixth launch the Pegasus satellite carried had a small cryostat with solid hydrogen [Reference 3]. Solid hydrogen transforms directly from solid to gas extremely quickly if vacuum is lost and, due to the flammable nature of hydrogen, very special precautions were taken. A vent stack to get rid of the hydrogen had to be developed, and the gas vented (if necessary) through the Pegasus fairing and around the L-1011 to burn harmlessly in the air. In addition, inert purge gas had to be carried for this vent stack, so that the hydrogen would not freeze the air in the vent stack and form a plug. The launch was successful, and the satellite placed into orbit with its full load of hydrogen intact.

Dual Orbits - 1999

The two payloads on board Pegasus' twenty seventh launch could not be placed in the same orbit. In addition, the forward payload was a "Thompson spinner" (spinning end over end, rather than axially). To perform the mission, the rocket was required to reach 4 different orbits. HAPS and the Pegasus control system performed flawlessly, reaching the first orbit accurately, putting the forward satellite into its Thompson spin, stabilizing the rocket after forward satellite separation, separating the adapter cone (the aft payload was a load-bearing design), performing a Hohlmann transfer of the aft satellite to its required orbit and separating it over the designated ground station, and then deorbiting itself before venting all the propellants and gases to render the final stage inert.

Equatorial Operations – 2000

The most recent Pegasus launch was from Kwajalein, a United States-leased island in the Territory of the Marshall's Islands. By launching from Kwajalein, Pegasus was able to reach an orbital inclination of 2 degrees, and the satellite was optimized for deep space observations.

2. Pegasus Present

2.1 *Successful Operations*

Pegasus has launched 16 consecutive successful space missions and is working very hard to extend that streak, as a launch vehicle is only as good as its last launch. In the commercial launch vehicle market, failures cause lost contracts and reduced new business opportunities. There are two reasons why the program has achieved this level of success - people and processes.

People make the rocket. The hardware is designed, assembled, tested, qualified, and missions performed by personnel who are dedicated to putting hardware into space. Pegasus shares a manpower pool with Taurus and Minotaur, letting the technicians and engineers stay current on common hardware, while enjoying an occasional change of task. The manpower base is stable, and many of the people who work on the rockets today were there when the lessons of success were learned years ago. The Taurus and Minotaur programs have both reaped the success of Pegasus lessons learned - neither has ever experienced a failure.

The Pegasus program learned the hard way on its ninth launch that people are only as good as the processes that guide them. This means the training, certification, work instructions, and launch procedures are taken extremely seriously. ISO 9001 codifies this process - write what you do, do what you write. The Pegasus personnel live the Orbital quality motto - "We Launch Quality".

2.2 *Continued Reliability Upgrades*

Pegasus continuously updates hardware on the rocket. When a battery failure caused the failure of the DPAF-equipped Pegasus on flight fourteen, all batteries of that design were changed. When a better flight computer was available, Pegasus upgraded, increasing software processing capability and telemetry processing. The program is upgrading to a third generation inertial navigation system; one that is fully capable of flying with GPS-aided navigation. Pegasus is evaluating the proper time to phase in the Minotaur integrated avionics hardware. Reliability and value added are the focus of the Pegasus engineers. 16 consecutive successes qualify the Pegasus launch vehicle for the highest value NASA payloads and the Pegasus program takes this qualification very seriously.

3. Pegasus Future

3.1 Acantara

In two years, Pegasus plans to perform a mission from another remote launch site, Alcantara, in Brazil. This site will be the fourth continent, sixth telemetry downlink and range safety system, and eighth runway used for Pegasus operations. The site will extend the capacity of the Pegasus space launch system even further. Alcantara is closer to the equator than any other land-based space launch range in the world, and has a wide range of launch azimuths available. Pegasus will be able to reach any posigrade, and virtually all retrograde inclinations directly, with approved, direct flight trajectories.

3.2 Launch on Demand

With a satellite specifically designed for a rapid launch, and some modifications to the rocket, time from approval to launch to orbit may be shortened even more. The program is investigating scenarios to utilize fully the flexibility and rapid response that an aircraft-based system best serves.

3.3 Autonomous Rendezvous

Pegasus has been chosen for NASA's Space Launch Initiative to prove autonomous rendezvous – the Demonstration of Autonomous Rendezvous Technology. Such technology will be required for the second generation reusable launch vehicle to remove dependence on man-interactive guidance.

3.4 Alternate Access to Station

In an Alternate Access to Station study for NASA, Pegasus has investigated providing a quick response capability for getting supplies to the International Space Station (ISS). Pegasus can meet the ISS orbit by moving its drop point significantly by flying the aircraft to a different location. This gives Pegasus more than one launch opportunity to an ISS rendezvous orbit per day. The mass it can carry is not great, but does support the majority of line-replaceable hardware items on the ISS.

References

1. Goddard Space Flight Center, National Air and Space Administration: *Pegasus-Pegsat Payload Environments Data Report*, December 1990
2. Justin Ray/Florida Today: *Florida Today Space Online*, <http://www.flatoday.com/space/explore/stories/1997/042197d.htm>, April 21, 1997
3. SpaceViews: *SpaceViews News*, <http://spaceviews.com/1999/03/04b.html>, March 4, 1999

Small Satellites in a Large Launch Vehicle Marketplace:

NASA's Strategic Effort to Provide Space Transportation

J. C. Schafer, NASA Space Access, Office of Space Flight, NASA Headquarters, Washington DC 20546, USA

e-mail: John.Schafer-1@ksc.nasa.gov

Abstract

As NASA enters the new millennium, the planned expendable launch rate is as robust as it has ever been in history. Included in this robust launch rate is NASA's increased use of small satellite technology. This paper illustrates NASA's near term requirements and long-term projections of launch services for small class satellites.

We identify NASA's challenges of providing space transportation for small satellites in an environment where both the defense and commercial market trends are towards larger satellites and larger launch vehicles. In order to meet these challenges, we explore NASA's strategic effort to meet the needs for small satellite space transportation such as secondary payload manifesting, multiple manifesting, and strengthening alliances and aggregating launch requirements with other small satellite users.

1. Overview of NASA's Space Access Goals and Objectives

1.1 National Space Policy

NASA's Space Access goals and objectives are rooted in the President's overarching 1996 National Space Policy and are further refined through NASA's strategic plans and Enterprise strategic plans. According to the National Space Policy, the goals of the US space program are to:

- Enhance knowledge of the Earth, the solar system and the Universe through human and robotic explorations
- Strengthen and maintain the national security of the United States
- Enhance the economic competitiveness, and scientific and technical capabilities of the United States
- Encourage State, local and private sector investment in, and use of, space technologies
- Promote international cooperation to further US domestic, national security, and foreign policies.

M. Rycroft and N. Crosby (eds.), Smaller Satellites: Bigger Business?, 239–247.

1.2 NASA Strategic Plan

The National Space Policy designates NASA as the lead agency for research and development in civil space activities. As the lead agency for research and development of civil space activities, NASA's mission is to:

- Advance and communicate scientific knowledge and understanding of the Earth, the solar system, and the Universe
- Advance human exploration, use, and development of space
- Research, develop, verify, and transfer advanced aeronautics and space technologies.

To achieve the overall NASA mission, five Enterprises have been established with specific, focused goals and objectives. These are the:

- Space Science Enterprise
- Earth Science Enterprise
- Biological and Physical Research Enterprise
- Human Exploration and Development of Space Enterprise
- Aerospace Technology Enterprise.

In carrying out the science objectives, each of the Enterprises has numerous programs dedicated to their science objectives. Each of these programs has various transportation requirements that may include:

- Launch Services (ELV)
- Space Shuttle
- International Cooperative
- Secondary Payload
- Sub-Orbital Launch
- Sounding Rocket
- Aircraft
- Balloon
- Earth Based
- Other.

1.3 Space Access Goals and Objectives

NASA's Office of Space Access provides launch services to the science community via a service/customer relationship. The Office of Space Access is organized such that it acts as the service organization to the science Enterprises to provide launch services. NASA's Office of Space Access goals and objectives are:

- Provide safe, reliable, on-time, and cost effective launch services that meet or exceed customer requirements
- Structure government technical oversight to maximize probability of mission success within constrained resources
- Establish contractual mechanisms to enable access to the fullest range of available launch services.

Primary responsibilities of the Office of Space Access include chairing the Flight Planning Board, which forms the launch manifest, and establishing launch service related policy. NASA's Office of Space Access launch services historically implied Expendable Launch services only, but more recently included Space Shuttle capabilities as well. The Office of Space Access customers are the individual science enterprises within NASA. The Earth Science and Space Science Enterprises have been the predominant users of NASA acquired launch services on ELV's. The Human Exploration and Development of Space Enterprise and the International Space Station, as well as the Biological and Physical Research Enterprise, are the predominant customer of the Space Shuttle Space Transportation System. Small payload customers also take advantage of the Space Shuttle's capabilities utilizing the Shuttle's small payloads carrier systems, which include the Hitchhiker, Getaway Specials (GAS), and Space Experiment Module (SEM), as well as mid-deck lockers on the Space Shuttle.

According to the 1990 Launch Services Purchases Act, NASA must use competitively procured commercial launch services to launch US Government payloads. Further, the commercial launch services must use launch vehicles manufactured in the US Exceptions are limited, but include International Cooperative Agreements with other nations or Presidential Approval for non-US Acquired Services.

In order to meet the launch needs of the five Enterprises the Office of Space Access, through the Program Office at the lead center for ELV at Kennedy Space Center, has established contracts with qualified US ELV Launch Services Providers. The Launch Service Providers on contract to NASA provide a complete range of launch capabilities from small secondary payloads on a Pegasus or Delta II class launch vehicle to the largest of the future EELV launch vehicles. Launches requiring the Space Shuttle Space Transportation System are managed through the lead center Johnson Space Center, while the small payloads are managed through the Goddard Space Center's Shuttle Small Payloads Project.

2. NASA Launch Requirements

During most of the 1990's, NASA's ELV launch rate averaged about 4 to 5 launches per year for all payload classes. Starting in 1999, NASA's launch rate has more than doubled to an average of 8 to 10 launches per year with some years surging to 12 per year.

In the same timeframe, dedicated launches for small payloads and secondary payloads sometimes surged to as many as 2 to 3 per year, but averaged fewer than one per year. Starting in 1999, the launch rate of small and secondary spacecraft has nearly tripled. This is due to the science Enterprises' commitment to the smaller spacecraft and the maturation of the "Faster, Better, Cheaper" philosophy that has enabled more frequent launches within fiscal constraints.

Today, the current NASA manifest shows a yearly planned launch rate for small and secondary payloads nearly equal to the entire yearly launch rate in the early to mid 1990's (see Fig. 1).

The NASA manifest indicates an average of 2 to 3 small-class launches per year. The secondary payload manifest grew from less than one per year to an average of two per year in the 1998-2001 timeframe. Beyond 2002, the secondary payload launch rate again drops to one or fewer per year, due to fiscal constraints.

Because of NASA's unique science requirements, dedicated, medium sized launches account for nearly half of NASA's planned ELV launches over the course of the next decade. NASA's aggregate small and secondary payloads launched on ELV's account for approximately one third of NASA's launches over the same time period (see Fig. 2). By definition, the secondary payload requirements do not result in an additional launch.

3. NASA Effort to Meet Small Payload Launch Requirements

In order to meet NASA's needs for small payload launches as well as its unique science requirements, the Office of Space Access has established a wide range of launch capabilities.

3.1 Space Shuttle Small Payload Capability

Included in the Space Shuttle capability is the Shuttle Small Payloads Project (SSPP) organization that designs, develops, tests, integrates and flies a group of small payload carrier systems for the Space Shuttle. These carriers - the Hitchhiker, Getaway Specials (GAS), and Space Experiment Module (SEM) - support small payloads supplied by NASA, other US government agencies, universities, high schools, domestic commercial customers, and foreign nationals and governments. These carriers can support payloads that range in size between 50 lbs (23 kg) and 4000 lbs (2270 kg). The small payloads are manifested according to science priority and available opportunities. During the International Space Station assembly period, small payload manifest opportunities will be limited.

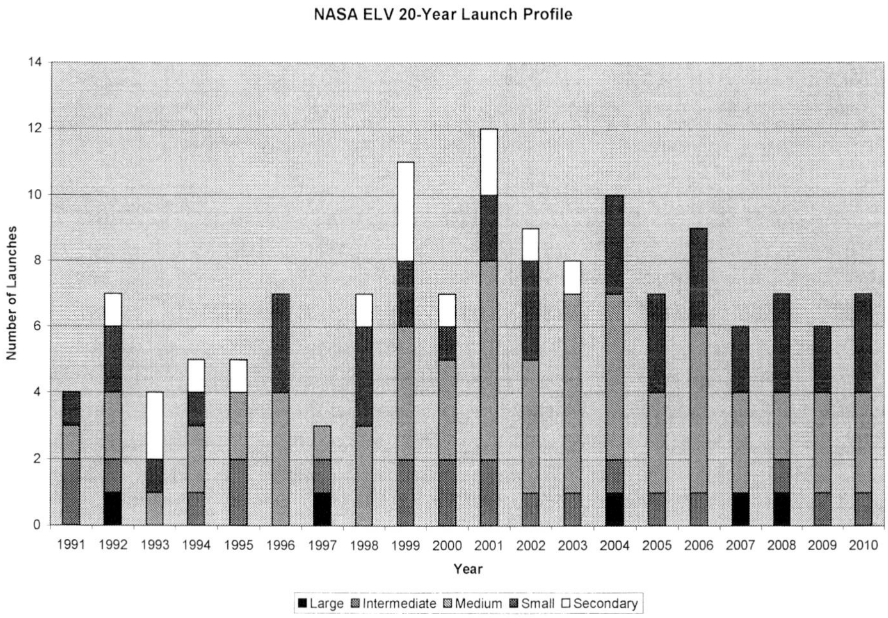

Figure 1. NASA ELV 20-Year Launch Service Profile [Reference 2]

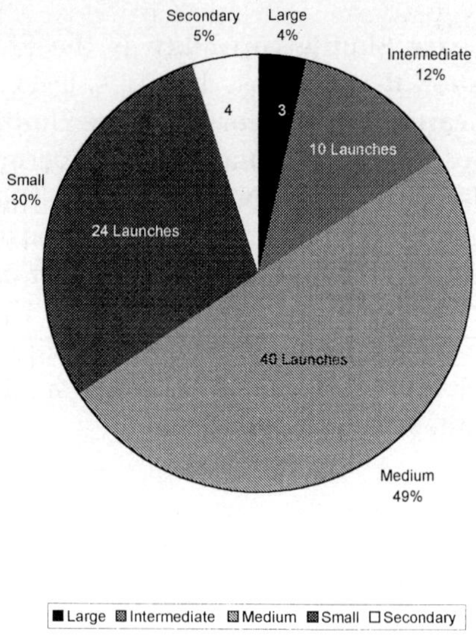

Figure 2. NASA 10-Year ELV Launch Requirements by Payload Class [Reference 2]

3.2 Dedicated Expendable Launch Vehicle (ELV) for Small Payloads

Small payloads requiring dedicated launch vehicle capability utilize NASA Launch Services for small vehicles. NASA currently has commercial launch service contracts, which include launch services for the Pegasus, Taurus, and the Athena launch vehicles. As shown in Fig. 2 above, the small launch requirements for NASA are projected to comprise 17 percent of the total ELV payload launch demand over the next decade. Small class launch vehicle demand is projected to comprise approximately 33 percent of the total ELV launch demand (excluding the secondary payloads, which do not result in additional launches; see Fig. 1 for historical and projected launch demand for all classes of launch vehicles).

3.3 NASA ELV Co-Manifest and Secondary Payloads

Launch vehicle secondary and co-manifest capabilities on ELV for NASA include the Pegasus, Taurus, and Delta II. In addition, secondary payload capability is currently in development on new Evolved Expendable Launch

Vehicles (Delta IV and Atlas V), which are currently in development. NASA continuously evaluates planned or future primary payload launches and compares them with potential secondary payloads in order to identify possible matches. NASA maintains regular dialogue with the USAF to take advantage of secondary payload opportunities. Co-manifest and secondary opportunities with commercial payload launch services are addressed on a case-by-case basis. The technical evaluation includes identifying a primary payload whose technical requirements are compatible with that of a co-manifest or secondary payload (e.g., mass, orbit, performance margin, etc.). Once a potential match is found which addresses the technical issues, then other considerations are evaluated which include issues such as cost, schedule, and risk.

3.4 International Cooperative Agreement

According to the President's 1996 National Space Policy, the US Government (including NASA) may enter into cooperative agreements with international partners to provide launch services. This may include instruments of opportunity on a foreign spacecraft, secondary payload launch, co-manifested payload, or primary payloads.

4. Small Payload ELV Launch Challenges

Due to forecast flat or declining commercial market demand for small launches, NASA is concerned about the commercial availability of domestic, commercially available, dedicated, small Expendable Launch Vehicle (ELV) capability. This capability is especially important to NASA's Access to Space customers, primarily the Earth Science and Space Science Enterprises. The decline in demand for small payload launch capability could impact NASA Enterprise customers in terms of launch and mission cost, mission and payload complement, launch availability, launch flexibility, and ultimately how NASA implements its strategic plans.

4.1 Small Launch Vehicle Market Trends

Consistent with the National Space Transportation Policy and Launch Services Purchases Act, the US Government is required to purchase launch services commercially from domestic sources to the maximum extent practical. The availability of those domestic launch services is dependent on the aggregate launch demand from all the market sectors (civil, military, and commercial space transportation). With the relatively small launch demand in any single market sector, all the market sectors are dependant on the launch demand of the others.

The predicted robust commercial launch demand materialized only briefly and is currently flat. The near term outlook is not optimistic. In 1998 the worldwide commercial market surpassed the Government market for the first time. However, in 2000, the trend reversed with 75% of domestic launches being Government sponsored. Reduced demand results from many factors, such as:

- Recent LEO industry bankruptcies and financial difficulties
- Inexpensive international ICBM-based small launchers dominating the market
- Worldwide oversupply of launch vehicles
- Terrestrial-based systems more flexible and growing faster than expected.

Launch demand is even more pessimistic for the small launch capability. Current commercial and DOD market trend is towards larger spacecraft and heavier launch capability. NASA's requirements are predominately for small (and medium) sized scientific payloads and launch systems. The trend towards larger launch systems leaves NASA ELV in a market niche (small and medium), diverging from the predominant commercial and military market for larger systems. According to latest manifest projections, NASA is now the largest user of small, and soon medium, ELV launch capability (over 80% of current NASA ELV launch demand is for small and medium class vehicles).

NASA demand alone is not sufficient to sustain an indigenous competitive commercial small launch capability. With the DOD and commercial launch trend towards larger launch vehicles, much concern exists within NASA and the space communities of the future viability of small launch capability.

4.2 NASA Effort to Maintain Small Launch Capability

Access to small, dedicated, commercially available launch services has become a priority within NASA as well as with the USAF Space Test Program. NASA seeks to aggregate its Enterprise requirements and work with the USAF Space Test Program (STP) in order to maintain US small class launch space capability. Increased use of cooperative missions with international partners, where NASA contributes the small US launch capability, is also under study. There is also an ongoing effort investigating alternative methods of access to space to the International Space Station with contingent launch capability. While this investigation is not limited to small launch systems, it could provide additional launch opportunities for small launch systems.

Because of the past and future contributions of small launch systems to NASA as well as the US space program, NASA is confident that it can overcome the recent difficulties in the small launch sector and is determined to continue its success in launching small payloads.

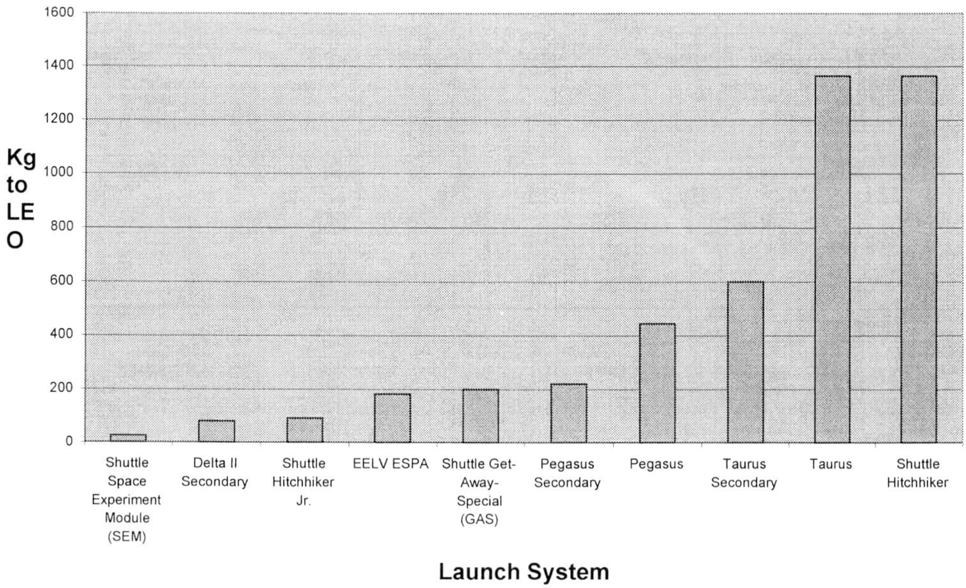

Figure 3. Capabilities of NASA's frequently used ELV and Space Shuttle small launch capability

References

1. Federal Aviation Administration's Associate Administrator for Commercial Space Transportation: *2001 Commercial Space Transportation Forecasts, May, 2001*
2. NASA Headquarters Space Access Office, *Flight Planning Board,* March 1, 2001

Vega : the European Initiative for Launching Small Satellites

R. Barbera, European Space Agency, ESRIN, P.O. box 64, 00044 Frascati, Italy

e-mail: romano.barbera@esa.int

Abstract

Vega is a small launcher under development by the European Space Agency (ESA). Its size and performance have been tailored to a design reference mission corresponding to the capability to launch a satellite of 1500 kg from the European Spaceport in Kourou (French Guyana) into a circular polar orbit at 700 km altitude. The possibility of other types of missions, including double launches, are included.

The development of Vega is part of an overall European launchers strategy devised during the year 2000, aimed at enhancing the European independent access to space by new launchers complementary to the Ariane family. In this perspective, Vega represents the smaller element of a possible family built up on the basis of common technology developments and components. During the last two and a half years, when the definition phase was carried out, several iterations have been performed, in particular in the selection of the first stage. The parameters playing a key role in the trade-off process were the overall development costs, the potential savings in the recurrent production costs, the increase of performance to adapt to the evolving market projections and the investment in new technology offering positive spin-offs in other ESA programmes.

The baseline configuration includes three solid-propulsion stages and an upper module with liquid propulsion for orbital manoevre and control. The first stage, called P80, features a 3 m diameter and carries about 80 tonnes of propellant accommodated in a new-technology filament-wound case. This stage adopts new technology solutions for thermal protection, for the nozzle and its actuators, as well as in the composition of the propellant grain. Due to its innovative content, this stage requires a longer development schedule, which drives the overall Vega schedule. The second stage makes use of the Zefiro motor, in a version carrying 23 tonnes of propellant: this motor is in an advanced stage of development with three successful full-scale firing tests. The third stage is a shorter and lighter (9 tonnes) version of the Zefiro motor.

The upper module, called AVUM for "Attitude and Vernier Upper Module", is composed of two separate elements to accommodate the liquid bi-propellant system for primary manoevres and the avionics modules with GNC, power supply, telemetry and safeguards.

1. Introduction to the VEGA Programme

The origins of the Vega programme go back to the early 1990's when studies were performed in several European countries to investigate the possibility to complement, in the lower payload class, the performance range offered by the Ariane family of launchers.

The Italian Space Agency (ASI) and Italian industry, in particular, were very active in developing concepts and starting pre-development work based on established know-how in solid propulsion. When the various configuration

M. Rycroft and N. Crosby (eds.), Smaller Satellites: Bigger Business?, 249–255.

options began to converge and the technical feasibility was confirmed, the investigations were extended to cost estimates and to market analysis.

As a conclusion of these preparatory activities, in February 1998, ASI proposed to develop a small launcher - in the meantime called Vega - in the framework of the European Space Agency (ESA), as a co-operative project with other European states. The main requirements defined then were the following:

- 800 kg payload launched into a Sun synchronous orbit (SSO) at 1200 km altitude
- launch from Kourou, French Guyana
- 2 m payload fairing diameter
- maximum synergy with Ariane developments
- low cost (less than US $ 20 million)
- first launch before early 2003.

The programme was adopted by ESA in June 1998 but the funding was limited to a Step 1, with the aim of getting the full approval by the European Ministers meeting in Brussels in May 1999. This milestone, however, was not met since it was not possible to obtain the agreement of France to participate in the programme. This situation gave rise to a period of political uncertainty and to a series of bilateral and unilateral negotiations aiming at finding an agreeable compromise. It is important, on the other hand, to record that during the same period the technical definition work continued without major disruptions. Development tests on the Zefiro motor were conducted with success.

The extension of the duration of the Step 1 provided the opportunity for revisiting and updating the market analysis, considering the evolution of potential customers and of competitors. The result of this exercise indicated the need to refocus the reference mission towards Earth-observation payloads, and to increase the performance up to the capability of launching a 1500 kg satellite into a 700 km polar orbit, whilst maintaining the cost objective. Several iterations were performed to optimise the configuration, the results of which are described later in this paper.

One of the major options selected after technical and programmatic trade-offs is represented by the adoption, as the first stage for Vega, of a new high-performance solid motor featuring several technology advances, in particular a filament-wound case (a case with a structure made of composite filament carefully wound) for size, propellant mass and internal pressure not available in Europe. This new motor, called P80FW in association with the 80 tonnes of propellant mass, will not only offer increased performance and lower production costs, but also pave the way to future applications of medium-size launchers complementary to Ariane 5 and in a new generation of boosters for Ariane 5 itself.

After about 2 years of definition and consolidation activities, the Vega configuration, including the new P80FW first stage, two additional solid stages (Zefiro 23 and Zefiro 9) as, respectively, second and third stages, and the upper module AVUM, was established and ready for development as an ESA programme. The formal funding was granted by the participating European States in December 2000, within a financial envelope of € 335 million. In parallel, the development of the P80 FW motor, was approved, with a budget of € 123 million.

2. Opportunities for the VEGA Launcher

The decision to develop a small launcher is a response to a Resolution in the Space Transportation Strategy adopted by the ESA Council in June 2000:

"completing, in the medium term, the range of launch services offered by the addition of European manufactured small and medium launcher, complementary to Ariane, consistent with diversified users' needs and relying on common element, such as stage, subsystems, technologies, production facilities and operational infrastructure, thereby increasing the European launcher industry's competitiveness".

Vega will also satisfy a potential market for launching small satellites identified in several forecasts. Thus NASA, has emphasised on "small missions" making use of low mass satellites and low capacity launch vehicles, and several European space agencies, especially the French and Italian agencies, will follow similar paths. The development of small satellite standard platforms, such as Minisat Proteus and PRIMA, has been initiated. It is expected that the availability of such standard platforms will attract several applications, allowing cost reductions and new projects.

From a technical point of view, the recent evolution of Earth observation technologies allows for a reduction of satellite mass. Optical and infrared detectors are now much smaller and, even in the field of radar observations, all weather surveillance can be performed using satellites having masses around 1 tonne. The Earth Observation programme in ESA comprises two main components - Earth Explorer, science driven missions, and Earth Watch for application missions. Both are based on multiple small missions instead of a single large satellite like Envisat. Small satellites are considered as a suitable alternative to traditional satellites, for either visible or radar military Earth observation missions. For scientific missions ESA is proposing a family of small satellites to demonstrate enabling technologies to be used for future larger missions (SMART for electric propulsion, etc.).

In the field of telecommunications, the number of systems relying on use of small satellite constellations is multiplying. Two possible types of mission are identified for small launchers. "Little LEO" constellations, which are dedicated to data transmission, store and forward services in real time and messaging applications, are based on satellites weighing some hundreds of kilograms. These satellites may be launched as a single or multiple payload by a small launcher. "Big LEO" constellations are based on satellites of about one tonne. For these systems (e.g. Globalstar, Iridium), spares management is not a trivial process, and the timely replacement of a failed spacecraft, typically in less than two months, allows a major saving. With constellations relying on a very large number of satellites, the market potential for small launchers could increase significantly due to the needs for such replacements, it being understood that the overall deployment strategy will rely mainly on medium and heavy launchers.

As result of different, independent assessments of the potential market for a European small launcher, it has been estimated that the number of European (and a few non-European) governmental missions, which will make use of a small launcher, will initially be of the order of two per year and may grow to four per year after 2005, with a total of 30-35 launches in the period 2004-2013.

The projections based on present requirements show that for little LEO constellation deployment, and for big LEO and broadband LEO satellite replacement, the market is still uncertain. Taking into account the rapid changes and the uncertainties as to how constellations will develop, the number of additional payloads deriving from the commercial applications is estimated to be one or two per year.

In the long term, an increase of the launch rate by up to 30 % may be envisaged as a consequence of the following possible events:

- an increase of the space budget of National Agencies
- confirmation of the small mission approach
- commercial success of constellations
- improvements in miniaturization technologies for small satellites.

In conclusion, combining the forecasts for the various categories of customers described above, the projected market for a European small launcher may be three to four launches per year, in the initial years starting from 2004. The total number of launches is expected to grow up to five or six launches per year, once the service is well established and the vehicle is marketed internationally, by an experienced organization such as Arianespace. This market represents a realistic baseline for Vega operations.

3. VEGA Small Launcher Objectives

On the basis of the identified European needs, particularly in the field of the radar satellite systems, and of the requirements emerging from the market survey, the required in-orbit capability for the reference mission is specified as:

- 1500 kg to a 700 km altitude, circular polar orbit
- Kourou launch site

In addition to the reference mission, Vega will be able to launch satellites into orbit for a wide range of missions and applications, for instance with a range of orbital inclinations from 5.2 degrees to SSO, altitude between 300 and 1500 km, and payload mass between 300 and 2500 kg.

Vega provides a minimum payload dynamic envelope, in a single launch configuration, defined by a cylindrical volume of 2.35 m diameter and 3.5 m height, plus an additional conical volume of 2.8 m (with a height of 2.8 m) growth potential studies are being carried out to investigate the design impact of increasing the length and/or diameter.

The payload (of special mass) is supported by a standard 937 mm diameter mechanical interface, the separation device being provided by the launcher.

The probability of a Vega launcher failing to inject the payloads in the specified orbit, due to failure or malfunction of any component, shall not exceed 0.02 (i.e. minimum reliability of 0.98, with a confidence level of 60%).

Consistent with the market projection and the offered performances the programme objective is to achieve the qualification launch of Vega by the end of 2005.

4. Summary

In its baseline configuration (see Fig. 1), Vega is designed as a single-body satellite launch vehicle composed of three solid-propulsion stages, and an additional liquid-propulsion upper module (AVUM) for attitude control, satellite release, and fairing for payload accommodation and protection.

Of the three solid stages, the most significant is the first one (P80 FW) which features advanced technologies, like a high-performance motor case in high-strength fibre, a low density thermal protection system, a new solid-propellant grain, a simplified design nozzle and lighter igniters.

The second and third stages with, respectively, 23 and 9 tonnes of solid propellant, are direct derivations from the Zefiro 16 motor already in an advanced stage qualification.

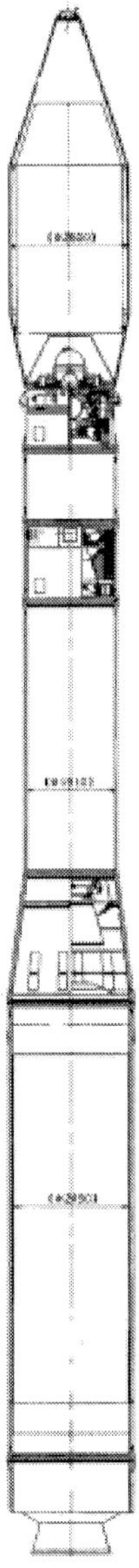

Figure 1. Diagram of the VEGA launcher

Overview of the Russian Launch Possibilities for Small Satellites

G. M. Webb, Commercial Space Technologies Ltd., 67 Shakespeare Rd, Hanwell, London W7 1LU, UK

e-mail: cst@cstltd.demon.co.uk

Abstract

Russian launchers, as well as those from Ukraine, and launching facilities in Kazakhstan, Plesetsk and elsewhere offer various unique facilities and features for the cost effective launching of small satellites. However, some of the characteristics and problems of dealing with Russia tend to prevent many potential customers from using the facilities and gaining from the advantages offered.

In this paper a description is given of all Russian and Ukrainian launchers currently extant. Their suitability and recent record for the launch of small satellites is also considered. Planned launchers and variations of current ones that are likely to be developed and used are also discussed. CST's experiences as a negotiator of launches and a supplier of logistical support for seven small satellites in Russia are reviewed, and new opportunities for future launches are described.

1. Introduction

From the launch of Sputnik-1 in 1957, the Soviet Union regarded space as a prime ground on which it could compete in the cold, chiefly ideological, war. The USA, the chief ideological opponent, responded by winning the key moon race, but, as a democracy, failed to produce any coherent space programme to fulfil the dreams of its public. The Soviet Union tried, and to a large extent succeeded, to substitute the glories of space achievement for the lack of consumer goods. Russian politicians and public remain, on the whole, space enthusiasts.

The cold war and its resulting trade restrictions produced gaps in Soviet satellite technology which, as a result, were quite "agricultural", short-lived and rather unreliable. This was compensated for by a launcher fleet that was mass produced, efficiently designed, technically advanced and very cheap to use. Therefore a policy evolved to make satellite systems cheaper, and cheaply replenished. As a consequence, in the years before the economic consequences of its command economy began to bite, it was performing ten times as many launches as the rest of the world put together. Even after the enormous contraction that has occurred in post-Soviet times, Russian space activity is still vast and multifarious. The plentifulness and cheapness of FSU launchers should not lead us think that they are technically primitive. Russian engine designs, the core of an efficient launcher design, are being adopted abroad, particularly the USA, a sure sign of their technically advanced status.

M. Rycroft and N. Crosby (eds.), Smaller Satellites: Bigger Business?, 257–265.

This paper is an attempt to summarize and assess the experience that Commercial Space Technologies Ltd (CST) has gathered during its launcher services brokerage activity for small satellites [Reference 1], which is part of its activity in Russia. Seven small satellites have been launched so far, and a similar number of projects are in progress. Small satellites as discussed here are defined as having a mass of up to 500 kg, i.e. covering the "micro" and "mini" ranges as defined by SSTL [Reference 2].

Russia is the most suitable space arena for launching small satellites for four main reasons.

- Russian space activity involves a large variety and quantity of launches to all the orbits required by small satellites (except equatorial)
- Due to the tough financial conditions in post Soviet Russia, prices are favourable and companies eager to please and be competitive
- The de-commissioning of a great many missiles has led to a flood of conversion launchers
- Also due to the mechanical robustness of Russian satellites a large number of possibilities for piggy-back launchers are available, and there is a willingness to adapt major national missions to this end.

In post-communist Russia most space organisations (and key personnel) correspond in major ways to their pre-1991 equivalents. The Russian Aerospace Agency (Rosaviakosmos) performs analogous functions to the old Ministry of General Machine Building but, of course, with much reduced financing. Some organisations, particularly the old "politicals" such as Glavkosmos, have faded. The master of the old field, the Korolev design bureau (now "Energia") needs patronage from Rosaviakosmos to function. Important facilities such as Yuzhnoe and the Baikonur cosmodrome are now in foreign countries. Some survive on continuing military needs and contracts (e.g., Makeev). Some, such as Khrunichev, which had the luck to have saleable commercial products, have burgeoned; Khrunichev has eclipsed even its central founding organisation, NPO-M. Others, such as TSNIIMash, retain exactly analogous positions.

Because of these analogous positions in the Russian space sector (as well as the similar position of the whole of the remaining military-industrial complex) the term CIS does not have the right feel when applied to space. A better term is Former Soviet Union (FSU), and this term will be used here.

2. General Features of Russian Launchers

The Soviet Union produced four families of launchers capable of launching medium and large satellites. These are relevant to small satellites only in the way of piggy-back launches.Of the four, only the Soyuz family (the first ever satellite launcher) and the Zenit family (one of the latest) have been used for piggy-back launches. Of the remaining two, the Energia/Buran system, even though it was the largest and most technically advanced overall, has been abandoned. The Proton system has not been relevant for small satellites (as piggy-backs) up to now, but may become so (see Section 7). Small launchers, conversions of missiles and projects relevant to small satellites in the micro and mini ranges are discussed in Sections 3, 4 and 5 respectively.

Russian launching systems had, and still mostly have, many common features making them easier and more economical to use, among which are:

- System approach to launcher, launch complex and operations design
- Long production runs planned at the outset (and usually achieved); not preceded by lowest cost prototype (as in STS, Ariane-5)
- Continuous and large (now mostly past) investment in engine design and development (use of efficient engines lowers the mass-fraction of the launcher)
- No-fuss approach, such as the use of railway transport, introduction of exotic fuels (e.g., cryogenic) only when absolutely necessary
- Standardisation of launchers, which are non mission-specific, which allows rapid turnaround/changes of payload at the range
- Very experienced launch crews (at least in the past)
- Some new small launchers are ex-missiles, with military specifications, and even better all weather capabilities than Soyuz or Proton (which were in themselves military designs)
- Convenience and cost savings associated with the universal adoption of horizontal integration and pre-launch testing.

3. The Future of Cosmos and Tsyklon

With the exception of the super-heavy Energia/Buran complex (from which Zenit is derived) and the unsuccessful N-1 manned lunar vehicle, all other Soviet Union launchers originated as weapon systems, even Soyuz and Proton, the origins of which are obscured by 40 years of their use as space launchers.

In the small launcher field the Cosmos and Tsyklon are seen as pure space launchers, despite their original pedigrees. After hundreds of launches each, this view is justified. Both launchers have achieved a very high reliability; up to now, they have been the small "workhorses" of the FSU launcher fleet. The Tsyklon launcher is one of the most advanced in the world, with completely automatic propellant loading. The Cosmos, although of much older design also has good capability and well established facilities. Both launchers can launch clusters of up to six satellites with masses of up to 1 tonne (Cosmos) or 3 tonnes (Tsyklon) [Reference 3].

However, in spite of this, both of these "workhorses" are likely to cease use in the near future. The production of Cosmos has been halted for sometime and its re-commencement will be prohibitively expensive. Tsyklon will be phased-out for Russian national launches (as a foreign launcher) so that, despite being technically advanced and highly reliable, it now depends on funding and the uncertainties of international political deals to continue. Cosmos is continuing to be used for Russian national programmes, so long as the stocks last.

4. Converted Missiles

Converted missiles offer cheap and reliable launches of small satellites until i) stocks run out, ii) they become old, or iii) the START (Strategic Arms Reduction Talks) treaty (or other disarmament agreements) inhibit their use. There are 3 basic types of converted missile launcher, depending on the launch method [Reference 3].

4.1 Silo Based ICBMs

Both currently deployed missiles, the SS18 and 19, are large and use storable liquid propellant in contrast to their US equivalents. Both can be readily and cheaply adapted into launchers and considerable stocks of each exist. Both have been successfully used for satellite launches and both are likely to be used in the future in several variants, resulting in two complete families of launchers. START Treaty provisions mean that the operation of both launcher families will be curtailed after 2007 (the Dnepr will probably be cancelled). Some basic facts are as follows:

SS-19 Family: Strela, Rokot K, and Eurokot are manufactured by the Khrunichev State Research and Production Space Centre. Commercial marketing companies are NPO-M and Eurokot, respectively; there are plenty available at present. They are affected by the START Treaty. Strela and Rokot have LEO capabilities of under 1 tonne and 1.5 tonnes, respectively, and the

Keynote Address

The Small Satellite Business: Contributions to, and from, Developing Countries

M. Othman, Office for Outer Space Affairs, United Nations, Wagramerstrasse 5, A-1220 Vienna, Austria

email: mazlan.othman@unvienna.org

Abstract

The most pressing need of the developing countries and, therefore, the most useful deed which we could undertake is the establishment of capacity in all areas of space activities. Although more engaging to address, technology transfer is just one component of the equation. Just as important is the necessity to build capacity in space science and technology applications, particularly to demonstrate clearly how small satellites can contribute to important decision making. While it is more profitable simply to sell technology, due consideration should be given to joint investment and financing for projects. Additionally, the issue of the sharing of intellectual property rights, as is deemed appropriate, has to be addressed seriously in order to allow the developing nations to advance incrementally.

Many developing countries have made, or are planning to make investments in small satellite programmes that could be harnessed to complement the space infrastructure in existence; in turn this could augment scientific knowledge on the environment, weather, disasters, etc.. The active involvement of such nations in the space enterprise will bring about a more universal recognition of the importance of space and of the benefits which it can bring to humankind.

1. Introduction

There is no denying that, in the realm of space, the divide between developed and developing countries is a yawning chasm. And while it is essential to concede that this divide does not go along traditional lines, since many countries considered economically developed fall under the "developing" category due to the absence of a space industry or a space programme, I would like to confine ensuing discussions to the definition of developing countries as widely understood within the United Nations context. In any case, in whichever way the word developing is applied, it is irrefutable that small satellites have a meaningful role to play in bridging the disparity.

The overall theme I will dissect in the next few minutes is that small satellite development programmes can, and should, bring about a situation where the cooperation between two parties is not a one-sided relationship. But, as in other things in life, a conscious effort to ensure a more equitable partnership needs to be instituted. This approach will guarantee that

M. Rycroft and N. Crosby (eds.), Smaller Satellites: Bigger Business?, 269–276.

developing countries will be able to contribute to the continued development of the small satellite industry.

I therefore first consider how the small satellite industry can make a meaningful contribution to developing countries within this equitable partnership context, and then discuss what the developing countries can do to advance the small satellite business.

2. Contribution of Small Satellite Development Programmes to Developing Countries

There are several ways in which small satellite development can assist in the catalysis, implementation and sustenance of a space programme in the developing countries. The literature abounds with examples of how this can be done. The question which we need to ask ourselves is: has it added value to development in a meaningful way? In my view, there are still gaps in the way we do things that we need to address, if we are truly concerned about making a real contribution. I will touch on three things.

First, we should be inordinately cognizant of the fact that, in developing countries, political subscription is especially important, and that there is an absolute need to show rapidly what is achievable in the programme and to demonstrate clearly substantive use of the space asset. There is no denying that space programmes in space faring nations are also under similar pressures, but one must admit that the magnitude is not the same. A particular project may be chopped but the space programme in that nation will in all likelihood survive. We can tackle this concern by embarking on an institutional and human capacity building agenda at the outset.

Secondly, there are financial issues that cannot be ignored. To address this, my thesis is that, it is appropriate for us to take an active role in identifying possible financing/funding mechanisms. Further it is advantageous to all of us if the space industry were to shift its marketing approach to embrace the concept of joint investment.

Thirdly, to ensure the sustainability of a programme, not only do we need to make space applications operational but also we have to make sure that the recipient of the technology is able to continue to solve emerging national problems using solutions offered by small satellites. To do this, local engineers and scientists must be able to progress to the next steps of satellite design. My argument is that, if we are serious in allowing them to achieve this, we need to surmount the issue of the protection of intellectual property rights.

I now explore the arguments to explain my case.

2.1 *Capacity Building*

There are two major pursuits in a capacity building agenda that can rapidly enhance the competence of developing countries in the space arena. One is to make technology transfer happen, and the other is to make space applications operational.

Most people subscribe to the belief that a small satellite programme lends itself well to an effective technology transfer to developing nations. The extensive use of commercial off-the-shelf products, the inherent requirement for small dedicated science and engineering teams, and the quick turn-around time from design to flight model are some of the commendable qualities suited to overcoming the constraints faced by these countries. We have good examples of past successes and we are, therefore, usually very satisfied and comfortable when we talk about technology transfer. There is no need for me to analyze this further.

Notwithstanding the excellent record which we have in technology transfer, have we unequivocally demonstrated the utility of small satellites in the day-to-day decision-making process, and in other activities? My view is we have not. To date, no nano- or micro-satellites in operation offer services which are similar to those offered by the LANDSAT or SPOT series of satellites. The Orbcom small satellite communications constellation today has claim to an only limited success. Many, if not most, micro-satellites do not have users beyond the institutions that built them.

My premise is that, we need to treat as top priority the requirement to ensure that small satellites become a useful, maybe even indispensable, component of the information system for a developing country in its decision-making processes. For this purpose, not only do we need to design small satellites that solve a real problem for that nation, but also that we have to put in place steps that lead to making space applications operational.

And how do we do this? By planning and executing holistic technology transfer and training programmes which amply address the issue of how the satellite data will be received, distributed and used by as large a user community as possible. Naturally, the developing nation has a major part to play in this aspect, but our role is important in providing leadership and direction. The world does not yet fully beholden to the utility of satellites for day-to-day operations. It is in the wider interest of satellite manufacturers, therefore, to take one step further in ensuring the optimum utility of the satellites delivered.

2.2 Financing and Joint Investment

In the numerous dialogues which the United Nations has had with developing nations on the obstacles that prevent them from fully benefiting from space science and technology applications, finance is one of the most cited parameters. It is in our extreme interest to know what that really means in the context of our industry. So let me share with you the responses that we have received. In most cases, they just need money to buy computers so that they can process satellite data; in some cases, they need financing to build receiving ground stations and, in a few cases, they assert that they do not have the resources to build their own satellite to serve their special needs. Let us look at the latter two reasons because they are most relevant to this Symposium, and let us see if there are ways in which we could contribute to surmounting the problems.

To help fulfil the need for either a small satellite ground station or the fabrication of a small satellite, our initial contribution could be in identifying, together with the developing countries, what funding mechanisms are available. Many of us know what can be done to access funding, but we sometimes miss the opportunity to highlight them. I recognize that in some instances there could be ethical issues; each case needs to be looked at according to its own merits and circumstances. But let us not ignore this as a tenable contribution.

The next step, which is how to obtain the funding, is more difficult and most crucial. In this case, we are in the position once again to assist the developing nation by providing the data and arguments needed to justify such funding. Who says that this sort of role can, and should only, be played by governments and organizations such as the United Nations? What we in the small satellite business have in our favour is that we can more easily make a case for either a ground receiving station or a small satellite, given the low costs of both of these in the context of our business. Our chances for success are therefore much greater.

I would now like to take you one step further into our discourse on fiscal matters and talk about joint investment. I can see that I could be questioned on the congruity of this subject. One party wants a product, another party sells it. End of story. There is no need for joint investment. My view is that, if we take this approach, it could also mean the end of business.

The latest books in the market on management, negotiations or marketing emphasize "win-win" outcomes, collaboration and partnership. I fully understand that this is a less than ideal world and one could still make an argument against these principles. But should we and need we in the small satellite business adopt this attitude?

We all agree that the small satellite business scenario is a rapidly changing one. One rather important aspect is that the technology evolves faster than we can cope with and we are constantly having to change the way we do things. I would like to assert here that there is special attribute of the small satellite enterprise which we have not fully grasped and exploited, and that is, small satellites are best suited for [and also best marketed as] small dedicated missions. The nature of such missions is they could be country specific, or application specific or payload specific. It is, therefore, extremely likely that it requires a fresh approach to mission design with much room for innovation and creativity. Now given this necessity for a change of approach in design, and given the inherent exigency to incorporate the latest technology for our own business survival, I think it is reasonable that this development cost is taken up by the satellite manufacturer. I would go one step still further and propose that we should endeavour to create a climate of partnership where in the process of design development, for instance, the participation of the developing country is actively allowed for and encouraged. Most of the developing nations who are starting up space programmes do so not only because they have certain needs to meet, but also because they already have some basic capability. It is up to us to fully utilize that basic capability and ensure it grows.

2.3 *Incremental Advancement in Technology*

I have mentioned before that, to enable the developing country to continue to solve emerging national problems using solutions offered by small satellites, local engineers need to have a real and demonstrated capability to move on to the next steps of satellite design and fabrication by building on whatever skills and knowledge which they have already acquired. This is all a part of the technology transfer process, you may say and, therefore, is a non-issue. I would like to assert that things are not as simple as we assume they are and that is why I am treating this separately.

The reason why a technology transfer programme is sometimes not what it appears to be is because most contracts have an intellectual property rights clause and these are sometimes worded in such a way to largely or completely prevent the technology recipient from using the technology, or even a part of it, in future satellites. In some cases, contracts preclude the transfer of information to a third party, which could sometimes be construed to include even educational institutions. Common sense will tell us that this situation will hamper progress and, in fact, engender continued dependence. I am not advocating that you give up your intellectual property rights entirely. Instead, I would like to appeal to our sense of what constitutes a genuine partnership and to think about, as well as embrace, the principles of such a relationship.

Once again, I would like to come back to the spirit of sharing that must prevail if we truly would like to make a contribution to the progress of developing nations. They, too, have an essential role to play in furthering the ideals and motivations of the small satellite enterprise.

3. Contributions from Developing Countries

Many developing countries have made, or are planning to make, investments in small satellite programmes. In this way, they inject the crucial catalyst that is needed to advance small satellite R&D and business, namely, finance. Apart from this, they are in a position to offer the services of a space infrastructure that will contribute to the scientific knowledge of the world. My final assertion is that the involvement of the developing nations in the space enterprise brings about a more universal recognition of the importance of space and the benefits which it brings to humankind.

Let me turn now to each of these aspects.

3.1 Financial Investment

Many developing nations have already invested in small satellite development programmes - some examples are Chile, Pakistan, Morocco, Thailand, Saudi Arabia and Malaysia. We will hear about South Africa's plans later on from Garth Milne. Many more are beginning to invest, including Algeria, Indonesia, Nigeria and Egypt. I suspect there are more in the pipeline, and there are situations which could be potential opportunities for the small satellite business.

We cannot deny that the investment by these countries serves to drive small satellite development. We, of course, can also make the observation that it is the nature of the small satellite enterprise that makes access to space more affordable and, therefore, more attractive to the developing nations. Hence this mutually serving scenario should be capitalized on by both the small satellite industry and developing nations.

3.2 *Scientific Contribution*

Having made an investment, the developing country is, of course, in a position to make further contributions. Most small satellites planned by the developing nations have specific uses to address their issues and problems but, as these objects circle the globe, they can easily serve a larger purpose. For this reason, developing countries can provide data on the global environment, weather, etc..

My view is, developing nations on their part should actively pursue this concept. If in our planning stages we contrive to fill the gaps in scientific knowledge (as is allowed by satellites), we could go a long way to making a significant contribution to the sum of scientific knowledge on questions that affect all humanity.

3.3 *Universal Acceptability of the Benefits of Space*

Space is by no means an easy arena to get into and this is true for all nations. Budget issues aside, questions still abound on the true motivations of desiring to be in space. But the fact that many peace-oriented developing economies, for example Malaysia, Thailand, Morocco, Nigeria, Chile, just to name a few, are now actively pursuing activities in space, speaks well for space. And the added fact that the entry point of most of these nations is small satellites, speaks volumes for the small satellite enterprise.

In other words, the participation of the developing countries is excellent PR for the small satellite undertaking and for the space industry as a whole. In addition, we should recognize the fact that one developing nation attracts the attention of another who sees an example that can be emulated. So the circle gets bigger. And this answers the question of this Symposium which is begging for an answer.

4. Conclusion

In suggesting ways in which we can contribute to activities in developing countries, I have asked that you go beyond technology transfer, to address the need to make space applications operational in these countries. I have appealed to you to provide leadership in identifying funding mechanisms and I hope to have touched on your big heartedness to seriously consider joint investment. I additionally urged that you be not too inflexible in the approach to the protection of your intellectual property rights.

I ask you these things because, more than being noble reasons, my view is they make good business sense. Because developing countries, with their past and intended investment, their potential contribution to scientific knowledge and their promotion of the universal acceptability of space for the benefit of humankind, so important to us all here, are part of the parameters driving our business. It is my unwavering belief that a smart partnership concept along the lines which I have outlined will forge a strong relationship that will foster equal participation and, which, for the small satellite enterprise, will ultimately mean bigger business.

Session 5

Broader Impacts and Implications

Session Chair:

S. Mostert, Sun Space and Information Systems (Pty) Ltd., S. Africa

SUNSAT to SunSpace : Meeting the Needs of Technology Transfer and Commercialisation

G. W. Milne, J. J. du Plessis, J. A. Koekemoer, S. Mostert, A. Schoonwinkel, Electronic Systems Laboratory, Faculty of Engineering, University of Stellenbosch, Private Bag X1, Matieland 7602, Stellenbosch, South Africa

email: milne@ing.sun.ac.za, jjdp@ing.sun.ac.za, jkoekemr@ing.sun.ac.za, mostert@ing.sun.ac.za, schoonwi@ing.sun.ac.za

Abstract
South Africa's first satellite, SUNSAT, has led to a network of space activities in South Africa, including the formation of a new commercial micro-satellite organisation, Sun Space & Information Systems (Pty) Ltd, as a spin-off company of the University of Stellenbosch. The SUNSAT satellite was a graduate student-developed project, and included a NASA GPS receiver for occultation research, laser reflectors, magnetometers, star cameras, amateur radio communications, and a 15-m resolution, 3456 pixel, 3-band, stereo-capable push broom imager, and a live PAL TV downlink. As OSCAR-35 (SO-35), SUNSAT's amateur radio FM audio transponders provided excellent performance.

1. Introduction

Establishing and maintaining access to space is an undertaking that is particularly well fulfilled by micro and small satellites. South Africa's, and Africa's, first indigenously developed satellite, a 64 kg, research micro-satellite called SUNSAT, was launched into polar orbit on 23 February 1999 and demonstrated its major mission objectives over an operational period of 23 months [Reference 1,2].

The satellite developed at the University of Stellenbosch includes a 15 m ground spacing distance (GSD) pushbroom imager as the main payload. This imager is also in space on the Kitsat-3 satellite, and represents a significant improvement in performance for micro-satellites, providing co-registered, 3-band multi-spectral images from 800 km orbit height. The on-board amateur radio services were extended by software upload to include APRS, a rapidly spreading amateur radio protocol for short messaging, automatic position reporting, hazard mitigation and disaster management communications. This protocol is already implemented in some handsets, and will have a wide economic impact through its ability to monitor and track assets in a country.

An economic benefit and commercial interest is needed for the sustainable development and utilization of space. For countries that do not have a government program, or have a limited government program, a commercial spin-off is paramount to maintaining and developing the benefits made possible by space access. The formation of Sun Space and Information Systems (Pty) Ltd.

M. Rycroft and N. Crosby (eds.), Smaller Satellites: Bigger Business?, 279–287.

(SunSpace for short), as an affiliate company of the University of Stellenbosch, has established an economic mechanism to sustain the benefits from a start-up space programme. The company was formed to build on the success of the 15m GSD imager on SUNSAT and Kitsat-3, and to continue developments in remote sensing micro-satellites. By this extension beyond the academic environment in which SUNSAT was developed, SunSpace provides the opportunity for commercial contracting of space system development in a cost effective and sustainable manner. The company also provides career opportunities and stability for high-performing graduates from the academic program.

SunSpace is a key contributor to the South African government's effort to establish a national space programme to make the benefits from space available to the wider population. A significant increase in the positive impact on society is expected if the government establishes a "positive" regulatory framework which encourages the use of space technology for addressing the information needs of the people.

This paper examines the transition from an academic satellite development environment to a commercial environment where the needs of technology transfer, quality assurance, and commercialisation of small and micro-satellites can be effectively met.

2. SUNSAT Program

2.1 SUNSAT Origins

In the early 1990s in South Africa, space was considered to be the domain of the military, and the Greensat satellite was under development. During the political transition, this project slowed and was terminated in 1994, stopping ESA-standard professional space developments in the country.

Industry, however, had a need to increase the number of master degree qualified engineers emerging from universities who were able to contribute immediately in developments teams in their companies. This need motivated industry's support of students gaining research and development experience on a low-cost micro-satellite. SUNSAT resulted from this initiative, and involved over 100 students from its inception in 1992 till launch in February 1999; over 65 master's theses were completed.

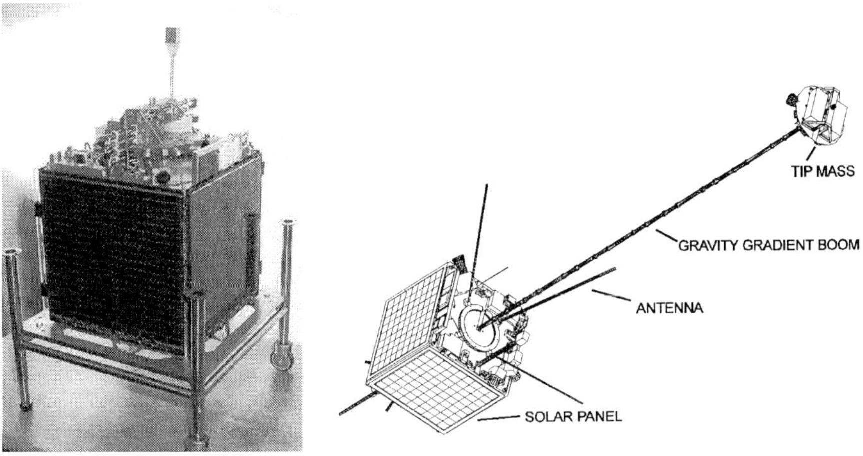

Figure 1. SUNSAT micro-satellite in handling frame prior to installation on launch vehicle (left), and the in-orbit configuration (right)

2.2 *SUNSAT's Big Step, High-risk, High Payoff Approach*

With the background of Greensat, industry only wished to support a serious project that would act as a technology demonstrator for a potentially marketable service. Micro-satellites being developed in 1992 were not exploiting the high resolution capabilities possible for 50 kg satellites, so SUNSAT's technology goal was set as demonstrating multi-spectral 15 m stereo imaging capability in a 50 kg class gravity-gradient micro-satellite (see Fig.1). For educational and outreach purposes a high quality and flexible amateur radio payload was defined and developed.

2.3 *Own Technology*

The SUNSAT program produced far more than a satellite. Funding bounds and a growing corps of students needing research and development projects provided the necessity and means to develop our own technology. This license-free situation has enabled products to be used in other international satellites, and is now permitting commercialisation of the investment.

2.4 *Test Facilities*

Vibration, modal survey, and electromagnetic compatibility (EMC) test equipment from the Greensat program is located within easy reach of Stellenbosch, and was used for major tests. These facilities have now become part of the Institute for Space and Software Applications. The Electronic Systems Laboratory [Reference 3, 4] also developed temperature and vacuum test facilities for ASAP-5 class satellites, enabling subsystem tests to be completed without personnel disruptions. Digital and microwave facilities are also available in the Department.

2.5 *Launch and Operation*

Funding restrictions required a sponsored launch, and NASA enabled the project to proceed by arranging a secondary payload launch with the Danish satellite Oersted in exchange for including a NASA JPL Turbo-Rogue GPS payload. The local time of Oersted's 650 km x 800 km polar orbit drifted an hour earlier every 70 days, which is undesirable for an imaging satellite, but provided the opportunity to demonstrate the technology. All launch vehicle and spacecraft engineering requirements set by NASA were satisfied.

SUNSAT operated for 23 months from launch, and endured a full-Sun and full-charge period when the orbit plane was normal to the Sun. Communications with SUNSAT ceased in January 2001. Failure of the redundant communications system is most likely due to debris impact, or a battery-induced power short circuit following the full-Sun heating.

2.6 *SUNSAT Highlights*

Satellite. The imager is housed in a rotating tube in the lower tray of the satellite, and is capable of left-right viewing and stereo viewing on the same pass, without requiring satellite attitude changes. Since the 3456-pixel linear CCD's are optically superimposed with less than 0.2 pixel misalignment, colour registration between bands is inherently assured. By suitable design and pre-qualification testing, no anomalies occurred with the imager during vibration tests or launch.

Imager and PAL TV camera. SUNSAT included a wide field of view PAL-standard TV camera that could be multiplexed into the wideband S-band downlink. Fig. 2 shows a section of the Western Cape of South Africa, viewed from 800 km, and a 48 x 46 pixel section of a 3456 pixel-wide multi-spectral image taken in Asia that includes the circular irrigator shown. The full resolution and false-colour multi-spectral information is not preserved in this reproduction, but is evident in images given in Reference 5.

Downlink. The 40-60Mb/s S-band QPSK modulator did not operate in orbit, preventing the satellite from delivering production quantities of high-resolution imagery. The fault tolerant architecture provided an alternative downlink which enabled the imager performance to be assessed.

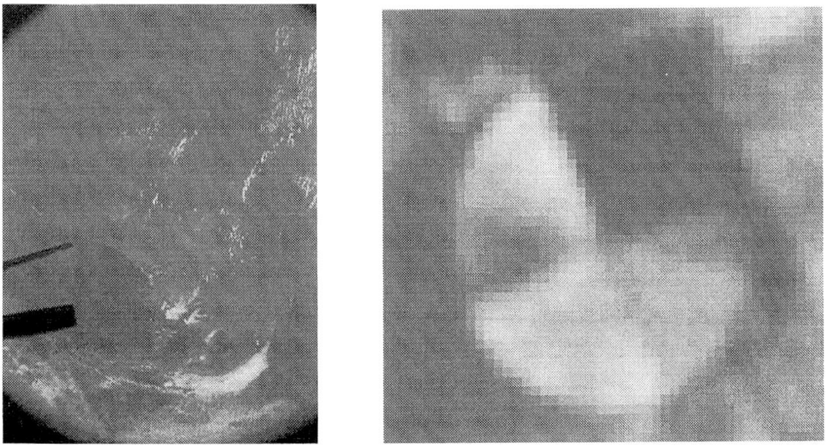

Figure 2. Section of the Western Cape of South Africa from a live TV downlink from SUNSAT (left), and 48 x 46 x 15m pixel segment from 3456 pixel-wide pushbroom image (right); the 3-band image is of an agricultural irrigator, and shows varying crop status in the four quadrants

Software. The complete suite of real-time software in SUNSAT, including the kernel and packet protocol stacks, was developed in-house. The on-board computers can be booted via interruptible RF links, and software uploads were done on a regular basis as new services were added.

SO-35 amateur payloads, Easy-sat. To encourage amateur radio satellite use, SA-AMSAT recommended that SUNSAT should include amateur payloads that were easy for beginners to use. As SO-35, SUNSAT provided hand-held cross-band transponding in mode B (UHF up, VHF down) and the inverse. It also included a simplex "Parrot" repeater, enabling 2m-only equipped amateurs to participate in satellite communications. Amateurs in Australia even made train-to-train contacts via SUNSAT. Audio recordings are available [Reference 6].

APRS introduction. Apart from being known as the best "Easy-sat", SUNSAT also pioneered the introduction of the APRS amateur short messaging service in space, including a demonstration of data collection from an automated amateur weather station in Antarctica.

3. Maintaining the Nation's Space Capability

The SUNSAT project has played a historical role in South Africa in completing the nation's passage into the club of satellite operating nations. It showed that:

- A large group of postgraduate students could successfully develop micro-satellite technology in a developing country
- A micro-satellite can perform in a multi-mission role, including 15m GSD multi spectral imaging, science payloads, and analogue and digital amateur radio services
- A first satellite can leap-frog the existing technology base in micro-satellite imaging, and introduce SPOT-2 like imaging into micro-satellite capabilities
- The amateur APRS communication service can be operated successfully from micro-satellites to pocket-carried transceivers.

SUNSAT capabilities that were not fully exploited in orbit included the 40-60 MB/s downlink, and the stereo capability. This is a consequence of the level of challenge in completing such a complex and capable satellite with a primarily student engineering force, supervised by academics who are not relieved of their normal teaching, thesis supervision, research, and administrative roles [Reference 3].

Overall, SUNSAT established a wide technology base, and a pool of satellite-trained manpower that largely remains in South Africa. It is also contributing to at least four overseas satellite projects, and has transferred satellite technology to over 100 black South Africans, thereby enabling South Africa to consider a next project in space.

From Big LEO Satellite Ventures to Smaller Satellite Systems: Lessons to be Learnt in an Apparently Evanescent Market

P. A. Salin, Center and Institute of Air and Space Law, McGill University, 3661 Peel Street, Montreal, Quebec, Canada H3A 1X1

e-mail: salin.patrick@uqam.ca.

Abstract

The Iridium venture had the privilege of introducing the international community to the novel concept of commercial mobile satellite communications. It fared well until competition demonstrated that alternative solutions could be developed as efficiently, with a faster and higher return on investment. This paper tries to identify a few lessons that can be learnt from the Big LEO adventure and to assess the possible evolution of other low Earth orbit satellite ventures that still intend to envelop the Earth with constellations of satellites. It also explores the development of smaller satellite systems that may be an interesting alternative.

1. Introduction

The Iridium experience was a *grande première* in the field of satellite constellations. It opened the way to others and, while it developed slowly, it also generated comments by incredulous observers that it would not work. Eventually, it did work, but the world was stunned to hear in the fall of 1999 that it finally did not work, commercially. It was ultimately salvaged for the sake of its maverick technical approach, as it seemed that such a mammoth investment should not be eradicated so abruptly. We also have seen the collapse, or withdrawal, of other satellite communications projects, including small satellite constellations. Therefore, we can hardly pinpoint the size or type of satellite itself in order to find the reason for these failures. We now try to have a global view on the Big Leo experience first, then to examine alternative solutions and, finally, to find the lessons that may be learnt.

2. The Big LEO Experience: From Rise to Fall, and Rise Again

2.1 *The Challenges that Iridium had to Meet until Its Official Failure*

Iridium wanted to be there first, and they were. The Iridium project sparked the imagination of the outer space community in the early 1990s, with its several specific challenges, which it met with varying degrees of success.

M. Rycroft and N. Crosby (eds.), Smaller Satellites: Bigger Business?, 289–298.

Technical challenges. The main challenge was to build, deploy and operate a space segment consisting of 66 satellites, with 11 in each of the six orbital planes. Figures such as the average cost of one satellite around US $15 million and the deployment of the fleet in over one year give an idea of the magnitude of the project. The architecture of the Iridium system was complex and led to several acclaimed technical breakthroughs in satellite performance. But the system was unsuccessful as its users complained about its poor quality of service and cumbersome handsets.

Regulatory challenges. Iridium did a good job of acquiring licenses and overcoming regulatory obstacle. Negotiations lasting three years were held with the US Federal Communications Commission (FCC), and Iridium obtained what it wanted in terms of frequency spectrum during the World Administrative Radio Conference (WARC) of 1992. By the end of 1998, it was reported to have received licenses from most of the countries where it planned to provide a service, about 90 priority countries, or 85% of its business plans. But other regulatory challenges had to be met in order to comply with ITU regulations.

Finance and marketing challenges. The newly formed Iridium Limited Liability (LLC) required financing in the US $ 3.5 billion range. It took advantage of the privatization of telecommunication companies and finally secured more than US $ 5 billion. But its marketing performance was disastrous. The satellite mobile telephony concept was conceived during the 1980s, at a time when no valid alternative existed in terms of global voice, data, fax and paging services. Iridium's Security Exchange Commission (SEC) filings estimated subscribers of 42 million by 2002. These figures never materialized, and it had only 10,300 customers in March 1999. With less than 10% of the US $ 30 million revenues anticipated, this was not enough to service a debt load of U $ 100 million every quarter. Iridium filed for Chapter 11 bankruptcy protection in August 1999, with the suspension of the trading of its shares on NASDAQ at a price of US $ 3 1/16, from a 1998 high of US $ 68.

2.2 *The Renaissance of Iridium, still a Valid Big LEO Project?*

While the threat of de-orbiting the Iridium fleet of 84 satellites was gradually being taken on board by Motorola, with complex legal and regulatory issues at stake, various plans to salvage the project were being elaborated and then dropped. On 12 December 2000, a new venture by the name of Iridium Satellite LLC announced it had purchased the operating assets of Iridium LLC and its subsidiaries for US $ 25 million. The approach by the new debt-free owner is to slash operating costs to their maximum. With monthly operating costs of about US $ 7 million, 10% of those of the former Iridium, the plans were to charge US $ 1.50 a minute compared to US $ 7 a minute previously. In addition, the scope of services was broadened to allow for the higher-speed transfer of data. We now need to see the financial outcome of this resurrection.

2.3 Experiences from other Failed or Changed LEO/MEO Projects

Globalstar: Iridium's historical challenger. Globalstar was a small satellite program, since the average mass of its satellites was in the 350 kg range. The concept of the 48-satellite constellation, the cost structure and pricing were different from those of Iridium. Backed by 11 foreign strategic partners, it also wanted to play a complementary role to regional cellular networks and planned to route its calls through the local public switched networks. Financial analysts looked upon it favorably, as it offered six times the capacity at half the cost of Iridium. The system was in orbit in November 1999, just after Iridium's default. Only 31,200 customers had subscribed to the service by late 2000. Debt payments were suspended in January 2001 and it was announced in early February 2001 that Loral Space would stop funding its affiliate, while losses for 2001 were announced to amount to US $ 3.8 billion. Similarly to Iridium, the Globalstar constellation was functioning from a technical point of view, but lacking customers. Restructuring is now awaited, while the service has been deemed viable since January 2001, including Internet access.

Amalgamation of several aborted projects. Odyssey, the third Big LEO project, was a front-runner and was licensed in 1995 for the launch of a 12-satellite MEO constellation. Its parents, TRW and Montreal-based Teleglobe, hoped to cover the Earth with a smaller constellation on a higher orbit and to provide much-needed phone services to remote areas of the globe. But they failed to obtain the necessary financing and the project was dropped in early 1998, while its main driving force, TRW, invested in ICO. As an offset of Inmarsat, ICO Global was boasted to be developed with a market-led rather than a technology-led approach. Yet, its sales projections collapsed in late 1999, before any satellites had been launched. ICO still had to cope with FCC regulatory problems and financing difficulties in gathering a missing US $ 1.7 billion. It was induced into financial rehabilitation by a US bankruptcy Court in October 1999, taken over and merged with Teledesic into a new ICO-Teledesic project. Subsequently, it was merged again, this time with Boeing's Ellipso venture, on 14 March 2001. The new group wants to develop a global voice-data-Internet communications satellite service and to reach all parts of the Earth at prices below those of the ailing Globalstar, i.e. between US $ 1 and US $ 3 per minute. It is still waiting for FCC approval, with a scheduled inauguration for 2005.

All Big LEOs are not dead. There are at least two other Big LEO satellites that are being operated with some success, AMSC-1, operated by American Mobile Satellite Corporation since 1995 and the very first Big LEO satellite of its own having been licensed by the FCC in 1989, and its twin, MSAT-1, operated in Canada by TMI since 1996 and also authorized by the FCC for North American coverage. Both spacecraft offer phone services for mobile and fixed users.

3. Alternative Solutions to These Total or Partial Failures

3.1 The Big Geostationary Orbit (GSO/GEO) Satellite Systems

GSO systems are the historical openers in the adventure of satellite communications and have developed the only profitable satellite market for the private sector so far. With the advent of LEO systems in the mid 1990s, a complex technical and operating costs debate started between the proponents of the two different orbits, GSOs versus LEOs. The novelty developed when it became known that regional GSO global mobile personal communications services (GMPCS) were scheduled to provide mobile services to specific geographic areas, thus competing with LEO GMPCS. This profusion of projects created tensions during several recent ITU Conferences as to the allocation of frequency spectrum for these systems as a consequence of the so-called "paper-satellite" issue. In addition, the market of satellite operators is now going through reorganization and the traditional boundaries between size, orbit and type of service no longer hold. Recent announcements to offer new broadband mobile satcom services, Internet Protocol (IP) applications, confirm the direct challenge to LEO small satellite services and the fact that the use of the GSO is not on a declining slope. Old and new operators are fueling what may be called an explosion of Internet access, all via the geostationary orbit.

3.2 The Little "Small Satellite" Systems

Here we consider the other small satellite systems, lighter non-geostationary (NGSO) systems as opposed to the bigger small satellites of Section 2. They include mini-satellites (100-500 kg), micro-satellites (10-100 kg), nano-satellites (1-10 kg) and pico-satellites (less than 1 kg).

Small satellite systems are already an old story. Small satellites have been used since the beginning of the space era, when many of the first US military satellites were microsats; the 24-small satellite Navstar GPS constellation is a good illustration. Microsats offer total costs that are lower than those of larger satellites and can be built and launched more quickly. Several factors helped with the development of smaller spacecraft, like the miniaturization of electronic components and the availability of smaller satellite terminals, factors that were compounded by the availability of lower-cost launchers and the belt-tightening impact of shrinking public budgets. Orbital Sciences first opened the private sector with its Orbcomm Global US $ 135 million project, a 26-small satellites LEO constellation, to provide two-way data communications to worldwide users of pocket-size terminals. Orbital also engineered small GSO satellites like Indostar to offer a 40-channel Direct To Home (DTH) television system. Unfortunately, Orbcomm was the last of the series of satellite constellation projects that eventually collapsed. It started service in December 1998 and was declared bankrupt in September 2000, without interrupting its communications services. Its customer base was deemed to be about 37,000 as of early February 2001. Its assets were finally auctioned in March 2001. Outside the US, France (with the Arsene and Cerise programs), Canada (SCISAT and MOST programs), and several other nations have developed smallsat programs mostly for scientific or military purposes.

The specific experience of SSTL. Surrey Satellite Technology Ltd. (SSTL) may have been first in anticipating and capitalizing in the early 1980s on the role that small satellites could play in all domains of space activity. It is now present in several sophisticated fields such as global disaster monitoring, Earth observations and Internet applications. Its approach was geared towards the non space faring nations by offering a smaller faster cheaper technology that it developed from its early UoSat micro-satellite missions. While the big-LEO constellations were applying for FCC licensing, SSTL was developing the concept of three-to-six satellite constellations at incredibly low prices, based on a three-steps approach - low cost, rapid response and advanced technology. It also maintained its independence from the aerospace establishment. By the end of 2000, SSTL had accumulated a heritage of 19 launched micro mini nano satellites and 9 microsat know-how transfer and training programs.

4. A Few Observations with the Benefice of Hindsight

4.1 *The "Low Cost" Concept*

Reducing the cost of a satellite project applies to spacecraft design and to the choice of a launch vehicle. Commercial smallsat systems want to recover their investment with client-paid revenues. "Largesat" projects may have overlooked the fact that being a NASA contractor is not the same as being a commercial entity.

Selecting a cost model may be risky but is a necessary task. Comparisons are useful, as they enable project managers to assess the class to which their project belongs. Naturally, the content of each project materializes where the project's overall cost stands within each range. We may say that a low-cost mission is one for which the life cycle cost is substantially less than the average cost of a standard mission of that type. Any later change in the program may increase its total financial burden.

Low-cost programs are easier to amend than high-cost programs. All projects strive to achieve their objectives on a cost basis. The failure of the Big LEO projects was spectacular not for the failure itself but for the magnitude of the figures involved, reaching several billions of US $. Then, the size of projects becomes a parameter in itself, since it has been shown elsewhere that the performance of very low-cost programs can be improved by adding relatively small amounts of funds, while the performance of high-cost programs can only be improved by large cost increases.

Distinction between "low cost" and "cost effective". Being "cost effective" often refers to conducting business as usual, implying that project managers are cost conscious in their everyday management of things, which may not be true. Being "low-cost" implies a more drastic approach; it may more easily yield a high performance result as soon as it reaches a level of optimal performance, which can be defined as a level of maximum performance per unit cost.

A "low-cost" mission implies sacrifice. While the failure rates seem to be about the same for low-cost and high-cost spacecraft, there seems to be a slight advantage to low-cost missions, because they are smaller, simpler and have fewer parts, larger margins and fewer failure modes. This apparently cannot be done without some sacrifice about what should be obtained, optimal performance, control and accountability, because of a smaller focus.

4.2 *The Financing of Satellite Projects*

After the costs issue, the financial issue is the complementary fundamental parameter of the success or the failure of commercial satellite ventures. The past five years have certainly taught us to distinguish between the "before Iridium" and "after Iridium" eras, which is not doing justice to Iridium, but it identifies a certain time, the Fall of 1999.

Before Iridium: up until the Fall of 1999. In 1993, PanAmSat was the first commercial satellite venture to raise large amounts of cash with high yield debt, opening the way to others. The first of its kind, Iridium did not have much difficulty in financing its development, benefitting from Motorola's first class

Small Satellites in Developing Countries - An Integral Part of National Development

F. D. Chizea, International Space University, Srasbourg Central campus, Parc d'Innovation, Boulevard Gonthier d'Andernach, Illkirch-Graffenstaden, France

e-mail: chizea@mss.isunet.edu

abstract

Space technology and access to space have been elusive for developing countries over the last half century, owing to the lack of awareness of policy /decision makers about the role of space technology in national development. Space technology was seen as very expensive and prestigious, meant only for the major industrialized countries, while the developing countries should focus on building their national economy and providing food, shelter and other social amenities for their ever-growing populations. In the last decade, the trend has changed, with many developing countries embracing space technology as one of the major ways of achieving sustainable development. This can be seen in countries like India, China, Morocco, South Africa Nigeria, Egypt, and South Korea, among others.

The present trend towards small satellites has also aided this transition because, apart from the small size, they tend to be cheaper to build and launch, with shorter development time, lower complexity, improved effectiveness and reduced operating costs. This in turn has made them more affordable and has opened up now avenues for the acquisition of satellite technology. This paper focuses on the various plans, programs and results of using small satellites in various developing countries, with special reference to Nigeria.

1. Introduction

Space technology, and the many practical benefits that can be derived from its utilization, will play a central role in national, regional and international economics and social development efforts. New applications of space technology are constantly being discovered and spin-offs from space technologies have led to advances in such diverse fields as medicine, material sciences and computers. Nevertheless, the high cost of participating in space activities has hindered the ability of many countries, particularly the developing countries, from taking full advantage of the many practical benefits that space offers for economic and social development.

The complexity of social, economic and political realities in developing countries continues to increase. This creates a need for better data and information as input for program development to remedy the situation, and the need for more enlightened decision-making. Indeed information provides for the current decision-making as well as a means for monitoring the outcome of past decisions. The need to improve the social and economic status of a rapidly growing population has also made the harnessing of raw materials and the intensification of agricultural production a high priority. At the same time

M. Rycroft and N. Crosby (eds.), Smaller Satellites: Bigger Business?, 299–306.

several development efforts have significantly modified the environment and induced ecological deterioration that is not reversible. The situation requires continuous monitoring and the processing of larger volumes of data into clear and simple forms for use by decision-makers.

It is generally recognized that satellite remote sensing provides valuable information for developmental planning and for environmental monitoring. However, with the present practices of data processing in the various big space faring nations (industrialized countries), the data are not often up to date or timely for many applications and can be very expensive. The absence of ground receiving stations in most developing countries coupled with the fact that many recently launched remote sensing satellites have no onboard recording systems has made it difficult to obtain potentially valuable information on the changing environment.

Urban and rural development programs require adequate telecommunications systems if they are to succeed. Experiences in other developing countries such as India and Indonesia have shown how satellite based communications systems have opened up the rural areas for development. Ineffective communications currently represents one of the greatest drawbacks to the socio-economic development of most developing countries. It is therefore important that participation in space technology must be defined in terms of this setting. Experience of developing countries, such as India, Brazil, China, Indonesia and Pakistan and of industrialized countries like USA, Russia, France, Japan, Spain, etc., have demonstrated inherent advantages from the introduction of national programs in space science and technology. These advantages have, for the developing countries, been most beneficial where vast areas are involved and the available resources are limited and meager.

Small satellites have considerably augmented the range of possible space missions, thereby reducing the cost of access to space for emerging countries. These satellites, though not a solution for all types of space mission, offer the possibility of performing ambitious science and technology applications to complement large space missions. It is to be noted that small satellites also arise as a result of electronic miniaturization, and the development of micro-technology for sensors and instruments. This has led to the integration of micro-electromechanical system (MEMS) with microelectronics for data processing, signal conditioning, power conditioning and communications, and may be defined as the application of specific integrated micro-instrument (ASIMP). Therefore, small satellites do not necessarily imply low technology but, on the contrary, very advanced technologies, offering larger payload mass fractions.

They offer valuable missions even with today's technology for science and applications as well as for education and training. The general advantage of small satellites, over larger ones, include:

- More frequent, and larger variety of, mission opportunities
- More rapid expansion of the technical knowledge base
- Greater involvement of local industries
- Greater diversification of potential users
- Reduced development time
- Reduced launch costs.

As a high technology initiative, the successful development of a small satellite project requires:

- Highly innovative technical staff
- Small, motivated teams
- Personal responsibility, rigor and quality
- Good team communications, close proximity between team members
- Well defined mission objectives/constraints
- Knowledgable use of modern components
- Layered, failure resilient system architecture
- Thorough testing of components and of the whole system
- Technically competent project management
- Short time (to prevent possible escalation of project objectives).

2. The Nigerian Experience

2.1 Overview

Nigeria is a country at the threshold of technological development and industrialization. It has a population of 88.5 million (1991 census) and a land area of 930,000 square kilometers of which about seventy five percent is estimated to be suitable for agriculture. The country is endowed with a wide variety of natural resources, which include solid minerals and hydrocarbons. Nigeria is a major agricultural producer; however, since the early 1970's there has been a shift towards petroleum as the main foreign exchange earner.

The provision of adequate food, clean drinking water, shelter, health care delivery, good roads and infrastructure for development, especially for rural dwellers who constitute about eighty percent of the population, has remained the prime objective of the three tiers of government in Nigeria. The government has introduced various programs at one time or another in the past, which are aimed at the realization of its objectives for the rapid development of the country. These approaches view rural development as an integrated process designed to raise the quality of life of the rural and urban population on a self-sustaining basis.

The general feeling among Nigerians is that the idea of space science and technology is beyond us as a developing country. This feeling is partly due to the lack of interest and faith in our scientists and engineers, and partly due to the belief that the problems are mundane and Earth-bound, and thus that the solutions to those problems must be sought on the ground with classical and non-space based technology. It is now known that space platforms are very good vantage points from which to collect the data necessary to solve many Earth-bound problems. There is thus a unique potential for Earth-oriented space effort especially in telecommunications, broadcasting, meteorology and remote sensing. The developing country can maximally harness these potentials in a cost-effective way.

Training and education in space technology started in Nigeria with the advent of a number of space science related activities in various institutions in the 1950s. These appear to have concentrated on the undergraduate and graduate students in physics, geology, electrical and electronic engineering, surveying, urban and regional planning and a few other basic and applied sciences. There has been training for technicians and technologists in surveying, especially in remote sensing application areas such as cartography. Evidence of training for many other categories of personnel is not available. There is no deliberate policy of integration of space science issues into the curriculum of primary and secondary schools in the country.

Despite the many years of involvement in space science and technology, the institutions where these programs are based are unable to keep up with the pace of emerging trends in these areas. This arises due to severe funding problems in these institutions. In most cases, the initial take-off grant appears to have been the only outstanding grant ever received by the institution for its space program. This meant that, over the years, even the few facilities acquired could no longer be maintained. There are also the problems of motivation for researchers who find themselves becoming idle in their later years, with less and less involvement in active research. This results in decreased levels of training received by students as the teaching aids and research facilities necessary to demonstrate pertinent issues are lacking.

In order to correct the situation, interaction between scientists and policy/decision makers have been going on over the years. These interactions became symbolic of the country's inability to march forward with others in space science and technology; it is a clear demonstration of why the country has not succeeded in translating the well-articulated proposals and plans by scientists into coordinated national programs. Top government functionaries who are not scientists but who must approve the funds often are not aware of the full benefits of space technology in national development. Thus the recurring problem in implementing space programs is always how to convince bureaucrats to approve and release funds for the relevant programs. Some enlightened Nigerians blame scientists for not doing enough to educate policy /decision makers and those controlling the approval and release of funds, in spite of the fact that scientists have been in continuous contact and lobbying with governments in Nigeria since the late 1950s.

The establishment of the African Regional Centre for Space Science and Technology Education (ARCSSTE) at the Obafemi Awolowo University, as an affiliate of the United Nations Office for Outer Space Affairs, is a welcome development for the enhancement of basic space science training and research activities in Nigeria specifically and Africa generally. The basic space curriculum of the ARCSSTE, includes: an introductory course on astronomy and astrophysics, the astrophysics of the solar system, the physics of stars, galactic physics, cosmology as well as technologies for probing outer space. The pragmatic initiation of research and training in these issues will be of help to other centres in Nigeria and Africa, with interest in science and technology. In this regard, ARSSTE is expected to be a major rallying point for all participating countries in the continent, on issues concerning science and technology in general, and space science in particular.

2.2 *New Developments and Future Directions*

In April 1999 the Nigerian government established the National Space Research and Development Agency (NASRDA) to consolidate all the space science and technology related activities in order that they might have greater impact on the development efforts in Nigeria. The immediate objectives of NASRD include the following:

- To develop indigenous capabilities for research and development and applications in the major areas of space science and technology
- To use these capabilities as tools for
  (i) Natural resources management and infrastructure development
  (ii) Environmental monitoring and sustainable development
- To develop an effective and efficient communications system
- To train Nigerians in the acquisition and application of modern technology.

The major issues of concern include the lack of capacity in satellite technology even though the country has had training and research programs in satellite based technologies over two decades. Therefore, capacity building in satellite technology is a major focus. The preparation for Nigerian space science and technology programs through the development of its own satellite in the shortest possible time is being targeted together with practical demonstration of the satellite technology. To this end building of a pilot and small satellite, including various potential small satellite applications is envisaged.

2.3 *Economic and Social Benefits*

In spite of the fact that Nigeria has a large land area that is agriculturally very productive, a vast array of natural resources and high manpower, the country lacks reliable vital national data such as a good population estimate as well as the location and quantities of its natural resources. There is also a dearth of information to the general public, and especially to scientists, engineers, policy makers and planners on the current issues in space science and technology and their potential benefits to the economic and social development of the country. This implies that the country is yet to take advantage of the social and economic benefit from its vast natural resources, as well as many applicable technologies and databases available internationally to plan its development programs.

Nigeria's policy makers and planners do not seem to aware of the immense socio-economic benefits derivable from space science and technology. Government agencies, which were set up to inform and create awareness on important human and material development issues, have not created the required enlightenment on the utilization of space science and technology for national socio-economic development of the common man in Nigeria. For example there are no up-to-date topographic maps of many parts of the country.

It is most welcome that some ministries and extra-ministerial units have started using space technology for mineral exploration and for weather forecasts. These have assisted scientists, policy makers, farmers and other sectors requiring such information to improve the performance of their activities. It is a major concern that these activities are not carried out sustainably, but in such an ad-hoc manner that they are not likely to give the data generated the spatial coverage and continuity they require for effective application. It is also of concern that several other units do not follow such examples in their activities and services.

Most countries of the world have opened up their rural areas using satellite communication for education, health care and food production. It is of great concern that Nigeria is not even aware of some of these benefits of space technology, hence the lack of plans and policy objectives to follow this global trend. In line with the above, it is also important that other important applications of space science and technology, especially for crime monitoring and prevention and the management of rural and urban traffic planning are recognized as important pathway for socio-economic development in Nigeria.

In recognition of the need to acquire its own satellite, the Nigerian government signed a contract with Surry Satellite Technology Limited (SSTL), UK, in the year 2000 to build her first satellites (NIGERIASAT) and a ground receiving station in the country. Apart from building NIGERIASAT, Surrey would provide satellite technology know-how training for ten Nigerian engineers who would pioneer the building of subsequent satellites in the country. NIGERIASAT is expected to have a mass of 50 kg, carry payloads for disaster monitoring, and fly in constellation with four other satellites of the same payload to be built by Surry for South Korea, Morocco, China and Indonesia. The project is expected to take-off soon.

3. Conclusion

Small satellites are one of the most valuable tools to initiate and develop indigenous space capability, Apart from their low cost which makes them affordable by almost all countries, they offer the opportunity for developing countries to use the knowledge and skills of their citizens in different aspects of space science and technology with a view to using them for various socio-economic activities. These satellites offer low cost data and accessibility via direct reception to low cost receiving stations, which can be operated by users. However, there is still a lot of training required to gain expertise in transforming these data into useful information.

Geographic information is essential for proper decision-making, and so the development of small satellite programs with the above tasks is highly recommended. Joint development, construction and operation of a variety of small satellites offers opportunities for a suitable project to enable space research, technology demonstration and related applications in communications and Earth observations to generate the information for required national planning.

It is also important that adequate steps should be taken to educate decision-makers on all the benefits which can be derived from space technology and its applications, and the value of using such information for national planning.

References

1. National Agency for Science and Engineering Infrastructure, Nigeria: ***National Policy on Space and Science and Technology,*** Draft Document, 1993
2. Sundararamaiah, V.: ***Expectations and Experience of Developing Countries with Industry Partnerships,*** Proceedings of the UN/ESA-Symposium on Space Industry Cooperation with the Developing World, 1997
3. Delaye, M. and Contant, J.M.: *An Overview of Possibilities of Space Industry Cooperation with Developing Countries,* Proceedings of the UN/ESA Symposium on Space Industry Cooperation with Developing World, 1997
4. Perumal, A.: *The Role of Space Systems in an Integrated Management Plan for Sustainable Development,* Proceedings of UN Workshop on Space Technology as a Cost-Effective Tool to Improve Infrastructures in Developing Countries ,1997
5. Jia- Jun, Y. and Chun-Fa, L.: *Space Technology Applications: The Cost-Effective Alternative for Developing Countries,* Proceedings of UN Workshop on Space Technology as a Cost-Effective Tool to Improve Infrastructures in Developing Countries, 1997
6. National Space Research and Development Agency: *National Paper of Nigeria,* paper for the Third UN Conference on the Exploration and Peaceful Uses of Outer Space, 1999
7. <http://www.nigeria.com>
8. <http://www.africana.com/index>

Prospects for Involvement in Space Activities in Developing Countries via Small Satellites

N. Dhanji, International Space University, Strasbourg Central Campus, Parc d'Innovation, Boulevard Gonthier d'Andernach, 67400 Illkirch-Graffenstaden, France

e-mail: dhanji@mss.isunet.edu

Abstract

Due to the large capital investment and high risk associated with space activities, developing countries have not yet been able to benefit from the use of space technology to provide valuable resource and environmental data. However, with the advent of small satellite technology, the prospects for involvement in space activities by developing countries have greatly increased. Currently, developing countries face significant hurdles in implementing space technology and extending their technical capabilities; this is largely attributed to the lack of awareness on the part of governments and the user community of the benefits derived from space technology. The information acquired from small satellites would enable developing countries to better manage their resources, thereby providing an effective tool for achieving sustainable development.

1. Introduction

At the dawning of the new millennium, the major space faring nations of the world are expanding the frontiers of outer space in an effort to build the largest scientific and industrial research platform, the International Space Station. While these nations strive to achieve new technological heights, developing nations are still struggling to achieve sustainable development.

For decades, the major space faring nations have taken advantage of space technology to provide timely, accurate information about their resources and environment, enabling these countries to develop rapidly. The information acquired from space-based systems has enabled the advanced nations to better manage their resources, respond to a rapidly changing environment, monitor the global environment so as to ensure national security, and enhance technological development within local industries.

Developing countries face the pressing challenge of pursuing resource development with due consideration of the mid-term and long-term effects of such environmental and technical development. To some of these countries, the challenge could turn into a dilemma if the exploitation of their resources is mismanaged in their pursuit of a respectable level of development.

M. Rycroft and N. Crosby (eds.), Smaller Satellites: Bigger Business?, 307–314.

In order to provide viable solutions to the problems of developing countries, it is important to examine the special needs of these countries and assess the feasibility of meeting these needs using space technology. Most developing countries are faced with problems of poverty, scarce resources, widespread malnutrition and poor health, low agricultural productivity, a low standard of living and a high rate of illiteracy.

Although space technology offers developing countries an effective tool in achieving sustainable development, these countries face significant constraints in implementing the technology. An understanding of these constraints is necessary in order to develop clear guidelines for implementing space technology in developing countries.

2. Constraints in the Implementation and Application of Space Technology in Developing Countries

The large capital investment and high risk associated with space activity are factors that have ruled out the participation of developing countries in the field of space technology. As a result, developing countries have placed little priority on developing their own technical capabilities in space. Currently, the major constraints faced by developing countries in the implementation and application of space technology include:

- The lack of awareness and understanding, on the part of governments and the user community, of the benefits of space technology
- The lack of technical expertise, and of adequately skilled and experienced manpower
- The lack of clear national strategies and policies for the collection and use of resource and environmental information
- Mandates which do not reflect the need to collect and process data for generating information, as well as the sharing of such information with other organizations
- The lack of appropriate legislation that reflects the specific needs of the country and the requirements for meeting those needs in terms of financial, technical, material and human resources
- The lack of appropriate national institutions to plan the adoption and adaptation of space technology programs to meet the national objectives and priority areas for which the technology will be applied.

In order to implement space technology in developing countries, it is necessary to address these constraints by developing a strategy to overcome the hurdles and set up the framework for increased involvement in space activity. A general strategy for implementing space technology in developing countries may consist of the following objectives:

- Convincing the governments of developing countries of the benefits derived from space technology, thus encouraging them to develop their technical capabilities
- Developing appropriate policy that reflects the specific needs of the country, and incorporating space technology as an inherent part of the requirements for meeting those needs
- Developing a National Space Agency that will provide the framework from which all space-related activity within a country can be controlled and executed
- Developing partnerships amongst neighboring countries so as to benefit from any cost-sharing and risk-sharing opportunities in meeting common needs
- Implementing training programs for nationals.

3. The Benefits of Space Technology for Developing Countries

The primary objective of space technology applications is to provide relevant and timely information to local communities about their natural resources and environmental conditions. Satellite technology offers repetitive global coverage, allowing continuous environmental and resource monitoring. The data acquired from space-based systems, when organized and processed, provide information that can assist in critical decision-making regarding developmental issues. Such Earth observation applications include:

- Land cover characterization
- Natural resource inventory and management
- Agricultural planning and crop production
- Environmental monitoring and impact assessment
- Timely provision of climatic and meteorological information
- Physical infrastructure and utilities planning and management
- Urban and rural planning
- Natural and human-induced disaster assessment.

Geographic Information Systems (GIS) may be used to create value-added products such as Digital Elevation Maps, for simulation or modeling purposes, and Flood Plain Maps for disaster management and land insurance purposes. GIS combined with remote sensing technology provides an effective tool for managing natural resources and assisting decision makers to define optimal strategies in the use of these resources to achieve sustainable development. Examples of how some developing countries have used remote sensing technology and Geographic Information Systems to enhance their development include:

- Efficient managing of natural resources to increase food production for highly populated areas in China [Reference 1]
- Improving urban infrastructures in the Philippines [Reference 2]
- Producing a digital geographic database for land use and vegetation cover of the African continent [Reference 3]
- Water quality monitoring and freshwater resources in South Africa [Reference 4]
- Selecting sites for industries to minimize land and water pollution in Pakistan [Reference 5]
- Monitoring of the coastal environment of Morocco [Reference 6]
- Providing low-cost distance education services in rural and remote areas in Korea [Reference 7].

Satellite imagery from SPOT, Landsat and ERS is widely available on the market. However, in evaluating the benefit of implementing a national space program, governments of developing countries should lend consideration to the cumulative cost of obtaining such data, the possible lack of availability of information that is relevant to developing countries, and the continued reliance on outside sources for information. The development of local space-based capabilities not only provides long-term cost benefits but also provides developing countries with a tool for local sustainable development.

Another major application for space technology is satellite telecommunications, providing distance education and cultural programs, access to information in rural and inaccessible areas to combat natural disasters, tele-medicine and telephone services to remote areas, and tracking and monitoring of wildlife. The ubiquitous nature of satellites eliminates the need for expensive ground infrastructure and also plays a key role in eliminating the "last mile" problem.

4. The Role of Small Satellites in Achieving Sustainable Development

In the past, developing countries have been unable to benefit from the use of space technology to provide valuable resource and environmental data due to the large capital investment and high risk associated with space activities. However, with the advent of small satellite technology, the prospects for developing countries becoming involved in space activities have greatly increased. Small satellite technology offers developing countries an economically viable, technically effective and environmentally benign tool for achieving sustainable development. In addition, small satellites play an important strategic role in gaining affordable access to space.

As small satellite technology offers a low-cost, low development time option for increased space activity, countries previously unable to pursue space activities now have a greater opportunity to enter the space arena. Today, small satellites are creating exciting opportunities for business and research while at the same time responding to constrained budgets.

The short developmental time of small satellites has helped to decrease the time to scientific return, enabling researchers, governments, university students and businesses to reap the benefits of satellite technology within a few years rather than decades. Small missions are further characterized by a large amount of flexibility and response to new scientific opportunities. In addition, the shorter development periods of small missions keep the overall costs to a minimum.

In addition to serving as a tool for sustainable development, small satellites offer developing countries the opportunity to acquire rapid training in the field of space technology. Small satellite programs offer undergraduate and graduate students opportunities to experience the complex nature of scientific research in all its various stages from conception, realization and analysis to publication. Furthermore, students gain technical experience in building, operating and controlling small satellites. Having an average developmental time of about two years, small satellites make it possible for students to benefit fully from such projects during the course of their studies.

Small satellites offer opportunities for international cooperation between space faring and non space faring nations in facilitating technology transfer. Developing countries can take advantage of the many training courses in satellite technology offered by space faring nations as well as by the UN and other organizations.

Surrey Satellite Technology Ltd. (SSTL) in the UK, a pioneer in the development of small satellite technology, has collaborated with various developing countries to enable them to develop their own small satellite programs and to facilitate technology transfer [Reference 8]. The Centre for Space Science and Technology Education in Asia and the Pacific has been developed by the UN Program on Space Applications to address the training and education needs of that region. It is the first of several such centres planned to be established under the UN Program on Space Applications [Reference 9]. The International Space University (ISU) in France offers a multidisciplinary graduate program in space studies; the program provides comprehensive knowledge of the technical, legal, political, business and scientific aspects of space activities [Reference 10].

With the further advance of technology, largely driven by miniaturization, there are strong indications that small satellites could reach unprecedented levels of performance and reliability in the near future, while at the same time maintaining mission costs at acceptably low levels. Therefore, small satellites are well adapted to developing countries having limited space expenditures.

5. Space Policy for Developing Countries

Understanding a country's developmental needs and the role that space technology can play in meeting those needs is the first important step in implementing space programs to enhance the development of a particular country. However, formalizing the process by developing a national space policy provides the necessary framework for coordinating and enhancing the use of space technology applications within the country.

The National Space Policy of a developing country should aim to [Reference 9]:

- Acquire affordable access to space for the purpose of achieving sustainable development and human well-being through the use of small satellite technology
- Give priority to areas of space applications in which direct results could be achieved and benefits accrued to the common people
- Utilize the knowledge and technology gained by other advanced countries rather than trying to "reinvent the wheel"

- Provide education and training in space science and technology to develop self-reliance in the field
- Facilitate industry involvement so as to benefit from cost-sharing opportunities while strengthening ties between industry and government, as well as promoting technical development and business opportunities within the country
- Develop partnerships amongst neighboring countries having common needs, thereby strengthening political ties and taking advantage of risk-sharing and cost-sharing opportunities
- Ensure that the space policy is not easily thrown off course by changing governments.

In order to gain the support of decision makers, emphasis should be put on how the information acquired from small satellites will achieve the specified objectives and not on the technology itself. Taking, for example, the objective of mitigating disaster, the policy may read: "the use of remote sensing technology to develop flood plain maps for disaster mitigation in regions susceptible to floods".

The space policy of a developing country should be dependent upon its technological and financial capabilities as well as its maturity in mastering space technology applications. Therefore, it is important for the policy to consider the ramifications of various space technology applications in terms of their running costs and social impacts.

6. Conclusions

The struggle of developing countries to achieve sustainable development has been made easier with the advent of small satellite technology. Small satellite technology has enabled countries like India and Brazil to enhance their development so much so that they can hardly be categorized as third world countries anymore. The ability of these countries to compete economically on a global level has been made possible by having local access to vital information about their resources and environmental conditions via satellite technology.

Although small satellite technology offers developing countries a viable means of improving their technical capabilities and gaining affordable access to space, these countries are still dependent upon the major space faring nations of the world to provide technology transfer and launching facilities. Therefore, it is crucial for developing countries to build strong ties with the advanced nations so as to obtain the necessary skills and access to space.

In addition, building alliances with other developing countries will help to unite the developmental effort, thereby providing a stronger presence of developing countries in the global economy. Finally, careful consideration must be paid to the ramifications of technology development within developing countries in terms of their social and economic impact.

References

1. Jing Li: *Efficient Management of Natural Resources to Increase Food Production for Highly Populated Areas,* UN/ESA/EC Symposium on Space Technology Applications for the Benefit of Developing Countries, September 9-12, 1996
2. Solis, J.G.: *The Potential of Space Technology for Improving Urban Infrastructures,* UN/ESA/EC Symposium on Space Technology Applications for the Benefit of Developing Countries, September 9-12, 1996
3. Latham, J.: *AFRICOVER – Producing a Digital Geographic Database for Land Use and Vegetation Cover of the African Continent,* UN/ESA/EC Symposium on Space Technology Applications for the Benefit of Developing Countries, September 9-12, 1996
4. Howman, A. and van Vliet, H.: *The Use of Remote Sensing and GIS for Water Quality Monitoring and Freshwater Resources Management in South Africa,* UN/ESA/EC Symposium on Space Technology Applications for the Benefit of Developing Countries, September 9-12, 1996
5. Ishaq Mirza, M.: *Using Satellite Remote Sensing Data for Appropriate Site Selection of Large Scale Industries to Minimize Land and Water Pollution,* UN/ESA/EC Symposium on Space Technology Applications for the Benefit of Developing Countries, September 9-12, 1996
6. Abdellaoui, R.: *The Use of Remote Sensing Data for Monitoring the Coastal Environment,* UN/ESA/EC Symposium on Space Technology Applications for the Benefit of Developing Countries, September 9-12, 1996
7. Jong Chung, S.: *Space Systems in Support of Low-Cost Distance Education Services in Rural and Remote Areas,* UN/ESA/EC Symposium on Space Technology Applications for the Benefit of Developing Countries, September 9-12, 1996
8. <http://www.sstl.co.uk>
9. Committee on the Peaceful Uses of Outer Space: Report on the United Nations/International Astronautical Federation Workshop on Expanding the User Community of Space Technology Applications in Developing Countries, Melbourne, Australia, September 24-27, 1998
10. <http://mss.isunet.edu>

A Microspacecraft Platform for Missions in Earth Orbit and Beyond LEO

P. Eckart, S. Angelucci, L. Appolloni, M. Canales, E. Igenbergs, A. Opris, K. Pauly, F. Schlerka, F. Schwertfirm, W. Seefelder, A. Stamminger, Division of Astronautics, Technische Universität München, Boltzmannstrasse 15, 85748 Garching, Germany

email lunarsat@lrt.mw.tum.de

H. Baier, Chair of Lightweight Structures, Technische Universität München, Germany

E. Freidl, Daimler-Chrysler Aerospace, Ottobrunn, Germany

M. Kesselmann, OHB System GmbH, Bremen, Germany

L. Tarabini, GMV, Madrid, Spain

A. Straub, Universität der Bundeswehr (German Forces), Neubiberg, Germany

Abstract

This paper presents an overview of mission opportunities for microspacecraft with high Δv capabilities of up to about 1,500 m/s, especially for auxiliary launches from the Ariane 5 ASAP. The proposed concept is extremely flexible and can be adapted to specific mission requirements, thus permitting low-cost missions to Earth orbit, as well as to the Moon, Mars, and selected asteroids. A particular focus of this paper is the modularity of the proposed concept, including the bipropellant propulsion system needed, the lightweight structure, and the 3-axis attitude determination and control system (ADCS).

The proposed microspacecraft platform is derived from the LunarSat concept, which has been developed by a team led by the Technische Universität München, Germany, and Surrey Satellite Technology Ltd., UK. The LunarSat spacecraft and mission are presented as an example of the proposed concept, designed by young engineers, scientists, and students from around Europe, with support from numerous institutions and the space industry.

1. Introduction to the LunarSat Mission

Microsatellites have recently become a viable alternative for a variety of space applications. This is due to new mission concepts, as well as major advances in microelectronics and the increasing availability of commercial off-the-shelf (COTS) components.

The concept presented here is derived from the LunarSat concept [Reference 1], to be sent into an orbit around the Moon to perform scientific investigations concerning the lunar environment and its characteristics. It could be launched as an auxiliary payload on an Ariane 5 ASAP platform, have a

M. Rycroft and N. Crosby (eds.), Smaller Satellites: Bigger Business?, 315–320.

mass of less than 120 kg in GTO, and orbit the Moon in a highly elliptical polar orbit with its perilune above the lunar South pole area.

A microsatellite, i.e., a spacecraft with a mass of only about 100 kg can have a maximum Δv of about 1500 m/s, using conventional propulsion technology. Therefore, their scope of missions from GTO is limited to:

- Selected Earth orbits
- Lagrangian points of the Earth-Moon system
- Lunar orbit [Reference 2]
- Mars fly-by
- Near-Earth object fly-by.

The lunar orbit mission case is presented here. It was planned that the LunarSat Operations Team will be located at ESOC and supported by ESOC staff. During the launch and early orbit phase (LEOP), which will be kept as short as possible (3 days), the contact with the spacecraft will be almost continuous. After injection into Lunar Transfer Orbit, the spacecraft will be monitored during the passes, but operations will occur off-line to a large extent, using regularly uplinked time-tagged schedules.

2. Spacecraft Structure and Configuration

The maximum allowed spacecraft size and mass, as defined in the Ariane 5 ASAP user manual, are 600 x 600 x 800 mm^3 and 120 kg, respectively. Consequently, the main considerations in selecting a shape for such a microspacecraft are:

- Packaging considerations, i.e., to provide enough volume to contain the subsystem components and to fit within the fairing envelope
- Structural considerations, i.e., compatibility with the payload and launch vehicle mechanical interfaces and efficient, in-line structural load paths between payload and launch-vehicle.

The most important criterion for selection of the spacecraft shape is that it must be able to contain the largest packaged components. Because of the desired velocity capability of up to 1500 m/s, such a microspacecraft requires about 60 % of its total wet mass for the propulsion system. This means that the design process is mainly ruled by considerations regarding the required tank volume that has to fit inside the ASAP envelope.

The proposed microspacecraft is divided into two main segments:

- Payload Bay, which contains all payload and sensors, plus Telemetry Tracking and Command (TT&C) and On Board Data Handling (OBDH) and parts of the Electrical Power System (EPS) subsystem of the spacecraft
- Service Bay, which contains the main thrusters, the propellant tanks and all propulsion elements. The proposed system uses four main thrusters with a thrust level of 22N each, using Nitrogen Tetroxide (NTO) and hydrazine. Four 1N hydrazine thrusters are used for attitude control, along with three reaction wheels. Also situated in the service bay are the Li-ion batteries and the laser gyros.

Several configurations have been analyzed for the propulsion system, for which five tanks (two for the required fuel, two for the oxidizer and one pressure vessel) are required.

The resulting baseline design uses a symmetrical tank configuration of four propellant tanks (2+2), all mounted on the same level. The pressure tank is mounted below these four tanks, on the geometric center axis. The propellant tanks are located on the tank panel (Middle Tank Plate - MTP) of the propulsion module. This configuration is shown in Fig 1.

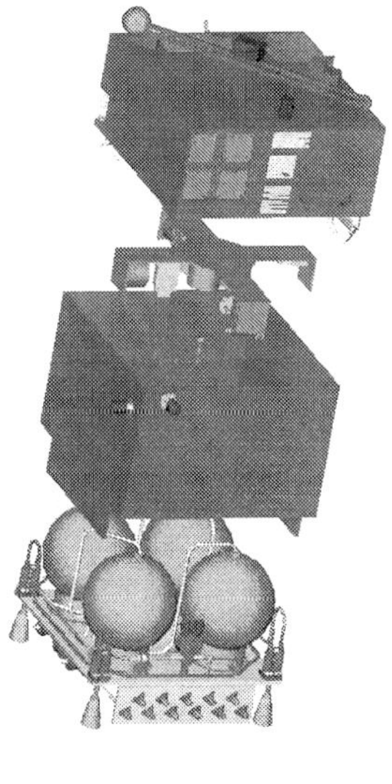

Figure 1. Baseline spacecraft configuration [Reference 3]

3. Propulsion System Design

After conducting a trade-off regarding the influence of thrust and main engine mass on gravity losses during the injection firings, a four engine configuration (used in pulsing mode) has been selected [Reference 4]. Specifically, for LunarSat the injection firing from GTO into LTO was investigated as a reference because its Δv of about 565 m/s is most decisive concerning the gravity-losses.

The driving design aspect of the propulsion system is cost-effectiveness. This goal is achieved by three means:

- Using COTS components, in order to avoid developing costs
- Integrating the propulsion system mainly on one device to shorten assembly time and to gain a high modularity and flexibility in production.
- Establishing a modular construction system of the whole spacecraft for adapting the main design easily to variable mission profiles.

4. Attitude Determination and Control System (ADCS)

The main task of the proposed microspacecraft ADCS is to determine and control the spacecraft attitude, i.e., to stabilize the spacecraft, and to orient it in the desired directions during the mission, despite the external and internal forces acting on it. Specifically, for the LunarSat mission the following tasks and analyses have been conducted during the course of the ADCS design:

- Operational ADCS modes have been identified
- Impact of the environment on the ADCS have been described
- Subsystem constraints have been identified
- Hardware components and the subsystem architecture have been selected
- Power and mass budgets have been provided
- Interfaces with other subsystems and with the payload have been identified
- Required ADCS algorithms and control strategies have been identified
- Implementation of determination and control algorithms is in progress.

Due to the frequently required re-pointings, the complex mission goals, and the inertial characteristics of the spacecraft, a 3-axis stabilisation technique has been selected for the LunarSat spacecraft. The ACDS hardware that is foreseen consists of:

- 4 fiber optic laser gyros
- 1 Sun sensor and 1 star sensor
- 4 solar cells (broad sun sensor)
- 3 reaction wheels
- 4 attitude thrusters.

The main characteristics regarding the LunarSat control strategy is that it is not designed for an Earth orbit mission. Therefore:

- It has to accomplish different main engine firings (increasing the difficulty of the mission)
- It has to accomplish tasks that are completely different (such as inertial pointings, imaging, etc.)
- LunarSat will be launched with an ASAP (Ariane Structure for Auxiliary Payload) platform of the Ariane 5
- LunarSat is a micro-satellite that implies it has a restricted mass budget. (It is not yet proven that micro-satellites can conduct interplanetary missions).

All these aspects drive the control of LunarSat spacecraft towards a 3 axis-stabilized solution, which permits stable and accurate maneuvers, depending on sensors and actuators.

One of the most critical phases for the ADCS is control during the orbital insertion firings. The inertial characteristics of the satellite change dramatically when propellant is consumed. Also, the position of the center of mass will change, due to the non-symmetrical configuration of the spacecraft structure. This shift will act as a disturbance torque that has to be counteracted in a short time.

Different control algorithms have been developed and implemented using the Matlab/Simulink™ platform:

- Detumble
- Control During Firing
- Inertial Pointing
- Spot Pointing
- Sun Pointing.

The firing phases algorithms require the main engines and the attitude thrusters as actuators. For inertial Sun, Earth and Moon pointings reaction wheels are used. Slewings are mainly accomplished with the reaction wheels.

5. Conclusions

The microspacecraft platform presented in this paper may be used to reach selected Earth orbits, the Lagrangian points of the Earth-Moon system, or lunar orbit, and to achieve a Mars or Near-Earth object fly-by from GTO. The proposed design is based on the development of the LunarSat spacecraft and uses mainly COTS components. It thus provides a low-cost platform for certain solar system exploration missions.

It needs to be pointed out, however, that this type of spacecraft is significantly more expensive than a conventional microspacecraft due to the required propulsion system and the resulting increased complexity. Also, the operations costs are typically much higher than for Earth orbiting missions. These aspects lead to the fact that a typical microspacecraft mission beyond LEO may cost million, including launch, platform, payloads, and operations. Nevertheless, the designers of this concept strongly believe that it will open the door for future low-cost microspacecraft exploration missions.

References

1. Eckart, P et al.: *LunarSat, the European Microorbiter Mission Results of Phase B and Project Status*, DGLR Jahrbuch, 1999
2. Seefelder, W.: *Transfer Trajectories for the LunarSat Mission*, DGLR Jahrbuch, 1999
3. Schlerka, F.: *LunarSat Configuration and Database Developement*, Master's Thesis Division of Astronautics, Technische Universität München, Germany 2000
4. Kesselman, M.: *Calculations of the g-Losses with an ESOC Software Tool*, Division of Astronautics, Technische Universität München, Germany 1998

The Brazilian Microsatellite Development Strategy: an Assessment of the Scientific Applications Satellite (SACI) Programme

E. J. Costa Filho, A. T. Furtado, State University of Campinas (UNICAMP), Geociences Institute, Scientific and Technological Policy Department, P. O. Box 6152, Code 13083-970, Campinas -SP, Brazil

e-mail: costafi@ige.unicamp.br;furtado@ige.unicamp.br

Abstract

At the end of 1994 the Brazilian National Space Research Institute (INPE) started the development of a scientific microsatellite programme, called the SACI Programme. This satellite was based on the experience acquired during the execution of the Complete Brazilian Space Mission (MECB)

The SACI Programme was sponsored by the Federal Financing Agency for Studies and Projects (FINEP). Brazilian universities also participated, developing some of the scientific experiments. The choice of the experiments was made under the responsibility of the Brazilian Academy of Sciences (ABC). The scientific experiments selected were PLASMEX (study of plasma bubbles), MAGNEX (geomagnetic experiments), PHOTOEX (airglow photometer) and ORCAS (solar and anomalous cosmic rays).

1. Introduction

Several countries have increasing interests in the space sector, especially in the satellite field. There exists, nowadays, several classes of satellites such as telecommunications, meteorological, remote sensing, scientific, etc.. We are witnessing the technological contrast between conventional satellites and new generations of microsatellites [References 1,2]. If on the one hand, conventional satellite technology uses expensive and proven technology to increase reliability, on the other hand, microsatellites use new technologies to reduce costs and to increase their performance.

Since the beginning of the 1990's, INPE was concerned with improving the scientific satellite area [Reference 3]. It was a "common sense" at that time that this field was unexplored, and that it was a very important area for microgravity research. However, the institute suffered budgetary constraints at the time, that made the project unfeasible in the short term. The INPE solution to this problem was to obtain resources from FINEP, as a grant of US $ 5.8 million for the satellites. (The SACI-1 and 2 cost US $ 4.6 and 1.2 million, respectively.)

The SACI is a low Earth orbit (LEO) satellite for scientific and technological purposes. The total mass is about 60 kg, 47% of which is destined for scientific experiments. It is important to note that all the subsystems of the satellite were

M. Rycroft and N. Crosby (eds.), Smaller Satellites: Bigger Business?, 321–327.

designed aiming at multiple missions. The satellite bus can meet the requirements of different payloads without important modifications in the subsystems.

The choice of the experiments was made under the responsibility of the Brazilian Academy of Sciences (ABC), which issued an Opportunity Announcement to the scientific community to use space for their experiments. The experiments approved were measurements of the Earth's magnetic field (MAGNEX), a study of equatorial plasma bubbles (PLASMEX), measurements of photochemical emissions in high atmosphere (PHOTOEX), and the observation of solar and anomalous cosmic rays (ORCAS). SACI-1 was launched in October 1999 using a Long March 4 as a *piggyback* of CBERS (China-Brazil Earth Resources Satellite), and SACI –2 in December of the same year using a Brazilian satellite launch vehicle (VLS). Unfortunately, both satellites did not work. SACI-1 could not send information to the ground stations, because of a technical failure in the on-board computer system and the second one was lost because of a launcher failure.

2. The SACI Programme: Main Features of the Scientific Satellite

SACI is a scientific satellite that includes new technologies in its five subsystems (power supply, attitude and control, thermal control, onboard computer and communications). Furthermore, the SACI satellites were conceived to use the INPE ground stations which the Institute developed during the 1980´s. In Table 1, the main features of the satellite are given.

Mass of 60 kg, including 28 kg payload mass
Development time about 28 months
Spin stabilization with geomagnetic control
1° pointing accuracy
18 months expected lifetime
Payload power requirement of 30 W
Passive thermal control
Modular concept
Simple technical solutions

Table 1. Main features from SACI satellite

3. The Choice of Experiments

In 1994, the ABC started the process of choosing experiments for the SACI programme. The scientific and technological objectives of the programme were to increase:

- Competition for experiments on board the spacecraft
- Possibilities of international cooperation and participation in the present and future satellite projects [Reference 5].

The creation of new space activities in the satellite field through the participation of the Brazilian scientific community stimulates new knowledge and new fields, and decentralises the development and the benefits resulting from the programme

On 8th of September 1995, the ABC selected four experiments to constitute the SACI payload, as shown in Table 2.

Experiment	Objective	Principal Investigator	Institution
PLASMEX	Study of equatorial plasma bubbles	Dr. M. A. Abdu	INPE
PHOTOEX	Measure the intensity of terrestrial airglow emissions	Dr H. Takahashi	INPE
ORCAS	Measure solar and anomalous cosmic rays fluxes in the magnetosphere	Dr. A Turtelli	UNICAMP
MAGNEX	Measure the Earth's magnetic field	Dr. N. B. Trivedi	INPE

Table 2. Experiments selected for the SACI programme

The total cost for the scientific payload was US $ 800,000. The managers of the programme kept this cost low as a result of cooperation and the use of already developed rocket experiments The main features of the experiments are described in the following subsections.

3.1 *Study of Equatorial Plasma Bubbles (PLASMEX)*

The main objective of this experiment is to investigate the generation, development and decay of plasma bubbles in the ionosphere, especially above Brazil. This research was intended to elucidate the influence of the bubbles and their associated plasma turbulence on several space applications systems (remote sensing with radar, space geodesy, trans-ionosphere telecommunications, etc.)[Reference 6].

Several aspects of these phenomena such as their dependence on ionospheric environmental and electrodynamic conditions are still not well studied. More detailed knowledge of such aspects are fundamental to improve the predictability of the occurrence of these bubbles.

To accomplish this the following instruments were used [Reference 5]:

- A high frequency capacitance probe to measure the plasma density
- A Langmuir probe to measure the electron density and the spatial spectra of the plasma irregularities
- An electron temperature probe to measure the electron temperature in the ionosphere plasma

3.2 *Measure the Intensity of the Terrestrial Airglow Emission (PHOTOEX)*

This experiment has the objective of measuring the intensity of the terrestrial airglow emissions of oxygen, OI 557.7 nm, OI 630.0 nm, and hydroxyl OH (8,3) emissions. The photometer system is composed of 4 sensors to measure in 4 distinct wavelength regions. The photometer was designed and manufactured by a technical team at INPE. This experiment has been conducted by INPE with the cooperation of the Federal University of Rio Grande do Norte (UFRN), The Federal University of Paraíba (UFPB), the Institute of Space and Aeronautics Science (ISAS), and the National Institute of Polar Research (NIPR), both from Japan.

3.3 *Solar and Anomalous Cosmic Ray Fluxes (ORCAS)*

In interplanetary space, heavy nuclei of He and Ne with energies below 50 MeV/amu, called Anomalous Cosmic Rays (ACR), were discovered in 1970-1973 by the space probes IMP, Pioneer 10 and 11. The main objective of the telescope ORCAS was to measure the ACR fluxes from C to Ne, "trapped" in the magnetosphere, by using solid-state detectors and by identifying the time and the direction of particle arrivals. The ORCAS will allow studies of the distribution and of the dynamics of trapped particles for conditions of minimum solar activity, with intense fluxes.

Besides UNICAMP, where this experiment was developed, the Institute of Physical and Chemical Research (RIKEN) in Japan and the Naval Research Laboratory (NRL) in the USA cooperated.

3.4 *Earth's Magnetic Field (MAGNEX)*

The geomagnetic field controls the movement of charged particles in the space around the Earth and protects the planet from the direct incidence of the solar wind. Simultaneous measurements of the geomagnetic field on the Earth's surface and in the near space plasma are essential to the study of geomagnetic phenomena. In the last 25 years electrical currents aligned with the geomagnetic field lines have been subject of important studies with satellite magnetometers.

The main objective of the experiments was the investigation of the geomagnetic field near the equator. The plasma electrodynamics, specially in the region of the South Atlantic Geomagnetic Anomaly (SAGA), would give important complementary information to the study of the geomagnetic effects.

4. The Framework of the Programme

The division of responsibilities was a favourable point in the SACI programme. Several INPE divisions were involved such as MIC - scientific microsatellites project, DMC -spatial mechanical and control division, DEA - electronic and aerospace division, DMF - manufacture department, CRN - regional centre of Natal, and LIT - integration and tests laboratory. The INPE staff working on the programme consisted of 2 persons full-time, 66 persons half time and about 20 persons involved between 20 to 30%. These staff worked on the development of the satellite, and most of them, worked sporadically on operations such as ambient tests, cleaning, painting, etc. .

Because INPE is a public institution they can only contract industries by a bidding process, which would be time consuming. Thus, all the industrial contracts were transferred to FUNCATE, established by INPE in 1993 when one of the main suppliers to INPE run out of business (the company named Esca). FUNCATE had 25 staff full time involved and 5 staff half time, besides the subcontractors. FUNCATE contracted some institutions to develop, with INPE, subsystems for the satellite; these institutions are listed in Table 3.

Systems	Institution
On-Board Computer	Federal University of Ceará (UFC)
S-Band Antennas	Aeronautic Institute of Technology (ITA)
Thermal Control	Federal University of Santa Catarina (UFSC)
Structure and Control	Technological Research Institute (IPT)
Protocol of communication and solar sensors	University of São Paulo (USP)

Table 3. SACI satellite - systems contracted outside INPE

After the project phase, the foundation subcontracted a total of nine companies for the supply of many components. The contracted Institutions are listed in Table 4.

Suppliers	Components
LEG	Mechanic Structures
Intercircuits	Press Circuits
MEQ	Electric Components
CPqD	Communication Components
IPT	Structure
FIBRAFORTE	Electric Projects and Design
DIGICON	Power Supply and Antenna
EQE	Electric Components
Microeletrônica	Electronic Components

Table 4. FUNCATE industrial subcontractors

This institutional arrangement provided for technology transfer for the Brazilian suppliers, and a transfer of critical knowledge for the universities and institutions involved.

5. Conclusions

This paper has shown the main features of the SACI Programme. The external financing to INPE guaranteed a flow of resources in spite of the budget constraints during the 1990's. Although neither satellites are working, some important lessons were learnt by the main institutional actors:

- A co-operative philosophy was developed between INPE and other scientific institutions with space research capabilities
- The programme has approximate the ABC to the INPE, inducing other collaboration such as the French – Brazilian Microsatellite Programme

- The INPE Microsatellite area has its own budget, indicating the importance of this area inside the institute
- The participation of the Brazilian industrial firms and of the scientific community are positive aspects of this programme.

Some of the developments carried out in the SACI programme could be employed in other satellite projects at INPE. The experience acquired in the SACI satellite programme has contributed to change the Brazilian space activities, which became more open to external actors (national and international scientific community and local high-tech industry).

References

1. Fleeter, R. and Warner, R.: Design of low-cost spacecraft, *Space Mission Analysis and Design*, 2nd edition, Larson, W. J. and Wertz, J. R. (editors), chapter 22 Torrance, CA, USA ,1992
2. Neri, J . A. et al: The Brazilian Scientific Microsatellite SACI - 1, *Acta Astronautica, Vol.* 39, pp 707-709, 1996
3. Sobral, J. H. A. et al: The Brazilian Scientific Satellite, *Acta Astronautica, Vol.* 32, pp. 675-682, 1994
4. Neri, J. A. et al: SACI - 1, *A Cost Effective Microsatellite Bus for Multimission Payloads*. International Conference on Small Satellites: Mission and Technology, Madrid pp. 9-13 , 1996
5. Neri, J . A. and Fonseca, I. M.: *The Brazilian Scientific Microsatellite SACI-1*, 4th International Symposium, Small Satellites Systems and Services, Antibes. Juan les Pins, France, September 14-18, 1998
6. INPE: Scientific Applications Satellite, Technical Report 1, *Satélite de Aplicações Científicas SACI-1*, São José dos Campos, Brazil, March 1995

Micro/Nanosatellites as an Enabler of the New Information World

T. J. Grant, Atos Origin Nederland B.V., P.O. Box 1444, NL-3430 BK Nieuwegein, The Netherlands

e-mail: Tim.Grant@atosorigin.com

Abstract

This paper compares the old and new information worlds and the old and new space worlds. I argue that only by embracing the "small is better" approach can the space world take part in the new information world. I look not only at how micro/nanosatellites can be an enabler of the new information world, but also at what the information world has to teach the space world. The progression in the old information world from mainframe through mini to PC must be mirrored in terms of satellites. It must become possible to develop satellites in one or two years, rather than decades, and to set up satellite production lines in small countries or - better still - in developing countries. If university students can build and operate small satellites, then every enterprise can have its own constellation. The space world must become a network of constellations, a Grid of micro/nanosatellites.

1. Introduction

In 1949, the IBM founder Thomas Watson predicted that the world would need just five computers. In a half-century we have seen this prediction swept aside by the introduction of 45 million PCs. At the close of the 20th century, the old information world had brought us laptop and palmtop computing, electronic mail, the World Wide Web, two generations of terrestrial mobile telephones, the first generation of satellite telephones, the hype around dotcom companies, Business-to-Consumer systems, Enterprise Resource Planning, and the global threat of the millennium bug. The contribution of the space world was limited to launching large telecommunications satellites into geostationary orbit. Only the Global Positioning System and a handful of small satellites gave a hint of what the new information world could look like.

With the advent of the 21st century has come the new information world. In just 12 months the millennium bug proved to be a damp squib, the dotcom bubble burst, and the world narrowly avoided a downpour of decommissioned Iridium satellites. The e-gurus now tell us that salvation no longer lies in Business-to-Consumer systems (B2C), nor even in Business-to-Business (B2B), but in Business-to-Employee (B2E) systems. Web-enabled devices will dwarf the number of PCs. Enterprise Resource Planning (ERP) is going out of fashion, and once again we hear that programming will become obsolete.

M. Rycroft and N. Crosby (eds.), Smaller Satellites: Bigger Business?, 329–336.

Suddenly, the space world seems to have woken up. All the talk is about constellations, satellites flying in formation, swarms of mobile robots on other planets, and extending the Internet across the solar system. The European Space Agency gets together with the European Union's Fifth Framework Programme to develop the second-generation Internet, the Data Grid, to share databases containing petabytes of information [Reference 1]. Space and information technologies seem to have joined hands at last.

This paper compares old and new information worlds with old and new space worlds. The question is whether the space world can keep up with the new information world. I argue that only by embracing the "small is better" approach can the space world take part in the new information world. I look not only at how micro/nanosatellites can be an enabler of the new information world, but also at what the information world has to teach the space world. The progression in the old information world from mainframe through mini to PC must be mirrored in terms of satellites. It must become possible to develop satellites in one or two years, rather than decades, and to set up satellite production lines in small countries or - better still - in developing countries. If university students can build and operate small satellites, then every enterprise can have its own constellation. The space world must become a network of constellations, a Grid of micro/nanosatellites.

2. Old and New Information Worlds

2.1 *Old Information World*

Just half a century ago databases consisted of shelves holding paper dossiers. "Computer" was a term that referred to people who performed calculations according to a set procedure. If they were lucky, they might be provided with calculating machines. Companies such as Rolodex and International Business Machines were established to provide the necessary office infrastructure.

The stored-program computing machine was invented out of necessity during World War II. In the subsequent half-century we have seen computers progress from mainframes, through mini-computers, to PCs, and now to palmtops. Even the experts were unable to predict the growth in the computer market. In terms of processors, the number of devices with embedded computing power will shortly outnumber the PCs. The key process at work here is miniaturisation. By reducing the size of computers, it has been possible to reduce their cost. With cost reduction comes a widening of the market.

The development of the computer was preceded by the development of telecommunications. This started in the 19th century with the telegraph, progressed through the analogue telephone, and has now progressed to digital telecommunications networks of global scale carrying multimedia data.

Developments in computing and in telecommunications are converging. The Internet, e-mail and the World Wide Web (WWW) have been built on top of the global telecommunications networks, linking the majority of the world's computers with one another.

2.2 *New Information World*

Around the turn of the 21st century a new information world emerged. The new information world is characterised by the rational reconstruction of networks of computers. Whereas in the old information world the fundamental unit was the computer, in the new information world the unit is the network. For example, in defence the latest fashion is to talk about "network-centric" warfare [Reference 2]. Network-centric thinking also lies at the heart of the so-called "New Economy" [Reference 3]. Entrepreneurs believed that money was to be made by reaching out to the millions of consumers who had purchased a PC and linked it to the Internet for personal use. The "dotcom" company was born, initially aimed at the B2C market. Supposedly, market share was the Holy Grail, profitability was unimportant, and a variety of business models was possible [Reference 4].

At the very start of the 21st century one might think that the New Economy, or the new information world of which it is part, is merely hype. The ERP industry hit a bump in the road with Baan's demise. At the same time, the telecommunications industry came unstuck when governments applied market thinking in auctioning third-generation mobile telecommunications licences.

The factor common to these events is that the new information world had to re-learn the lesson that making a profit is essential to commercial survival. Nevertheless, this lesson does not invalid network-centric thinking. On the contrary, there are quiet successes brewing that truly exploit the advantage of the network. For example, 600 cryptologists were able to crack a supposedly-uncrackable encryption code by working together over the Internet. SETI@Home exploits the underutilized processing power of its 3 million volunteer users to scan the skies for extraterrestrial life [Reference 5]. The Philanthropic Peer-to-Peer Program, hosted by Intel and developed by United Devices, the National Foundation for Cancer Research and the University of

Oxford, is designed to find cancer cures. Napster, Gnutella, Aimster, AudioGalaxy and others enable users to share files over the Internet.

What these successes have in common is that they use the Internet for peer-to-peer (P2P) computing. P2P computing is the sharing of computer resources and services by direct exchange between computer-based systems [Reference 6]. These resources and services include the exchange of information, processing cycles, cache storage, and disk storage for files. (By contrast, B2C, B2B and B2E models are hierarchical, with information exchange being indirect.) Computers participating in a P2P architecture can act as both clients and servers, assuming whatever role is most efficient for the network. P2P computing allows organisations to leverage their collective power to benefit the entire enterprise.

P2P computing works best when a very large number of computers are connected together. Business applications provide:

- *Collaboration:* P2P computing empowers geographically-separated individuals and teams to collaborate on projects, increasing productivity
- *Edge services:* edge services move data closer to the point where they are consumed, acting as a network caching mechanism, speeding up the session and saving money by eliminating the need for local storage
- *Distributed computing:* businesses with large-scale computer processing needs can use idle processing power and disk space, with faster completion times and lower costs
- *Intelligent agents:* computing networks can dynamically work together using intelligent agents.

The new information world - and P2P computing in particular - supports concurrent design and engineering, collaborative processing of large data-sets, enables collaboration between geographically-separated users, and makes more efficient use of information and communications technology resources. The keys lie in the P2P architecture and the (very) large numbers of participants, each contributing a small part of the overall process.

3. Old and New Space Worlds

As in the information world, one can distinguish old and new space worlds. The old space world is characterised by institutional investment and governance, by long development time-scales, by a flat market and by a steady increase in spacecraft complexity and size. Old space missions are primarily scientific, or are related to the International Space Station, or are to be found in geostationary orbit.

By contrast, the new space world is robust and growing at 15 % per annum [Reference 7]. Features include commercial investment and governance, spacecraft production lines, LEO, MEO or highly elliptic orbits, and architectures based on constellations.

Like the new information world, the new space world has had to re-learn the lesson that profit is essential to survival, as evidenced by the financial difficulties of Iridium, ICO and Globalstar. Profit can be gained by increasing revenue, by reducing costs, or by a combination of the two. Revenue increase can be sought in a premium market or in a high-volume market. Space telecommunications has successfully exploited both means. By contrast, other missions, such as remote sensing, have yet to find the key to revenue increase.

The other side to profitability is cost reduction. Again, there are several ways to achieve this. The Global Positioning System (GPS) shows one way. GPS can be regarded as a success on the scale of space telecommunications, to the extent that cheap GPS receivers are sold to ramblers and mountain bikers. The success of GPS can be attributed to the fact that it is a space-based service that is essentially free to all users. The space segment has been funded by the US military. Only the cost of a simple terminal device has to be borne by the user. Thanks to the stimulus of the Persian Gulf war, when terminal devices were manufactured in their thousands for military purposes, the cost of a GPS receiver is comparable to that of other consumer electronics products. In short, one way to reduce costs is to get the military to pay for the space segment.

The GPS also offers a second clue. Both the space and ground segments of GPS consist of large numbers of small objects, paralleling one of the key aspects of the new information world. The parallel is complete when we observe that the GPS system, less its control segment, is a peer-to-peer system. Each GPS receiver receives signals from multiple GPS satellites, and each GPS satellite broadcasts its signals to hundreds or thousands of receivers.

A third way of reducing cost can be borrowed from the new information world. In common with other speakers [References 8, 9, 10, 11, 12], I argue that miniaturisation - in this case of spacecraft - leads to cost reduction. Micro- and nano-satellites can be built by universities and developing countries, as evidenced by Stanford University's SAPPHIRE, OPAL and EMERALD spacecraft [Reference 13], and, above all, by the emergence of Surrey Satellite Technology, Ltd., from the University of Surrey [Reference 14]. SUNSAT from the University of Stellenbosch, South Africa, is the prime example of a small satellite built by a university in a developing country [Reference 15]. But I differ from other speakers in arguing that the new space world is characterised by large numbers of small satellites in peer-to-peer architectures.

4. Enterprise Satellites

Enterprise satellites are examples of the new space world. An enterprise satellite is a satellite or constellation owned by a large commercial organisation as a part of its infrastructure. By the nature of spacecraft, the corporation will almost certainly be a multi-national or an extended enterprise (i.e. the organisation together with its suppliers and clients). The advantages of company ownership are privacy, security, no need to share capacity with other users, and the freedom to relocate the satellite or constellation as required.

In the NEDSAT concept study [Reference 16] three Dutch companies took part in presenting an enterprise satellite system based on a constellation of small satellites. The study aim was to establish whether there was a market for such systems. The illustrative mission was store-and-forward messaging, with on-board applications data processing. Example applications could be the monitoring and control of widely-dispersed assets, such as oil and gas pipelines, coffee farms, large-scale land reclamation projects, or tracking and tracing mobile objects (e.g. vehicles, containers, bags). The study results showed that there were niche markets where an enterprise satellite system could address needs that were not covered by existing systems. The key features of such markets were identified. In the Planning for Autonomous Smallsat Constellations project, Atos Origin demonstrated that a NEDSAT constellation could perform its own mission planning and scheduling by peer-to-peer methods [Reference 17].

5. Enabling the New Information World

The new information world and the new space world have roles to play in enabling each other. The new information world's role is in providing modular, component-based software for on-board autonomy and in providing low-cost (e.g. PC-based), e-enabled control and user segments [Reference 18] for managing and sharing small satellites and constellations. The new space world is foreseen as the provider of the broadband, space-based Internet. In addition, the new space world can beneficially adopt e-business models, leading to a shortening of the "time to market" for space systems, and peer-to-peer, highly distributed architectures, like the forthcoming DataGrid.

The new space world would consist of large numbers of small satellites, produced to standard, interoperable designs on production lines in developing countries. The satellites would be grouped into constellations, each owned by a commercial enterprise with a global reach. The constellations would be linked in a peer-to-peer network, reflecting the enterprises' value-chain relationships. With inter-satellite links and substantial on-board processing power and data storage capacity, the network of constellations would form a space-based DataGrid.

References

1. CERN: *HEPCCC Grid Initiative - The Data Grid Project*, <http://www.eu-datagrid.org/>, CNR, Italy, June 5, 2001
2. US DoD C4ISR Cooperative Research Program: *Network-Centric Warfare*, <http://www.dodccrp.org/ncw.htm>. US DoD, USA, June 6, 2001
3. Kelly, K.: *New Rules for the New Economy: 10 radical strategies for a connected world*, Viking Penguin, New York, NY, USA, 1998
4. Rappa, M.: *Managing the Digital Enterprise - Business Models on the Web*, <http://ecommerce.ncsu.edu/topics/models/models.html>. North Carolina State University, Raleigh, NC, USA, June 5, 2001
5. SETI@Home: *Search for Extraterrestrial Intelligence at Home*, <http://setiathome.ssl.berkeley.edu/>. University of California at Berkeley, CA, USA, June 5, 2001
6. Peer-to-Peer Working Group: *Peer-to-Peer Working Group - Home*, <http://www.peer-to-peerwg.org/>. Peer-to-Peer Working Group, Portland, OR, USA, May 18, 2001
7. Charest, P.: *Satellite Procurement: A commercial perspective*, Lockheed Martin Commercial Space Systems, presented at the Space Business and Management workshop, ISU Summer School Programme, Suranaree University of Technology, Thailand, July 23, 1999
8. Sweeting, M.: *Micro/Nanosatellites - The New World*, presented at the 6th ISU Symposium, Strasbourg, France, May 21-23, 2001
9. Lewin, A.W.: *Market Growth Potential for Small Satellites: An Industry View*, presented at the 6th ISU Symposium, Strasbourg, France, May 21-23, 2001
10. Overright, J.: *Nanosatellite Science Applications*, presented at the 6th ISU Symposium, Strasbourg, France, May 21-23, 2001
11. Fleeter, R.: *The Trajectory to Commercial Microspace*, presented at the 6th ISU Symposium, Strasbourg, France, May 21-23, 2001
12. Doll, B.: *Future Satellite Business is Small Satellite Business*, presented at the 6th ISU Symposium, Strasbourg, France, May 21-23, 2001
13. SSDL: *Stanford: Space Systems Development Laboratory*, <http://ssdl.stanford.edu/aa/>, Stanford University, Palo Alto, CA, USA, June 6, 2001
14. SSTL: *SSTL*, <http://www.sstl.co.uk/>. Surrey Satellite Technology, Ltd., Guildford, UK, June 6, 2001
15. ESL: *SUNSAT Main Page*, <http://sunsat.ee.sun.ac.za/>. Electronic Systems Laboratory, University of Stellenbosch, Stellenbosch, South Africa, June 6, 2001.
16. Bradford Engineering B.V., PNO Management Consultants B.V., and Origin Nederland B.V.: *NEDSAT - Down-to-Earth Communications*, Report number NEDSAT-0001, July 6, 1998
17. Grant, T. J., and Broeils, A. H. *Distributed Planning in Constellations of Autonomous Spacecraft*, presented at the 5th International Symposium on Artificial Intelligence, Robotics and Automation in Space (i-SAIRAS 99), June 1-3, 1999, ESTEC, Noordwijk, The Netherlands. ESA SP-440, pp. 575-580, 1999
18. Grant, T. J., and Geurts, P. W. M. (2000). *Web-Based Command & Control: The millennium experience*, presented at the 2nd People in Control (PIC'01) conference, UMIST, Manchester, UK, June 19-21, 2001

Report on Panel Discussion 5:

Broader Impacts and Implications

K. Murphy, V. Rusyn, A. Volp, International Space University, Strasbourg Central Campus, Parc d'Innovation, Boulevard Gonthier d'Andernach, 67400 Illkirch-Graffenstaden, France

e-mail: murphy@mss.isunet.edu, rusyn@mss.isunet.edu, volp@mss.isunet.edu

Panel Chair: S. Mostert, Sun Space and Information Systems (Pty) Ltd., South Africa

Panel Members:

F. Chizea, International Space University, on leave from Nigerian Space Agency
E. Costa Filho, State University of Campinas (UNICAMP), Brazil
T. Grant, Atos Origin Netherland b.v., The Netherlands
M. Othman, Office of Outer Space Affairs, United Nations, Austria
P. Salin, Institute of Air and Space Law, McGill University, Canada
F. Schwertfirm, Technische Universität München, Germany

This panel discussion involved questions and answers on topics related to the role of small satellites for developing countries, and their potential.

As **F. Chizea** pointed out in his presentation, from the user perspective it is important to have the correct balance between the cost of a satellite and the cost of the ground segment. Both costs have to be considered when establishing new satellite services.

T. Grant pointed out that, from the consumer's viewpoint, the satellite itself is not important; it is just a tool to perform a service. The user has a need for a specific service, e.g., for telecommunications, not for a particular satellite architecture. Users are aware of the cost of the equipment which they use in-house; they are not so exposed to the cost of the satellite.

The sometimes used analogy between the evolution of satellite architecture and of the computer was criticized because of the aesthetic difference between having a tangible piece of hardware in front of the user and a dematerialized satellite in orbit. More important was the existence of user requirements and a market potential for satellite services. On the other hand, it was pointed out that

M. Rycroft and N. Crosby (eds.), Smaller Satellites: Bigger Business?, 337–338.

the PC started as a concept, a Personal Computer, without any idea of what the market was going to become.

It was noted that developing countries require help to establish their satellite programmes. The problem is often not that they lack money, but that the priority lies elsewhere, for example in military programmes. The lack of awareness of what satellites can contribute to a developing country is more fundamental. If money is available for a space project, how is that actually converted into benefits for the country?

Even though small satellites can be very useful for disaster monitoring, or for mapping erosion or oil spillage, the awareness amongst politicians of space contributions to a nation's needs is limited. Small satellites are a direct measure of achievement for the scientists and engineers involved. **M. Othman** pointed out that training is the key for space successes in developing countries, where remote sensing images have many users.

The question was raised as to whether small satellite projects are truly producing the educated engineers, as is claimed. **F. Schwertfirm** confirmed that almost all students involved in the LunarSat programme end up working in the German or European space industry. **M. Othman** reflected upon her experience in Malaysia, and informed the audience that a considerable number of people with experience of space are active in the Ministry of Science there. As an example, students develop complex skills of communication and, even if they apply these later to terrestrial systems, their training is of definite benefit to their country.

Session 6, Part 1

Debate: The Business Case for Small Satellites has yet to be made

Session Chair:

K. Doetsch, International Space University, France

Report on Panel Discussion 6, Part 1

Debate: The Business Case for Small Satellites has yet to be made

P. Milani, M. Rycroft, International Space University, Strasbourg Central Campus, Parc d'Innovation, Boulevard Gonthier d'Andernach, 67400 Illkirch-Graffenstaden, France

e-mail: milani@mss.isunet.edu, rycroft@isu.isunet.edu

Session Moderator: K. Doetsch, International Space University, France

Panel Members:

For: **R. Fleeter**, AeroAstro Corporation, USA
R. Parkinson, Astrium Ltd., UK

Against: **M. Sweeting**, Surrey Satellite Technology Ltd., UK
G. Milne, Sun Space and Information Systems (Pty) Ltd., SUNSAT, South Africa

K. Doetsch outlined the importance of the Symposium as a means to create an atmosphere where the experts share their knowledge and experience in the development of small satellites with people without expertise or without knowledge. They come up with recurrent themes and different points of view, and also bring new information about small satellites. This process can either confuse or reach a consensus, and also offer permanent knowledge through the written proceedings of the Symposium.

To initiate the debate, **K. Doetsch** asked a fundamental question: "Will small satellite systems become a significant part of the information age infrastructure? If so, when and how?"

M. Sweeting spoke of his vision of the young, whose interest in science and technology was stimulated more by space activities than by their immediate environment. **G. Milne** agreed that children could achieve much once they had been shown how to participate.

R. Fleeter agreed, but asked the question: "Who owns the entrepreneurship of space?". Noone. New space ideas could grow, hugely. However, a satellite is only a small part of an operating system. The important

M. Rycroft and N. Crosby (eds.), Smaller Satellites: Bigger Business?, 341–343.

concept is of a longterm business that does not demand continuous investments; that has still not been found. **R. Parkinson** defended the idea that small satellites will make a difference if they make money and if they do what individuals wish them to do. He mentioned one of the segments of the small satellite market - the developing world; space can directly reach many, many small communities there and make new sources of information available. He asked the question: "Can small satellites make a difference to the evolution of these societies, bringing up their level of knowledge, information, and standards of living?"

From the audience **J. Burke** made the comment that individuals can indeed do something about their intolerable environment. Space activities can lead to a cleaner environment and to energy saving.

K. Doetsch asked the MSS students to give their views, as young and new entrants to this field, on these applications, and the rationale for moving ahead with the development of new satellites. **V. Denis** (France) pondered the students' need to see the spirit of the entrepreneur. The new space business should come from a new and small company as happened for the development of the personal computer. **J. Hair** (USA) stated that the spacecraft is not the end product; it is necessary to improve the applications of small satellites. As well as more focus on smaller satellite programmes, to design for small costs was important, as was a reduced programme time to be achieved by choosing cheaper, easier, faster and autonomous technologies.

R. Parkinson pointed out that the big difference between computers and spacecraft is that, with a personal computer (PC), you can do anything that you want, but that a small satellite is designed for a specific application. **M. Sweeting** added that they now share the same way of development. Small satellites are still at the very beginning, as the computer was 40 years ago. In 20 years time, he sees microsatellites as "parts", being reconfigurable in orbit to do anything that an individual wants done. **G. Milne** considered possible problems if small satellites become very numerous, like frequency management and space debris issues.

From the audience came the request for a consideration about how the space industries can take advantage of the militarisation of space, both defensive and offensive. **R. Fleeter** observed that there are business opportunities in the developing world for remote sensing satellites, for defence, for locating regions of standing water (and hence malaria hazards), and for student projects.

J. Burke posed a question about the possibility of the fusion of the cellular phone, Internet and satellite industries through the development of some system architecture, once all these industries are driven by the same microtechnology developments? **R. Parkinson** stated his concern that such a completely integrated system architecture means that someone would have to be in charge of the system. We need to guarantee that individual options remain open, with competition in the marketplace being based on performance rather than on requirements.

In conclusion, all four panel members basically agreed that small satellites do have a big future.

Session 6, Part 2

Concluding Panel Discussion: Highlights and Lessons Learned in all Sessions

Session Chair:

K. Doetsch, International Space University, France

Report on Panel Discussion 6, Part 2

Concluding Panel Discussion: Highlights and Lessons Learned in all Sessions

P. Milani, M. Rycroft, International Space University, Strasbourg Central Campus, Parc d'Innovation, Boulevard Gonthier d'Andernach, 67400 Illkirch-Graffenstaden, France

e-mail: milani@mss.isunet.edu, rycroft@isu.isunet.edu

Panel Chair : K. Doetsch, International Space University, France

Panel Members:

J. Blamont, CNES, France
J. van der Ha, Consultant
D. Hernandez, CNES, France
S. Mostert, Sun Space and Information Systems, SUNSAT, South Africa
K. Murphy, MSS6 Student Panel Representative
J. Schafer, NASA HQ, USA
R. Zee, University of Toronto Institute for Aerospace Studies, Canada

K. Doetsch convened the concluding panel discussion, asking each panel member for his top-level comments about the message to be taken home from the Symposium.

D. Hernandez summarized his own view of the Symposium, making it clear that there are important niches for small satellites in the telecommunications area, technology demonstration and qualification. Small satellites are a growing "club", driven by the participation of young professionals and students.

J. Blamont had an ambiguous feeling. But so long as the performance of components aboard small satellites grows because of technological progress it will be possible to market small satellites to compete in performance with large satellite systems. In this way the small satellite business will increase.

J. Shafer observed that the reduction in size and weight of satellites stimulates the launch industry and further stimulates investment in the development of small satellites. And the reduction in satellite size helps large satellites to become more cost-effective. The small and large satellite businesses are therefore complementary to each other.

M. Rycroft and N. Crosby (eds.), Smaller Satellites: Bigger Business?, 347–348.

R. Zee considered that the future for nanotechnology and small spacecraft is related to widespread terrestrial applications, since large investments are needed. New applications should address the added value in order to stimulate investment.

J. van der Ha stressed the point that small satellites are seen as an alternative way of doing things. It is necessary to focus on the improvement of technology, but also to realise the customers' needs for services as well as a company's ability to be competitive in the overall marketplace. Small satellites are now getting bigger, and bigger ones smaller.

S. Mostert suggested two things: the use of small satellites for proven new applications and services for humankind, and space programmes to stimulate young and bright people to develop their skills and to become a means to realise their entrepreneurial spirits.

K. Murphy expressed the MSS students' perception of a lack of commercial focus during the presentations. In order to increase public awareness, the aerospace sector needs to show how space can bring benefits. Ideas like the creation of an International Space Chamber of Commerce and an International Bank for Space Development, developed by the Space Generation Forum at the United Nations UNISPACE 3 conference, should increase international cooperation and business. The students also wanted to know what are the visions of the leaders in the aerospace industry today which, through their work, they may achieve.

K. Doetsch thanked the speakers for their useful thoughts and the MSS students for their views. He opened the discussion to public comments.

A member of the audience came with view that, if the satellite providers develop systems that can offer easy interactions with the final users, there is a change which makes them more "present" in public life. Through their project students could provide a focus in the development of new applications and services for the end users. **J. van der Ha** believed that, first, there should be a statement of what the customer wants, a particular service, in which small satellites can play an important part. Performance satisfying customers' needs was crucial.

K. Doetsch closed the Symposium by thanking the panelists for their views and the attendees for their participation.

Posters

Mission Model Considerations using the NEC Advanced Small Satellite Bus

J. Aoyama, K Murata, T Ono, K. Todome, M. Kawaguchi, M. Adachi, NEC, Space System Division, 4035 Ikebe, Tuzuki, Yokohama, Japan

e-mail: j-aoyama@ax.jp.nec.com

Abstract

In recent years, the trend of satellite system design is to become more flexible and efficient by miniaturization and cutting the costs of the satellites. In the field of Earth observations, high accuracy, agile and timely observations are prominent as mission requirements. In the field of satellite communications/broadcasting, the customer desiring operations with sovereignty is increasing. NEC has developed 500 kg class small satellites, OICETS for on-orbit validation of LEO-GEO optical inter-satellite communications and MDS-1 for on-orbit evaluation of commercial-off-the-shelf parts as well as small scientific satellites.

In this paper, an advanced small satellite bus now under development is described. Then, the feasibility of a small GEO broadcasting satellite and an Earth observation satellite using this advanced small satellite is discussed.

1. Advanced Small Satellite Bus System

The development target of the Advanced Small Satellite Bus system is as follows:

- Equivalent functional performance, reliability and redundant system constitution to those of medium and big satellites
- Common system for GEO and LEO
- Introduction of Plug and Play concept to satisfy the requirements of missions specific functional performance
- Piggyback compatible satellite of the Ariane 5 ASAP class

Five key technologies are as follows:

(1) High performance integrated data processing technology and Plug and Play technology

A high performance on-board MPU of the 64 bits 100 MIPS class and a highly reliable CRAFT-type redundant processor system are now being developed. The high performance processor realizes the integration of satellite control functions such as satellite control-data processing, attitude control and navigation, and solar array drive, thermal and power control. Besides a Plug and Play type IEEE 1394 data bus and a remote terminal chip set in the user unit are now under development. They lead to the construction of a small, light

M. Rycroft and N. Crosby (eds.), Smaller Satellites: Bigger Business?, 351–354.

weight and integrated high performance data processing system meeting various mission requirements flexibly.

(2) Plug and Play Technology of onboard software using object-oriented language

Object-oriented programming is now being utilized to meet various satellite performance requirements using JAVA. Then, the efficient development of onboard satellite control software is enabled at reasonable costs by realization of the Plug and Play facility of the onboard satellite control software.

(3) Integrated attitude sensor technology

Attitude sensors are integrated into star sensors having the ability for all sky attitude determination.

(4) Generalization of the propulsion system

The propulsion system for the GEO and LEO systems are identical using a bi-propellant 10N thruster. Apogee firing for a GEO satellite is performed by four thrusters at the bottom of the main body firing simultaneously. Thus, the use of a propellant with high Isp for a single propulsion system for GEO/LEO keeps the design simple and the weight to a minimum.

(5) High performance and minimum mass of power supply and solar array panel

A lithium ion battery is now being developed with a capacity of three times that of a conventional, commercial-off-the-shelf battery. It helps to lighten the power system. Besides, the solar array panel mass becomes two thirds that of a conventional panel by adopting a multi- junction cell as the solar array cell.

2. Satellite Mission Model

Mission models for a small GEO broadcasting satellite (BS) and a small Earth observation satellite compatible with the Ariane 5 piggyback ASAP minisat class have been studied using this advanced small satellite bus shown in Fig. 1.

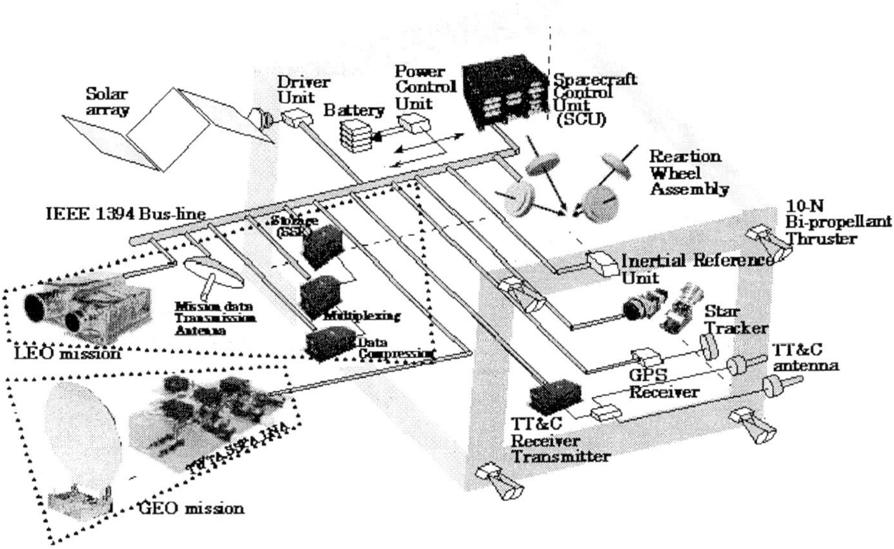

Figure 1. Schematic diagram of the advanced satellite bus

(1) Small BS mission model

A small broadcasting satellite (BS) with two 150 W-class transponders broadcasting several digital channels is proposed. The weight of such an advanced GEO small satellite bus is estimated to be 120 kg. The payload weight including two Traveling Wave Tube Amplifiers (TWTA's) is about 37 kg. The propellant weight for a GEO satellite of 5-year lifetime launched by Ariane is 144 kg. The total mass is therefore 300 kg. Thus, it satisfies the minisat weight criterion for a piggyback payload ASAP of Ariane 5, being 300 kg. Here, the solar array power is to be 800 W and the battery capacity 1300 Wh.

(2) Small Earth observation satellite mission model

A polar orbit ocean color observation satellite, with 7 bands in the near infrared, is suggested. The dry weight of the Earth observation satellite bus is 175 kg. The payload consisting of a radiometer, mission data processing and recording system, and mission data transmitter has a mass of 77 kg. The total dry weight is therefore 252 kg. The propellant mass for a 5-year lifetime is 14 kg. It satisfies the minisat weight of the ASAP, the mass being 288 kg. Here, the solar array power is 900 W and the battery capacity 600 Wh.

3. Summary

The development of an advanced small satellite bus has been discussed briefly. Moreover, its feasibility is confirmed by the study of a mission model. Substantiation of the advanced small satellite bus is now progressing for both a broadcasting satellite and an Earth (ocean) observation satellite.

SCALE: A Scaleable Wire Boom System

L. A. Bylander, O. H. Carlstrom, T. S. R. Christenson, F. G. Olsson, Royal Institute of Technology (KTH), Alfvén Laboratory, SE-100 44 Stockholm, Sweden

e-mail: bylander@plasma.kth.se, carlstrom@plasma.kth.se, christenson@plasma.kth.se, olsson@plasma.kth.se

Abstract

The SCALE wire boom is a compact wire boom unit designed mainly for scientific spin-stabilized satellites. The boom unit provides storage and deployment of a wire boom (length 5-50 m) and a probe (diameter up to about 100 mm). The wire is both a boom and a supplier of the electrical connections between the probe at the end of the wire and the experiment unit in the spacecraft body. The SCALE wire boom is unique because its volume and weight are scaleable with respect to the required boom length and probe diameter.

1. The Design

The boom wire, stored between two concentric cylinders, is fed using an electrical mechanism at the rear of the boom unit. This mechanism lifts the wire from between the cylinders and pushes it out along the centre axis of the two cylinders (see Fig. 1).

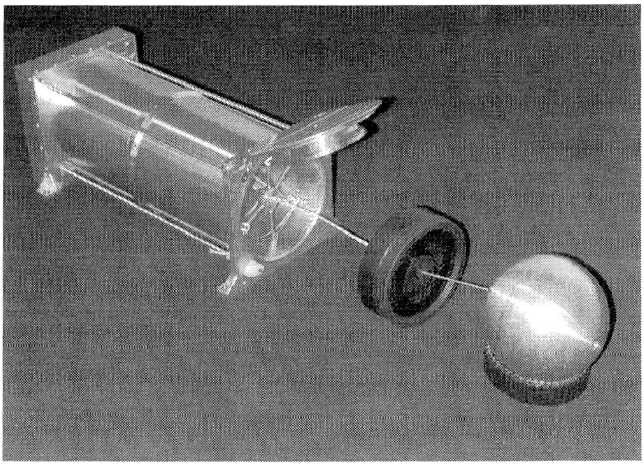

Figure 1. A partially deployed wire boom

The outer cylinder forms the body of the wire boom unit. The probe and the "hockey puck" are stored inside the inner cylinder. The probe at the end of the wire measures the potential of the surrounding plasma.

M. Rycroft and N. Crosby (eds.), Smaller Satellites: Bigger Business?, 355–356.

2. Design Advantages

The main advantages of the SCALE wire boom are that it is compact and scaleable. Parameters for three units of various sizes are given in Table 1.

Boom length (m)	Probe diameter (mm)	Dimensions (mm^3)	Volume (dm^3)	Weight (kg)
5	< 40	65*65*90	0.4	0.2
25	< 70	95*95*230	2.0	1.0
35	< 85	110*110*265	3.2	1.4

Table 1. Parameters of the scaleable wire boom system

Other advantages of the SCALE wire boom design are:

- The cylindrical shape of the boom unit is more favourable than that of a rectangular box in terms of mechanical stresses
- The volume of the complete system is minimized since the wire boom is stored around the measurement probe
- The electrical connection between the experiment unit on the satellite and the wire boom is made without slip rings
- It is easy both to deploy and retract the wire boom. Running the electrical motor in one direction deploys the wire boom and running it in the other direction retracts the wire boom.

3. Design Heritage

The wire boom system for the Astrid-2 satellite, a Swedish micro-satellite launched in 1998, was designed at the Alfvén Laboratory [Reference 1]. The wire is wound around the body of the satellite and the disadvantage of this boom system is that it is so tightly integrated with the spacecraft body. The lessons learned from the Astrid-2 boom system stimulated us to start the design of the SCALE wire boom system. The intention with the name SCALE is to point out that the boom design is scaleable. Details are available at http://www.plasma.kth.se/space/scale.html.

References

1. Hellman, H.: *Design of Wire Boom System for a Satellite*, Diploma thesis, Royal Institute of Technology, 1996

A Modular Design Concept For Small Satellites

L. A. Bylander, O. H. Carlström, T. S. R. Christenson, F. G. Olsson, Royal Institute of Technology (KTH), Alfvén Laboratory, SE-100 44 Stockholm, Sweden

e-mail: bylander@plasma.kth.se, carlstrom@plasma.kth.se, christenson@plasma.kth.se, olsson@plasma.kth.se

Abstract

The objectives of the new design concept (called Bylix) are to minimize the mass and power dissipation of the payload and the platform. The system design is modular to allow concurrent engineering throughout the design, integration and test phases, and to improve project management. The basis of the concept is a common "thin bus" for data and commands, power, frequency synchronization and time distribution, which requires a minimum of resources.

1. The Bylix Design Concept

What is Bylix? Bylix is an electrical system design concept for small scientific satellites. Bylix defines the functions of the software and hardware at the system level. To some extent low level software and hardware are also defined. The expression, small satellite, is mainly limited by the electrical performance of the bus. The design goal is a bus length of about 2 meters.

The system design features are:

- A common system bus
- A common real time clock to tag data
- Definitions of basic system functions
- Transparent to almost any number of satellites (i.e. 1 to 32 (TBD)).

The optimization criteria are:

- Maximum design modularity to reduce time to completion
- Maximum design re-usability to reduce time to completion
- Maximum testability during both the design phase and the integration phase to improve the quality
- Minimum power dissipation for the whole satellite
- Minimum mass and size of the electronics to improve the mechanical flexibility.

The optimization is achieved by:

M. Rycroft and N. Crosby (eds.), Smaller Satellites: Bigger Business?, 357–358.

- Selecting an optimum system design so that the platform and the payload electronics are minimized with respect to power dissipation and mass
- Selecting a design where each functional module is built on one printed circuit board to minimize the need for connectors and cables
- Selecting between implementing functions in hardware or software.

2. Example

The first task when a satellite is designed is to divide all functions needed into an appropriate number of units. Some functions are defined by the design concept (as indicated by *B* in the list below). The rule of thumb is one sensor per unit. Each item represents a physical unit:

- Power and frequency reference unit (*B*) generates stable power voltages (14 V and 3 V) and two frequencies (400 kHz and 1 Hz)
- Real time clock unit (*B*) distributes the correct time once per second
- Telemetry queue and transmitter unit (*B*) buffers all telemetry packages
- Command receiver unit (*B*) receives commands and distributes them to the correct receiving unit
- Attitude control unit controls the manoeuvre coils
- E-field unit measures the electric field from one probe, digitizes the value, stores a number of samples in a data package with a time stamp and sends the complete data packages to the telemetry queue unit
- B-field unit, a 3-axis fluxgate magnetometer.

3. Discussion

The driving force behind the Bylix design concept is to make it possible to build small scientifically comprehensive satellites at a low cost. Here small means about 10 kg, scientifically comprehensive means a typical plasma physics payload, and low cost is about US $ 3 million (comparable to the cost of the Swedish Space Corporation satellite Astrid-2). This task is expected to be accomplished by a modular design where each unit has few functions and can typically be implemented using one field programmable gate array and a few analogue devices mounted on one printed circuit board. This is contrary to making a few modules, with many functions implemented, using custom-designed devices manufactured using the micro electro mechanical systems technique. The use of the specification is open and a draft of the design concept is available at: http://www.plasma.kth.se/bylix/draft.pdf. Comments and/or contributions to the specification are welcomed.

NESS: Using a Microsatellite to Search for and Track Satellites and Asteroids

K. A. Carroll, Dynacon Enterprises Limited, 3565 Nashua Drive, Mississauga, Ontario, Canada L4V 1R1

e-mail: kac@dynacon.ca

A. R. Hildebrand, Department of Geology and Geophysics, University of Calgary, 2500 University Drive NW, Calgary, Alberta, Canada T2N 1N4

email hildebra@geo.ucalgary.ca

Abstract

The Near Earth Space Surveillance (NESS) mission concept will use a microsatellite platform to place a small optical telescope in low Earth orbit. This telescope will be used to search for, and track, two types of objects that are resident in space - asteroids and Earth-orbiting satellites.

1. NESS (Near Earth Space Surveillance) Mission Objectives

The NESS mission is proposed to serve the purposes of two separate space-tracking communities: satellite tracking and asteroid tracking. The mission will perform R&D in support of space-based satellite tracking programs, as well as search and follow-up tracking for Earth-approaching asteroids.

1.1 Satellite Tracking Objectives

Satellites are tracked for a variety of purposes, including provision of ephemeris information in support of mission operation functions, such as the pointing of high-gain ground-station antennas, and on-board estimation of magnetic field strength/direction for attitude estimation purposes. They can also support "space traffic control" functions and predict potential satellite collisions. NORAD uses satellite tracking data to distinguish satellites from ballistic missiles approaching North America. Because of the importance of this application, the U.S. Department of Defense's Space Command also maintains an accurate catalog of all trackable objects in Earth orbit. While satellites in low Earth orbit are efficiently tracked using ground-based radars, optical tracking has advantages for satellites in higher orbits (e.g., in geostationary orbit, or in Molniya orbits).

The Canadian Department of National Defence (DND), as part of Canada's contribution to NORAD, plans to develop an operational space-based satellite tracking system (SAPPHIRE). A microsatellite-based design similar to NESS is

M. Rycroft and N. Crosby (eds.), Smaller Satellites: Bigger Business?, 359–360.

one approach being considered. NESS objectives include testing various approaches to operating a space-based satellite tracking sensor, and collecting data on performance, image quality, noise, etc., to provide data that can be used to guide the design of the SAPPHIRE system. A key performance objective is to be able to detect satellites in geostationary orbit, as faint as magnitude 14.

1.2 Asteroid Tracking Objectives

Asteroid tracking is done for several reasons. Astronomers search to discover new asteroids, and use knowledge of their orbits to help develop models of solar system evolution and planetary formation. Follow-up tracking of asteroids is done in order to refine the accuracy of their orbit estimates, and to allow future close approaches with the Earth (i.e., good observing opportunities) to be predicted. Additional tracking allows a precise estimate to be made of the likelihood of a collision between an asteroid and the Earth. Looking ahead to future asteroid rendezvous and sample-return exploration missions, each new asteroid catalogued represents an additional potential mission target, and accurate orbit estimates are needed to enable exploration spacecraft to be navigated to their targets.

NESS objectives include searching for Earth-approaching asteroids, particularly those whose orbits are interior to that of the Earth's—the Aten-class asteroids, and the (hypothetical) Inner-Earth Objects (IEOs)—because these are difficult to observe from observatories on Earth, due to their angular proximity to the Sun. A key instrument performance objective is to be capable of detecting asteroids as faint as magnitude 20. A key operational performance objective is to be able to search the two regions "eastward" and "westward" of the Sun, between the solar elongation angles of 45 and 70 degrees, and ecliptic elevation angles of +/-20 degrees, once per month.

2. Outline System Design of NESS

The NESS mission objectives can be met using a CCD imager with a 0.85x0.85 degree field of view, 15-cm diameter optics, on a satellite platform stabilized to an accuracy of ~ 1 arc-second. A design for the NESS satellite has been based on the design of the 57 kg, 60x60x20 cm MOST space astronomy microsatellite, which is being developed by Dynacon, the University of Toronto and the University of British Columbia for the Canadian Space Agency. NESS will be able to re-use many elements of the MOST design, with some upgrades. In particular, the MOST attitude control system will be upgraded for use on NESS, to improve the accuracy by a factor of 10. Also, additional baffling will be needed for the telescope, to allow pointing as close as 45 degrees to the Sun.

Integrating Robustness Into ST-5 Nano-Satellite Development For Inexpensive Space Missions

G. Chen, Morgan State University, Department of Industrial, Manufacturing and Information Engineering, Baltimore, Maryland 21251, USA

E-mail: gchen@eng.morgan.edu

J. Oberright, NASA Goddard Space Flight Center, Systems Engineering Division, Greenbelt, Maryland 20771, USA

E-mail: john.oberright@gsfc.nasa.gov

Abstract

This paper presents an initial investigation into the cost reduction for multiple production runs of nano-satellites while maintaining high-quality standards. The objective is to reduce the recurring costs significantly. Nano-satellite design and manufacturing is an important issue in less expensive space missions. NASA has set strategic plans for technological development of small satellites, of which the Space Technology 5 nano-satellites development is one. It involves developing 10 to 20 kg-class spacecraft and operating them in multi-spacecraft science missions. The vision is to fly tens to hundreds of nano-satellites in a constellation to make multiple remote and in-situ measurements simultaneously in space. Thus, we can revolutionize the scientific investigations of key physical processes explored by NASA Space Science and Earth Science Enterprises.

1. Introduction

Satellites have become larger and more expensive than those in earlier time. The development span from concept to useful orbital operation usually consists of many years [Reference 1]. For timely validation of new ideas and technologies, it is important to explore novel design and manufacturing strategies for less expensive satellites. Indeed, the strategy should focus on high quality and low cost, taking the economic advantage of mass production. In short, small satellites, as complements to large ones, could provide scientists with the opportunity for affordable and timely access to space. It is not surprising that the demand for this has greatly increased over recent years. Concurrently, fascinating new technologies make it possible to produce lower-powered and lighter satellites [Reference 2].

NASA has set strategic plans for the technological development of small satellites for future space missions. Space Technology 5 (ST-5), one of NASA's new millennium programs, involves developing 10 to 20 kg-class spacecraft and operating them in multi-spacecraft science missions. Many nano-satellites as a constellation could make multiple remote and *in-situ* measurements simultaneously in space, thus revolutionizing the scientific investigations of key

M. Rycroft and N. Crosby (eds.), Smaller Satellites: Bigger Business?, 361–367.

physical processes explored by the NASA Space Science and Earth Science Enterprises. The cost projection is US $28.6 million for fabricating three pioneering nano-satellites [Reference 3]. Ideas and models will be recorded for future missions that will call for 100 similar spacecraft at a very low recurring cost. On a competitive basis, the NASA Goddard Space Flight Center has been assigned the responsibility of implementing the ST-5 Small Satellite Constellation Theme. This project will validate eight new technologies [Reference 4]. In addition, an important aspect of this small satellite project is to develop techniques to minimize the overall cost of production runs for future constellation missions.

2. Strategic Cost Reduction

Over recent years, the space and satellite industries have been confronted by the challenges of cost reduction while keeping high-quality manufacturing standards. More versatile, sophisticated and lighter satellites must be produced at lower cost. Satellite production has usually consisted of a unique, individual and one of a kind product. On the contrary, the mass production of satellites refers to some tens or hundreds. The Goddard ST-5 project considers the development of three modules as a pathfinder. Novel design and manufacturing strategies for such a development are based upon the benefits of mass production and nonlinear effects of learning curves. Incorporating new design concepts with the traditional technologies is also a focus in the effort towards accomplishing this goal. The objective is no longer merely to design and launch a successful satellite, but also to reduce its cost. Consequently, we should probe the area for a new methodology and new technological applications, rather than just look at cost reductions in a conventional way.

Robust design is a systematic approach to quality. Its ultimate goal is to build quality into products through good design. Robust design has three basic steps - system design, parameter design, and tolerance design [References 5, 6, 7, 8]. Similarly, robustness can be integrated into our cost-conscious nano-satellite design and manufacturing process. The total cost (TC) consists of a fixed nonrecurring cost and, a recurring cost given by

$$TC = a + bn \tag{1}$$

where a is the initial setup (or nonrecurring) cost to start production run; b is the variable cost, recurring for each unit of product; n is the number of products produced [Reference 9]. This model may be used for multiple productions of the same or similar satellites. Nonrecurring cost is the one–time expenditure to initialize the production. This cost can refer to that for conceptual design and development of satellites, manufacturing molds, durable equipment, tool and fixtures. In addition, due to the complexity of satellites and the nature of intensive high-tech skills, the recurring cost may be affected by a nonlinear "learning" curve for the first several satellites produced.

The learning curve was adapted from the historical observation that individuals carrying out repetitive tasks exhibit an improvement in performance as the task is performed a number of times. Empirical studies of this phenomenon [Reference 10] yielded three conclusions on which the current theory and practice is based upon:

- The time required to perform a task decreases as the task is repeated
- The amount of improvement decreases as more units are produced or, the shorter required time leads to the smaller improvement amount
- The rate of improvement has sufficient consistency to allow its use as a prediction tool.

The improvement has been found exists in the form of a constant percentage reduction in the time required over successively doubled quantities of units produced, as can be seen graphically in Fig 1. The mathematical model can be developed for this learning curve. We define $T_n = t_0 + t_n$ as the time required for the nth product (t_0 is a constant value or the basic time required to perform the task when the operator is very skillful, while t_n is the time spent in learning for the nth product). The derivative of T_n with respect to n (dT_n/dn) is the rate of change in the time required for each unit. In industrial application of learning curves, we assume that dT_n/dn exists. Its value is negative. The absolute value of dT_n/dn (improvement rate) should be proportional to the required learning time t_n (shorter t_n leads to smaller improvement). Thus, we have

$$\frac{dT_n}{dn} = \frac{dt_n}{dn} = -kt_n \tag{2}$$

where k is a constant value that reflects the percentage improvement with time. Integrating from Equation (2), we have:

$$t_n = t_1 e^{-k(n-1)} \tag{3}$$

Hence,

$$T_n = T_0 + t_n = T_0 + t_1 e^{-k(n-1)} \qquad (4)$$

where T_n is the total time needed to perform task n or to make the nth unit. Equation (4) can be used to estimate the effect of the nonlinear learning curve; this learning curve can also apply to the nonrecurring cost b in Equation (1).

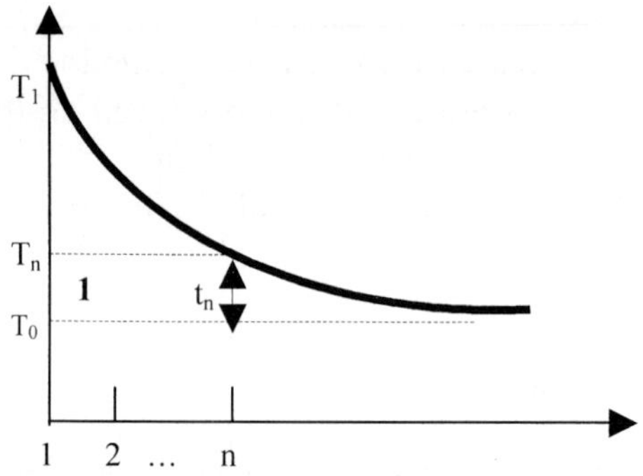

Figure 1. The Improvement Curve (or learning curve).

For total cost reduction, one of the most effective strategies is, apparently, to reduce the recurring cost b. This strategy pays off even at an expense of higher nonrecurring cost a. For instance, to produce and fly 100 similar nano-satellites, a reduction in the recurring cost can be reflected on the significantly reduced total cost. Examples include standardizing the manufacturing of structures and common parts or components, building up long-term, reliable partnerships with contract companies. We should try to reduce the recurring cost b to a smallest possible amount needed after several satellite production runs. For instance, the ST-5 project is a process of producing three pathfinder nano-satellites with a significant learning curve. The future production of other 100 similar nano-satellites will take place at a very low recurring cost. Thus, strategically, the cost of learning can be treated as a nonrecurring cost rather than as a recurring cost. That implies the use of the same labor resources for manufacturing, integration as well as testing, with long-term, dependable and trusted suppliers. Since standard production methods have lower costs, to reduce the recurring cost we should standardize as many parts and components as possible. For instance, main structures can be produced by a standard mold or standard process; the power supplies including the solar arrays can be produced as a standard module.

The duplicated and repetitive production will need enhanced quality assurance to minimize the deviations from the target performance. We should consider using the concept of robustness in the design and manufacturing process. By minimizing the deviations, the performance is robust or insensitive to the undesirable manufacturing variations as well as to predictable changes, such as a change in functional feature. We should implement quality management in the production. To do the job right first time is important. Doing jobs wrongly will always bring waste in production. That will further create loss and add significantly to the total cost.

3. Robust Design Applied To ST-5 Nanosat Trailblazer Project

The ST-5 nano-satellite constellation trailblazer project deals with the fabrication of three pioneering nano-satellites. It involves the development and investigation of 10 to 20 kg class spacecraft [Reference 4] and their operation in multi-spacecraft science missions. The conceptual pathfinder shown in Fig. 2 spacecraft platform is spin-stabilized, with body mounted solar arrays. The basic structure is an octagonal prism, selected for its application to volume efficient launch vehicle packaging for future nano-satellite missions. It is 20.0 cm in height by 41.6 cm in diameter across the flat faces (44.0 cm including the stowed magnetometer) and 45.0 cm across the vertices.

Robust design is cost-effective. Quality is built in through good design, rather than by enhancing on-line quality control activities. The application of robust design consists of three basic steps. The design criterion is the minimization of quality loss or of the variations in product performance. Taguchi defines quality as the loss imparted by the product to the society from the time the product is shipped. This is a more holistic viewpoint because the quality is related to the loss incurred, not only to the manufacturer at the time of production, but also to the consumer after the product is delivered. The quality activities must focus on reducing this loss. To integrate the robustness concept into the paradigmatic spacecraft design, we must define the goals in each stage of design and manufacturing. It is aimed at minimizing the total loss and production cost, for the life cycle of the spacecraft that includes idea generation, design and manufacturing, as well as the useful lifespan. For the ST-5 project, similarly, the entire process can be divided into three basic steps - conceptual design, detailed design and variation control.

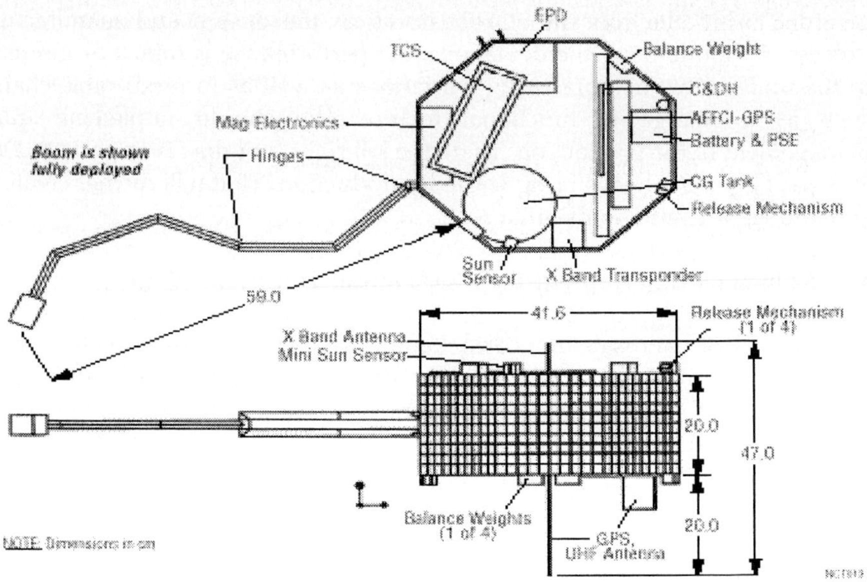

Figure 2. Diagram of the ST-5 nano-satellite, the dimensions being given in cm

4. Conclusions

Low-cost space missions are challenging issues confronting aerospace industries. Nano-satellites provide an opportunity of timely and affordable access to space for the validation of new ideas and new technologies, while validation experiments must be performed in the space environment. We have discussed the concept of robustness and the application of learning curves in the design and manufacturing optimization model of nano-satellites for large-quantity production at very low recurring costs.

Since we are dealing with multiple production runs (tens or hundreds) of satellites, optimal design and manufacturing techniques can lead to cost reductions while maintaining high-quality standards. Robustness can minimize the effects of variations in manufacturing processes, incoming materials, parts, components and subsystems, while having little impact on costs. Thus, we can achieve our goal of high quality and low production costs.

Acknowledgements

This research was supported by the NASA/ASEE Faculty Summer Research Fellowship. The authors want to thank James Andary, Systems Engineering Division Chief and the ST-5 Project Team at NASA Goddard for their collaboration.

References

1. Evans, B. (editor): *Satellite Communication Systems*, 3rd edition, Institution of Electrical Engineers, London, UK, 1999
2. Bertheux, P., Douillet, F. and Chiarini, J. C.: A low-cost multi-mission bus for small satellite applications, *Space Technology, Vol. 18,* pp. 171-174, 1998
3. NASA Goddard Space Flight Center, *Concept Study Report for the New Millennium Space Technology 5 Small Satellite Constellation Theme*, Goddard Space Flight Center, July 15, 1999
4. NASA Goddard Space Flight Center ST-5 Project Team, *ST-5 Systems Concept Review*, NASA, May 10-11, 2000
5. Kackar, R. N.: Off-line Quality Control, Parameter Design, and the Taguchi Method, *Journal of Quality Technology, Vol. 17,* 1985
6. Taguchi, G.: *Introduction to Quality Engineering,* Unipub/Kraus International Publications, White Plains, NY and American Supplier Institute Inc., Dearborn, MI, 1986
7. Taguchi, G.: "Quality engineering (Taguchi Methods) for the Development of Electronic Circuit Technology," *IEEE Transactions on Reliability, Vol. 44,* No. 2, 1995
8. Chen, G. and Kapur, K. C.: A Two-step Robust Design Procedure of Linear Dynamic Systems for Reducing Performance Variations, *International Journal of Reliability, Quality and Safety Engineering, Vol. 4,* No. 2, 1997.
9. Evans, J. R.: *Applied Production and Operations Management*, 4th edition, West Publishing Company, 1993
10. Modern Technologies Corporation: Defense Manufacturing Management - Guide for Program Managers, Defense Systems Management College, Fort Belvoir, VA, April 1989

Small Spacecraft for Space Weather Upstream Monitoring

N. B. Crosby, Mullard Space Science Laboratory, University College London, Holmbury St. Mary, Dorking, Surrey, GU6 8LP, UK

e-mail: nbc@mssl.ucl.ac.uk

O. C. St. Cyr, CUA/CPI, MS 682 NASA Goddard Space Flight Center, Greenbelt, MD 20771, USA

e-mail: stcyr@cua.edu

A. J. Coates, Mullard Space Science Laboratory, University College London, Holmbury St. Mary, Dorking, Surrey, GU6 8LP, UK

e-mail: ajc@mssl.ucl.ac.uk

M. J. Rycroft, International Space University, Strasbourg Central Campus, Parc d'Innovation, Boulevard Gonthier d'Andernach, 67400 Illkirch-Graffenstaden, France

e-mail: michael.j.rycroft@ukgateway.net

Abstract

Space weather is a multi-disciplinary subject that relies on dedicated spacecraft constantly monitoring the various relevant regions of space. The Sun is of special interest; continuous monitoring of definitive solar wind parameters upstream of the Earth is required, as are timely warnings of solar flares, coronal mass ejections and other extreme solar events. Measurements by instruments on small spacecraft would most cost-effectively provide continuous coverage of the various physical phenomena that cause undesirable space disturbances.

We discuss here a few examples of small spacecraft which have been proposed for such purposes, e.g., Space Weather Diamond, Triana, Geostorm, etc.. Emphasis is placed on the advantages of these spacecraft, and on how the observations could be used as inputs to space weather monitoring programs. We suggest that consideration be given to upstream monitoring using small satellites. Not only would the warnings given by these spacecraft be of importance to satellite operators and to utility companies, but also the observations would provide continuity of vital upstream monitoring information for space science research.

1. Why is Upstream Monitoring of Space Weather Important ?

Technological systems are becoming more sensitive to the environment, people are being sent into space on more of a routine basis, and society is growing more and more dependent on satellites. As the space environment can have harmful effects both on technological and biological systems (see Fig. 1), and perhaps even have effects on the Earth's weather and climate, it is essential

M. Rycroft and N. Crosby (eds.), Smaller Satellites: Bigger Business?, 369–372.

that it is monitored continuously, especially the Sun – the driver of our local space weather.

Spacecraft such as SOHO [Reference 1] and ACE [Reference 2], which orbit the Langrangian L1 point, 0.01 AU from the Earth on the Sun-Earth line, monitor the plasma, energetic charged particles and magnetic field characteristics at that location, providing an advance warning (about one hour) of geomagnetic storms that can overload power grids and disrupt radio communications on Earth, and present a health hazard to astronauts. But what will replace these spacecraft when they are no longer functional, and what of those space weather users that require additional lead-time? Smaller spacecraft monitoring the upstream solar wind conditions may help to provide the answer.

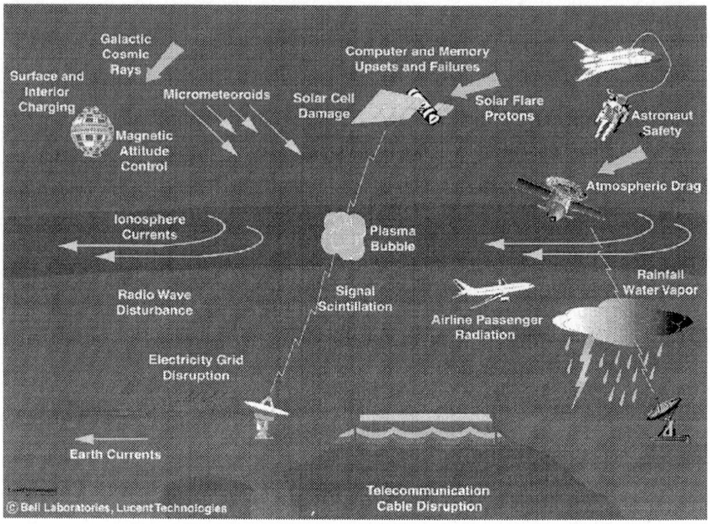

Figure 1. Overview of the user community for space weather products. Courtesy of Bell Laboratories, Lucent technology.

2. Examples of Small Spacecraft for Space Weather Studies

The spacecraft Triana [Reference 3] has been proposed to orbit the L1 Langrangian point. If launched it could help scientists construct more accurate models of Earth's climate and will make the first direct measurements of the radiant power emitted by the whole sunlit side of the Earth. The goal is to increase our understanding of how much of the Sun's energy is absorbed in the atmosphere and to improve our understanding of how global climate works as an integrated system.

Geostorm is a proposed mission that is still under review [Reference 4]. The spacecraft would use ultra-thin sails that take advantage of the push of sunlight to reach orbits closer to the Sun than those of ACE. Geostorm could provide data on an impending geomagnetic storm two to three hours before it reaches Earth. Such a mission probably would not be launched until the next solar cycle maximum.

Reporting plasma and magnetic field characteristics, the Space Weather Diamond offers a 10 times improvement of the warning time over similar platforms located at the L1 Lagrangian point, by monitoring solar wind conditions 0.10 AU upstream from the Earth [Reference 5]. There are four platforms phased into eccentric heliocentric orbits called 'distant retrograde orbits' which have the property that, from the perspective of a fixed Sun-Earth line, the spacecraft appear to orbit the Earth (see Fig. 2). The Space Weather Diamond offers the observer the possibility of separating spatial from temporal variations. This is important for studying the intrinsic properties and evolution of structures embedded in the solar wind. Especially, it should help in the understanding of the interaction of these structures with the terrestrial magnetosphere, highly important from the space weather forecaster point of view.

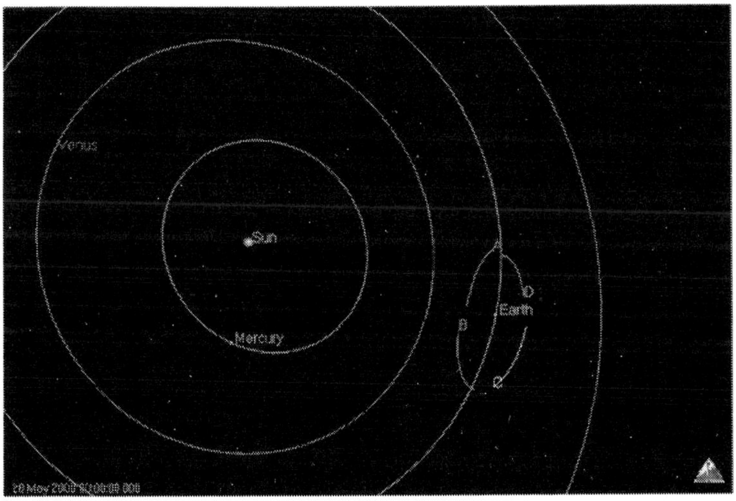

Figure 2. Perspective view of the Space Weather Diamond Orbit, populated with four spacecraft appearing to orbit the Earth [Reference 1].

3. Recommendations

It is suggested that simple mission concepts that do not require a constellation of > 4 spacecraft or exotic propulsion techniques should be considered as near-term solutions for space weather upstream monitoring. Instrumentation should provide *in situ* measurements of the solar wind plasma and the interplanetary magnetic field characteristics. Measurements from such missions would be beneficial both for space science research and for many practical space weather applications.

References

1. <http://sohowww.nascom.nasa.gov/>
2. <http://helios.gsfc.nasa.gov/ace/ace.html>
3. <http://triana.gsfc.nasa.gov/home/>
4. <http://science.msfc.nasa.gov/ssl/pad/solar/suess/Interstellar_Probe/ISP-MissionTech.html>
5. St. Cyr O.C., Mesarch M.A., Maldonado H.M., Folta D.C., Harper A.D., Davila J.M. and Fisher R.R.: Space Weather Diamond: a four spacecraft monitoring system, *Journal of Atmospheric and Solar-Terrestrial Physics, Vol. 62*, pp. 1251-1255, 2000

Nano Satellite Propulsion using FEEP Thrusters

C. P. Fernando, Cranfield University, College of Aeronautics, Cranfield, Bedfordshire, MK43 0AL, UK

e-mail: w.c.p.fernando.1998@cranfield.ac.uk

Abstract

A Field Emission Electric Propulsion (FEEP) system for nano-satellite propulsion is discussed. The feasibility of the technology is considered, taking into account both the nano-satellite and the FEEP thruster. In order to meet short-term nano-satellite applications, the emphasis is to understand performance capabilities of the total system in a LEO mission. This is achieved by considering the available technology, development and cost of launch for nano-satellite(s) as well as FEEP propulsion system requirements in order to meet those demands. FEEP propellant (Cesium and Rubidium) contamination is identified as one of the major problems with this propulsion system.

A FEEP thruster demonstration mission is considered here in order to validate this propulsion system for possible future missions. Both rotation and translation operational modes are identified for the first demonstrator mission. FEEP nano-satellite performance is demonstrated using different software tools such as Satellite Tool Kit (STK) and the AOCS Design Software (ADS) simulation programs from ESA.

1. Electric Propulsion for Micro-Nano-Satellites

Universities and other institutions are involved in micro-nano-satellite development with the incentive to reduce development and launch costs, to manufacture satellites in batches, and to launch them in series, possibly with a single launcher. To achieve this goal, standardisation of the satellite structure and optimum usage of the launcher payload adapter is essential.

Since the 1960's, electric propulsion has been used successfully for many categories of satellites [Reference 1]. The areas of application have been broad, from Sun pointing and attitude control to orbit raising. Initially, electrical thrusters were developed as a secondary propulsion system to provide attitude control and station keeping, mainly for geostationary satellites. However, with missions that require high pointing accuracy, it requires a well-controlled propulsion system (much smaller impulse bit). Space-borne "virtual platform" missions such as LISA [Reference 2] and Space Technology-3 (ST3) [Reference 3] are two examples that demand an accurate propulsion system. To accomplish "formation-flying" missions with high precision requirements, propulsion system like FEEP's are of interest.

1.1 Field Emission Electric Propulsion

FEEP technology has been developed within ESA, with Liquid Metal Field Ion (LMFI) sources for Electric Propulsion (EP) programs. At the end of the

M. Rycroft and N. Crosby (eds.), Smaller Satellites: Bigger Business?, 373–376.

1980's, FEEP was judged less competent than other electric propulsion concepts, especially ion thrusters and arc jets, mainly because of its high specific power. There was, however, renewed interest in FEEP at the beginning of the 1990's for high performance "drag-free" systems.

The main application for the FEEP thruster in the near future will be for fine-tuning with thrust levels in the μN range. Due to the low propellant consumption, the propellant tank could be integrated into the thruster itself, with the propellant being supplied to the thruster taking advantage of the capillary effect. In FEEP emitters, unlike most ion engines, ions are directly extracted from the liquid phase. High atomic weight, good wetting characteristics and low ionisation energy are the important characteristics in selecting a propellant [Reference 4].

The current FEEP emitter consists of two almost identical halves clamped together. The FEEP emitter/accelerator geometry facilitates the creation of the high electric field, which is essential for the production of ions, without an excessive potential gradient. As the propellant consumption is very low, the emitter's built in propellant reservoir is adequate for most applications. An external source of electrons (neutraliser) provides negative charges to maintain electrical neutrality of the thruster assembly.

As FEEP thrusters are "power demanding" (~ 60 W/mN) and require a very high electric fields (~ 10^9 V/m), the design and operation of the power conditioning unit (PCU) is complex. PCU's of low mass/size with small power consumption are required for FEEP-thruster applications in nano-satellites.

1.2 FEEP Propellants

Alkali metals are considered as the best FEEP thruster propellants; FEEP thrusters have been operated successfully using Caesium, Rubidium and Indium. Propellants with low melting points are advantageous, to reduce the power requirements for emitter heating (Cesium melts at 28.5 °C and Rubidium at 38.9 °C). A heater has to keep the propellant in liquid form during the mission but if the propellant temperature much exceeds its melting point, evaporation starts and the thruster operation degrades due to non-homogeneity in the flow. A temperature of 40 °C has been taken as the upper limit for Cesium for safe operation without significant propellant evaporation. The normal operating temperature for a Cesium FEEP thruster in a vacuum chamber is about 35 °C.

If exposed to the environment, Cesium reacts with atomic oxygen and water. When using Cesium for low Earth orbit missions, as well as during the pre-launch and launch scenarios, the emitter-slit may become obstructed, resulting in poor performance. It is thus a good idea to isolate the system using a protected capsule with inert gas, which can be purged, in space.

2. Suggested FEEP Demonstrator Mission

Complete validation of FEEP thrusters is impossible on ground as chamber tests have their limitations. A FEEP thruster should be tested in space prior to its planned use for more advanced missions. As a baseline for demonstration mission analysis, a nano-satellite model (250x250x250 mm, and approximately 10 kg mass) was selected. An average power level of 10 Watts was assumed to be available at all times. It is further assumed that the nano-satellite will be launched on top of a "master satellite" available for demonstration of the constellation. It is assumed that all communications are relayed through the mater satellite. It was further assumed that the launch vehicle would place the demonstration satellite in its final orbit. A short mission lifetime was selected, from a few orbits to a couple of days. The operational lifetime should be long enough to perform the demonstration, transmit the measured data and have confirmation of its reception by ground control.

The demonstrator mission will validate the FEEP thrusters, both in rotational and translational operations. Thrusters are employed to give angular momentum to the spacecraft in order to perform rotation. The rotational rates may be calculated using Sun or star sensors. In order to measure the capability of proportional thrust control, formation flying with respect to the master satellite, will be performed for a short period.

For the FEEP thruster demonstrator mission, the best possible mode of thruster operation is in torque mode. Rotation of the satellite is easy to perform and measurable with good accuracy. However it is thought that, prior to separation from the master satellite, the FEEP thrusters should be switched on and reduce their thrust to a reasonable level to enable the separation. This will allow a formation-flying scenario at the initial stage of the mission, as well as keep the demonstrator satellite in the vicinity of the master satellite.

3. Conclusions

The Field Emission Electric Propulsion (FEEP) concept will play an important role in future space missions as a fine pointing thruster to control attitude. This thruster could operate in continuous as well as pulse mode; either the emitter or accelerating electrode voltage can be pulsed to obtain the required pulses. According to ESTEC test results [Reference 5] with a 3-cm long emitter, impulse bits less than 10^{-8} Ns and minimum usable pulse length of 40 milliseconds were obtained. A demonstration mission was proposed.

References

1. Pollard J.E., Jackson D. E. and Marvin D. C. (editors): Electric Propulsion Flight Experience and Technology Readiness, pp. 7-8, (*AIAA 93-2221*). American Institute of Aeronautics and Astronautics, presented at the 29th joint propulsion conference and exhibit, June 28, 1993-June 30, 1993
2. Danzmann K. and Bender P.: LISA: Laser Interferometer Space Antenna., Horizon 2000 plus Forward, pp. iii, ESA, 96, December, 1995
3. Lau K., Colavita M., Blackwood G., Linfield R., Shao M. and Gallager D.: *The New Millennium Formation Flying Optical Interferometer Journal*, pp. 650-656. American Institute of Aeronautics and Astronautics, AIAA GN&C Conference, August, 1997
4. Wilson P.D., Hatt B. A. and Rogers T.: *FEEP Emitter Module Industrialisation (Phase-2)*, (ESTEC Contract No. 6764/86, Report). September 19, 1989
5. Valentian D., Bugeat J., Petit R., Bartoli C. and Rohden M. V.: *Application of Field Emission Thruster to Spacecraft Control and Orbit Positioning*, AIAA, 25th Joint Propulsion Conference, Monterey, CA, July 10, 1989-July 10, 1989

Angara and the Market for Non-Geostationary Orbit (NGSO) Satellites

E. Haase, R. Happersett, International Launch Services, 1660 International Drive, Suite 800, McLean, VA, 22102, USA

e-mail: ethan.e.haase@lmco.com, robert.e.happersett@lmco.com

Abstract

Small, non-Geostationary orbit (NGSO) satellites represent an important niche market within the satellite industry. Civil and commercial organizations have launched and will continue to develop these satellites for a variety of missions. In the near future, ILS will be introducing the Angara family of launch vehicles. Different versions of these vehicles to be manufactured at the Khrunichev State Research and Production Space Center in Russia will serve this market segment.

1. NGSO Market Overview

1.1 *Civil Non-Geostationary Orbit (NGSO) Satellites*

Civil agencies launch small NGSO satellites for several key mission types, including including remote sensing, scientific experiments, navigation, and technology demonstration. These types of satellites represent a robust niche market. Over the last five years, the civil NGSO sector has launched approximately 19 satellites per year in the 1-1000 kg range, with half of these satellites sized at 100 kg or less (determined from publicly available information on satellite mass for launches 1996-2000, researched by ILS). This level of launch activity is expected to continue over the next ten years.

1.2 *Commercial NGSO Satellites*

Commercial NGSO satellite systems have been planned for several types of communications markets, including voice, low data rate services, and broadband communications. Although the Iridium, Orbcomm, and Globalstar satellite constellations have been successfully launched and remain operational, each system has encountered difficulties in generating revenue. These difficulties have created a very challenging financial environment for other planned systems. Despite these concerns, however, additional companies remain committed to deploying constellations over the next few years.

Commercial satellite systems have also been deployed to provide remote sensing services. Several providers have been able to develop business plans and launch satellites. Although these systems do not require the large number of spacecraft used by commercial communications constellations, a steady

M. Rycroft and N. Crosby (eds.), Smaller Satellites: Bigger Business?, 377–378.

launch rate for these systems of a few satellites per year is expected over the next ten years.

2. The Angara Launch Vehicle Family

In the next few years, International Launch Services will begin offering launch services to address this segment of the market with the Angara family of launch vehicles. The Angara launch vehicle family is currently under development by the Khrunichev State Research and Production Space Center. Angara will provide configurations that address a wide range of launch requirements.

The Angara family will provide launch solutions that accommodate NGSO satellites. The Angara 1.1 and 1.2 vehicles could serve this marketplace through dedicated launches of NGSO satellites of up to 3,700 kg, or by launching clusters of smaller satellites.

The design of the Angara family represents an evolutionary approach, building on proven systems. Many Angara systems trace their heritage to currently operational vehicles. The first RD-170 and RD-180 engines used on the Zenit and Atlas vehicles. The Angara 1.1 Breeze upper stage and the payload fairing have flown on the Rockot launch vehicle. The RD-0124 upper stage engine and payload fairing on the Angara 1.2 have flown on the Soyuz vehicle.

Khrunichev have been developing the Angara family since 1994 to address international commercial and Russian domestic launch requirements. Currently, Khrunichev is conducting full-scale development, ground qualification and validation tests. The Angara 1.1 vehicle is targeted to enter operational service in late 2003. In 1999, International Launch Services was assigned the exclusive marketing and sales rights for the Angara family of launchers.

3. Summary

The NGSO market should remain an important niche market for satellite services. Although the number of systems launched will likely be smaller than projected a few years ago, commercial NGSO projects will continue to develop. In addition, the demand for small satellites to satisfy civil space projects is robust.

The WEOS (Whale Ecology Observation Satellite) System and University Education

T. Hayashi, Chiba Institute of Technology, 2-17-1 Tsudanuma, Narashino, Chiba 275-0016, Japan

e-mail: hayat@pf.it-chiba.ac.jp

Abstract

The Whale Ecology Observation Satellite System outlined here started with the submission of a plan prepared by students of the Chiba Institute of Technology under the author's guidance to the first Satellite Design Contest in Japan in 1992. Since then, efforts to realize the system have continued in the University, and WEOS is to be launched by a H2A rocket in the near future.

1. Introduction

Development of the Whale Ecology Observation Satellite (WEOS) system started in 1992 at the Chiba Institute of Technology when a mission was submitted to the first Satellite Design Contest in Japan. Around twenty students from many departments responded to the call made by K Watanabe, then a postgraduate student of the Electrical Engineering Department, and started to design a system of three segments, namely Space, Ocean, and Ground. They submitted a system design together with a mock-up model of both the satellite and the probe to be attached to a whale. The proposal won a prize of the Institute of Electronics, Information and Communications in Japan.

Based on the follow-on endeavors, the WEOS was selected as one of the piggyback payloads for a H2A vehicle in 1998, and this project obtained financial support from the Ministry of Education, Science and Culture in the same year as one of the Academic Frontier Research Programs. The WEOS is now scheduled for launch in the spring of 2002.

2. Outline of the System

A global study by means of satellites is the most appropriate means for revealing the ecology of whales. For this purpose a small satellite (50 kg) is to be put into a polar orbit; its attitude will be stabilized so as to keep the antennas facing the Earth's surface by means of a gravity gradient torque due to 3 m mast with a top weight [References 1, 2]. Onboard the satellite are a UHF command receiver, S-band transmitter for downlink telemetry, UHF phase lock loop receiver for receiving up-link from the probes on the whales, data processing unit with memory, and GPS receiver for localization of the probe.

M. Rycroft and N. Crosby (eds.), Smaller Satellites: Bigger Business?, 379–380.

A probe is tied by a length of cord to a small pin, which is attached to the blubber of the whale [Reference 3]. Instruments onboard the probe are a UHF telemetry transmitter, GPS receiver, solid state memory, and sensors for pressure, temperature, geomagnetic field, and acoustic signals.

Electric power is supplied from Li ion batteries, charged during the diving phase by electric power generated by the whale's motion. When the whale surfaces the sensor data stored in the memory of the probe are sent to the WEOS, together with the position data from the GPS receiver and with the identification code. The WEOS receives these data [Reference 4], and stores them in the onboard memory. When the WEOS passes over the ground operation center in the Chiba Institute of Technology, it sends a UHF command signal to the satellite and all whale ecological data are sent (S-band) to the ground station. Thus the migration patterns of individual whales and corresponding ecological data are obtained.

This system could also be applied to tracking smaller animals, or to constructing a search and rescue system.

3. Contributions of Students

Both undergraduate and postgraduate students have contributed through their theses to the cost effective WEOS program, not only by their laboratory work but also by their participation in the field tests.

The collaboration between students from different departments to construct one system provides them with a valuable training experience.

Acknowledgement

The author would like to express cordial thanks to Mrs. Masako Itoh for her work categorizing many theses and research activities.

References

1. Hayashi,T., et al.: *Whale Ecology Observation Satellite*, 9th Ann. Conf. on Small Satellites, Utah State University, 1995
2. Hayashi, T., et al.: *Whale Ecology Observation Satellite System*, 46th International Astronautical Congress, IAA-95-IAA.11.2.02, 1995
3. Hayashi,T.: Attachment of Probe for WEOS System, *J. Advanced Marine Science and Technology Society, Vol. 3*, pp. 23-30, 1997
4. Okamoto,Y., Hayashi,T., Hosokawa, S. and Yokoyama, K.: Positioning of the Signal Source Based on the Doppler Shift in Up-Link Signals Received by a Satellite, IAF-99-B-4.05, 1999

An Overview of the Technology Satellite of the 21st Century (TechSat 21) Program

M. Hughes, S. Norris, MicroSat Systems, Inc., 8130 Shaffer Parkway, Littleton Colorado, 80127, USA

e-mail: mhughes@itnes.com; snorris@itnes.com

Abstract

ITN Energy Systems, Inc., in partnership with the US Air Force Research Laboratory, has undertaken the task of developing the Technology Satellite of the 21st Century (TechSat 21). This program served as the foundation for ITN to create a separate operating company, MicroSat Systems, Inc. (MSI), on 2 April 2001 to provide small satellites and space qualified subsystems to commercial and government customers for a variety of different missions. Not only will this technology demonstration qualify new, lightweight and low cost subsystems that can be used in all areas of space application but also demonstrate the ability to use clusters of small satellites to replace large satellites, at significant cost savings.

1. TechSat 21: a Virtual Satellite

Government and commercial space missions have traditionally utilized large, expensive satellites, vulnerable to single point failures which result in the complete loss of the mission, and which require expensive launch services for access to space. Future military missions call for an increase in imaging (radar, optics, IR), weather science technology, Internet, and telecommunications capabilities at lower costs, and with increased levels of system robustness.

The purpose of the Technology Satellite of the 21st Century (TechSat 21) program is to demonstrate that high performance, inexpensive clusters of small satellites can replace such large satellites, with the additional benefits of increased mission performance, multiple mission utility, and less chance of hostile endangerment and loss of mission with a single point failure. This program will demonstrate close proximity formation flying, sparse aperture radar imaging, and the flexibility of modifiable cluster formations.

2. Technology Demonstration

MicroSat Systems Inc. (MSI) is taking promising technologies, developed by the Air Force Research Laboratory, and integrating them into TechSat 21 to validate their utility for distributed satellite constellations. The following enhanced technologies will support this validation:

- Thin Film Copper Indium Gallium Di-selenide (CIGS) Solar Arrays
- 3U PCI Avionics

M. Rycroft and N. Crosby (eds.), Smaller Satellites: Bigger Business?, 381–382.

- Pulse Plasma Micro-Thruster
- Low-Mass Phased Arrays
- Carrier Phase Differential GPS
- Li-Polymer Batteries
- RAID Disk Mass Storage
- RF MEM
- Autonomous formation flying, utilizing X-band inter-satellite link
- Composite structure and electronic boxes.

3. Future Applications and Benefits

The TechSat 21 concept enables a robust multi-mission capability, where different functions can be spread across the cluster. This distributed architecture is akin to the change from mainframe computers to a network of PC's. Application of the TechSat 21 concept is being explored for a number of military, civil and commercial customers, for applications including:

- Surveillance
- Global Communications
- Cooperative Aerospace Operations
- Commercial Remote Sensing
- Weather, Oil Spill, Surface Wind and Ice Detection
- Commercial Imagery and Precision Mapping.

Benefits of the TechSat 21 concept include on-demand operations, high reliability and a highly survivable system. If one satellite were to fail, the entire system is not jeopardized and the remaining satellites can be reconfigured to continue the mission. Additionally, these lightweight satellites can be launched on much smaller launch vehicles, further reducing the system costs. Independent estimates by the Aerospace Corporation predict a three-fold reduction in cost with three times the utility for these "Virtual Satellite" missions when compared to traditional satellite systems.

Vision of Future Space Systems Based on Recent MEMS and Nano-technology Developments

H. Kim, Korea Aerospace Research Institute, Satellite Division, P.O. Box 113, Yu-Song, Taejon, 305-600, Korea

e-mail: hkim@kari.re.kr

Abstract
MEMS (Micro Electro Mechanical Systems) and nano-technology have the potential and commercial prospect of revolutionizing current science and technological systems in space. Some considerations are discussed and some ideas presented.

1. Introduction

There is a widespread conception that information technology, nano-technology, bio-technology and environmental technology will play a dominating role in the 21st century. Satellites, space exploration systems and space transportation systems will become intertwined with these new technology-based industries in one way or another.

The factors that will change the shape of future space systems may be summarized as follows:

- Size: smaller and smaller nano-sats and micro space exploration systems
- Cost: affordable and adaptable, so that cheap access to space would be possible
- Risk: safer and smarter, so that lower risks can be attained by the combination of refined technologies, both conventional and new, and improved project management
- Life: varying, either prolonged or shortened
- Mission: diversified, flexible and requiring new launching systems.

2. The State of MEMS and Nano-technology

Even with the recent development and expansion in the intended application of these technologies, it is safe to say that they are still developing and not yet mature. The feasibility of adopting these technologies in "practical" space systems is being debated. On the other hand, some MEMS-based space components, MEMS-based RF switches, for example, have been already flown in space for some small and scientific missions. By the year 2008 or 2009, we will begin to witness conventional space systems being replaced by MEMS/nano-technology based space modules, subsystems, or even the entire system

M. Rycroft and N. Crosby (eds.), Smaller Satellites: Bigger Business?, 383–385.

[Reference 1]. This will include the micro sensors/actuators developed on a chip, on-chip electric/electronic devices such as the power source, and biologically-inspired computing systems, among others. In the long run, this will help to enable the highly autonomous GN&C system, self-healing, survivable electronics, avionics, and spacecraft [Reference 2].

3. Trends and Forecasts in the Satellite Industry

Such trends as the miniaturization and modularization of satellites, and manufacturing satellite buses as a commodity, have been seen in the satellite industries and in their supporting community such as research institutes and academia. As for the miniaturization of satellites, by conventional methods such as in the cases of OPAL, SNAP-1, or the latest MUNIN, can be viewed as a transitional phenomenon before the authentic development of new-concept nano-satellites, such as "Silicon Sat" [Reference 1], may prevail. It should be noted that the development of payloads in the micro/nano scale currently lies in the far less progressed state than the spacecraft bus. To make the miniaturized satellite in micro/nano or even smaller scale a more feasible choice for a commercial product in the space market, much more impetus should be given to the technology development for the payloads in this scale for their materialization in the practical sense.

4. Future of Space Systems

It is objectively foreseen that future space systems will be based on the fusion of old and new technologies, and the fusion of various new technologies. This would demonstrate expanded MEMS and nano-technology applications, changing of space systems' design and fabrication processes. A Virtual Reality-based facility and batch-processed micro fabrication may be established, and the nano-computer and nano-assembler realized in the long run.

5. Conclusion

Will a "Copernican conversion" in space systems or a drastic change of the paradigm in the space industry arrive by these new technologies? The current, overall consensus is that it will rather be a gradual revolution than a drastic change. It is also perceived that a more significant impact of these new technologies to space systems may lie in the potential enhancement of the performance of the existing space systems [Reference 3] rather than the embodiment of a totally new regime of space systems. It is expected that the effort to miniaturize space systems, and to refine these new technologies for their practical applications and the fusion with other new technologies in the coming years will be enormous

References

1. Helvajian, H., Robinson, E. Y.(editors): *Monograph 97-01, Micro- and Nano-technology for Space Systems,* Aerospace Press, 1997
2. Center for Integrated Space Microsystems, NASA/JPL: *1999 Annual Report*, 1999
3. Centre National D'Etudes Spatiales: *CNES Magazine,* No.9, June 2000

ROCKOT - a Competitive and Reliable Launch Vehicle for Small Satellites

M. Kinnersley, Y. Viertel, K. Schefold, EUROCKOT Launch Services GmbH, Flughafenallee 26, D-28199 Bremen, Germany

e-mail: eurockot@astrium-space.com

Abstract
The ROCKOT launcher, a three stage all liquid fuelled vehicle based on the SS-19 ICBM and flight-proven more than 140 times, is marketed to the LEO satellite industry by EUROCKOT Launch Services GmbH, Germany. This Joint-Venture is owned by Astrium GmbH, Germany, and Khrunichev State Research and Production Space Center, Russia. EUROCKOT offers commercial launch services to LEO with orbital inclinations between 48° up to Sun-Synchronous from two launch sites, Plesetsk and Baikonur. Inclined GTO and Earth escape missions can also be served with an additional commercial solid stage. Eurockot also addresses the mini- and micro-satellite market.

1. Launch Site and Infrastructure

EUROCKOT has made substantial capital investments in modernising and refurbishing the ground infrastructure at the launch site of Plesetsk, Russia (see Fig. 1) and enlarging the booster payload usable volume while retaining the impressive heritage of the vehicle. This resulted in a modern vehicle with a large usable payload volume and a unique, flat payload interface plane suitable for multiple payloads. Completely new state-of-the-art payload processing facilities, a mission control centre, lodging facilities and transport infrastructure have been built. For payload preparation, fuelling, integration and encapsulation a closed class 100 000 clean area (see Fig. 2) is provided.

Figure 1. Lift-off of Rockot from Plesetsk

M. Rycroft and N. Crosby (eds.), Smaller Satellites: Bigger Business?, 387–388.

2. Commercial Demonstration Flight

A Commercial Demonstration Flight (CDF) to commission the ROCKOT and the new dedicated EUROCKOT launch facilities at Plesetsk was undertaken on 16th May 2000. This simulated a commercial mission by successfully deploying two satellite simulators into polar orbit. The orbital parameters at the time of payload injection exceeded the values defined in ROCKOT's User's Guide, which are already impressive.

Figure 2. The Commercial Demonstration Flight was prepared in EUROCKOT state-of-the-art Cleanrooms

3. Launch a Piggy

With its program "Launch a Piggy", EUROCKOT Launch Services intends to attract operators of micro- and mini-satellites for shared launches on Rockot from Plesetsk. Groups of satellites can be multiply launched on either a standard or mission specific dispenser, with satellites typically of 50 to 150 kg class. EUROCKOT offers flight qualified separation systems including a clamp shell system with diameters of 300 mm or 600 mm or, alternatively, a pyro-bolt point attachment system. The environmental loads for piggyback satellites can be taken from Eurockot's User's Guide (see http://www.Eurockot.com). Technical Services for piggy-backs by EUROCKOT cover mission integration activities. The first launch LAP1 (Launch a Piggy 1) is planned for the near future and has already been reserved for a particular customer. Additionally a Sun-synchronous flight, LAP2 into a 700 to 900 km orbit is planned for the second quarter of 2002. Some payload capacity is still available for this launch.

Advanced Micro/Nanosatellite Attitude Control Systems (ACS)

V. J. Lappas, W. H. Steyn, C. I. Underwood, Surrey Space Centre, University of Surrey, Guildford, Surrey GU2 7XH, UK

e-mail: v.lappas@ee.surrey.ac.uk, h.steyn@ee.surrey.ac.uk, c.underwood@ee.surrey.ac.uk

Abstract

The next generation of small satellites is going to face a variety of different and more demanding mission scenarios from the past. Advanced, agile and high precision Earth and space Observation, satellite inspection, distributed platforms-constellations, satellite docking, miniature interplanetary probes are just some of the future missions to come. The present paper discusses the development of a new type of actuator for small satellites, a Control Moment Gyroscope (CMG).

1. Introduction

Attitude Control Systems (ACS) is an area critical for all types of spacecraft. Designing high performance ACS subsystems for high precision, agile mission scenarios of the future in the stringent physical size constraints of small satellites is not an easy task. Agility considerably increases the operational envelope and efficiency of spacecraft and substantially increases the return of earth and science mission data. Amongst other types of actuators (momentum wheels, reaction wheels), CMGs are by far the most capable actuator due to their torque amplification properties under low spacecraft angular rates.

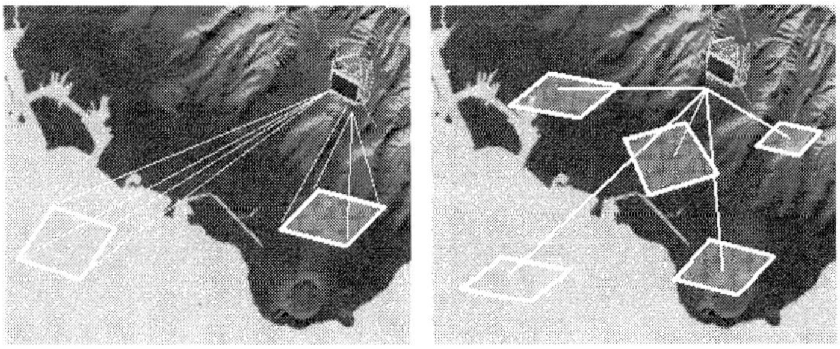

Figure 1. Agile Small Satellites will increase the return of earth and space mission data

M. Rycroft and N. Crosby (eds.), Smaller Satellites: Bigger Business?, 389–392.

2. Control Moment Gyroscopes (CMG)

The current standard slew maneuver rate for small satellites is in the range of 0.1-1^0/s. Fast slew maneuvers are those of the 3-8^0/s range. It has become evident that current ACS actuators (momentum wheels, reaction wheels) are not able to provide this agility on a very efficient way. Control Moment Gyros are actuators that can support such slew rates, mainly due to their torque amplification properties.

A CMG consists of two parts:

- The momentum wheel, which produces a large and constant angular momentum
- The gimbal motor (or set of gimbal motors), on which the momentum wheel is mounted, whose angular momentum vector is changed to the desired direction.

Torquing the gimbal results in a precessional torque that is normal both to the gimbal axis and to the momentum wheel spin axis.

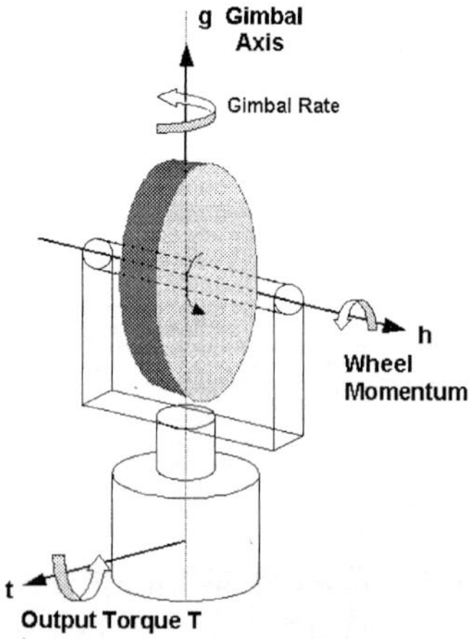

Figure 2. A Single Gimbal CMG

The total angular momentum h for a cluster of 4 CMG's is given by:

$$\mathbf{h} = \sum_{i=1}^{4} \mathbf{H}_i(\delta_i) = \begin{bmatrix} -c\beta\sin\delta_1 & -\cos\delta_2 & c\beta\sin\delta_3 & \cos\delta_4 \\ \cos\delta_1 & -c\beta\sin\delta_2 & -\cos\delta_3 & c\beta\sin\delta \\ s\beta\sin\delta_1 & s\beta\sin\delta_2 & s\beta\sin\delta_3 & s\beta\sin\delta_4 \end{bmatrix}$$

In order to get the maximum **h** we need a specific set of gimbal angles. From observation the maximum **h** is $\mathbf{h}_{\text{-max}} = h_0\,[3.15, 3.15, 3.26]^T$, where h_0 is the angular momentum of each CMG.

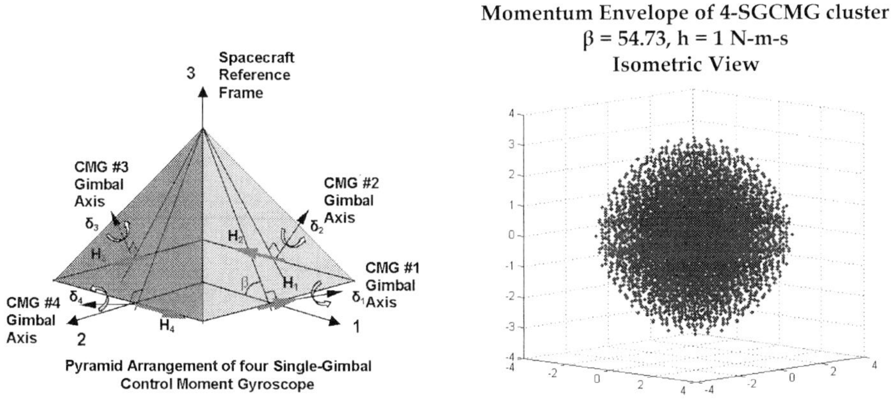

Figure 3. CMG cluster arrangement and momentum envelope

It becomes clear that the due to the gimbaling of the momentum wheel and cluster arrangement of the 4-CMG's torque amplification can be accomplished under low spacecraft angular rates.

3. Conclusions

CMG's have never been used in small spacecraft. Using an innovative design, a CMG actuator can be proved to be far superior to any actuator of its class. Combining this with the development of new algorithm techniques (controllers, singularity avoidance laws), this innovative and effective approach in attitude control for small satellites based on CMGs, can make low cost, miniature spacecraft more agile and versatile platforms than ever before.

4. Acknowledgements

The authors would like to thank all SSTL staff for their assistance in this project. Funding for the above work has been provided by Surrey Satellite Technology Ltd. (SSTL), the British National Space Centre (BNSC) and the Engineering Physical Sciences Research Council (EPSRC).

Small Satellite Development and Future Vision of NASDA

T. Nagai, M. Mokuno, National Space Development Agency of Japan, 2-1-1 Sengen, Tsukuba, Ibaraki, 305-8505 Japan

e-mail: Nagai.Takeshi@nasda.go.jp; Mokuno.Masaaki@nasda.go.jp;

Abstract

The National Space Development Agency of Japan (NASDA) has launched 40 satellites since its establishment in 1969. Meanwhile, the size of satellites has become larger due to enhanced mission requirements or enlarged space vehicles. The development period of large satellites tends to be prolonged due to complicated interfaces. However, the development, launch and demonstration of new space technologies are required in line with the rapid progress of science and technology. Therefore, NASDA has initiated the Mission Demonstration Satellite (MDS) series and other small satellites. MDS-1, the first satellite of the series, will be launched in 2001 by a H-IIA rocket, and the onboard mission candidates of MDS-3 are now under selection. NASDA has also developed an Optical Inter-orbit Communications Engineering Test Satellite (OICETS) and other small satellites. The advantages of small satellites are that the development period is short if the missions are few or simple, and that the mission can launch multiple satellites simultaneously. Therefore, while carrying on MDS-1 and other projects, NASDA is now studying an advanced LEO satellite for Earth observation, information technology, or technology demonstration, and will aim to realize a satellite system making the most of the advantages of small satellites in the new future.

1. Hopes for Small Satellites

While the National Space Development Agency of Japan (NASDA) has launched 40 satellites since its establishment in 1969, the satellites have become larger due to new mission requirements or enlarged space vehicles. Recent examples are the Advanced Earth Observing Satellite (ADEOS/3.5 tonnes) or the Engineering Test Satellite VII (ETS-VII/2.9 tonnes). The development period of a large satellite tends to be long due to complicated interfaces since it carries various pieces of equipment.

However, the development, launch and demonstration of new space technologies in a short time are required in line with the rapid progress of science and technology. Given the above, NASDA has initiated the Mission Demonstration Satellite (MDS) series, and has been working on its development. MDS-1, the first satellite of the series, will be launched in 2001 by a H-IIA rocket and the onboard mission candidates of MDS-3 are now under selection. Thus one of the hopes for small satellites is a timely demonstration owing to the short development time; another hope is for simultaneous launches of multiple satellites to form a constellation. For example, a communications or navigation system using a constellation of multiple satellites

M. Rycroft and N. Crosby (eds.), Smaller Satellites: Bigger Business?, 393–395.

has been proposed. In such a system, it is more economical to launch many satellites at once. In fact, some countries are planning, or conducting, simultaneous launches of more than three satellites for the purpose.

2. NASDA's Small Satellite Projects

NASDA is now developing two small satellites, the Mission Demonstration test Satellite-1 (MDS-1) [Reference 1] and the Optical Inter-orbit Communications Engineering Test Satellite (OICETS) [Reference 2]. MDS-1 is designed to verify the function of commercial parts in orbit, to verify minimization technology for components, and to measure space environmental data (e.g., radiation, etc.). MDS-1, a 480 kg spinning satellite, will be launched to the geostationary transfer orbit by a H-IIA rocket in 2001. OICETS is designed to conduct on-orbit experiments on optical inter-orbit communications such as tracking, acquisition and orientation technologies to the Advanced Relay and Technology Mission Satellite (ARTEMIS) of the European Space Agency (ESA). OICETS, a 570 kg three-axis stabilized satellite, will be launched to LEO by J-1 rocket in 2002. Also other satellites are now under development like Microsat-1 and Whale Ecology Observation Satellite (WEOS). Microsat-1 (70 kg) [Reference 3] will be launched in 2002 as a piggy-back satellite using a H-IIA vehicle. WEOS (approximately 50 kg) is now being developed by Chiba Institute of Technology.

3. Advanced LEO Satellite

Based on such experiences of satellite projects, NASDA is now working on an advanced LEO satellite that can be applied to future Earth observation, information technology, or technology demonstration missions. In these activities, NASDA is aiming at selecting the satellite mass from 500 kg to 2 tonnes, and shortening the development period from 5 to 3 years by adopting new satellite technologies or new system technologies. In the conceptual design of the satellite, the development period will be shortened by improving satellite integration, reducing the operational load by autonomy, and improving satellite function and performance by adopting advanced subsystem and component technology.

4. Conclusion

The advantages of small satellites are that the development period is short if the missions are few or simple, and simultaneously several satellites may be launched. Therefore, while carrying on with the MDS-1 and OICETS projects, NASDA is now studying an advanced LEO satellite, and will aim soon to realize a satellite system making the most of the advantages offered by small satellites.

References

1. Usui, M., et al.: MDS Project; New Challenge of Japanese Satellite Development, Paper IAF-00-U.1.06, presented at the 51st International Astronautical Congress, Rio de Janeiro, Brazil, Oct. 2000
2. Yamawaki, T., et al.: Development of LUCE (Laser Utilizing Communications Equipment) for OICETS, Paper IAF-00-M.2.05
3. Noda, A. : Future Vision of Micro/Nano Satellite in NASDA "μ-Lab Sat" and "TANPOPO (Dandelion)" ,Paper ISTS-2000-o-1-03v, presented at the International Space Technology and Science (ISTS) , Morioka, Japan, 2000, ISTS-2000-o-1-03v

The TAKO (Target Collaborativize) Flyer: a New Concept for Future Satellite Servicing

H. Nakanishi, K. Yoshida, Dept. of Aeronautics and Space Engineering, Graduate School of Engineering, Tohoku University, Aoba 01, Sendai 980-8579, Japan

e-mail: nakanisi@astro.mech.tohoku.ac.jp, yoshida@astro.mech.tohoku.ac.jp

Abstract

In order to keep the orbital environment safe and clean, free from collisions, the technology for satellite retrieval by robotic devices is essential. Rendezvous-docking and capture-berthing technologies have been demonstrated recently. However, the targets must have specific fixtures to be grasped and optical cue markers for vision-based control; such targets are called collaborative. On the contrary, most satellites in trouble are non-collaborative. The technology for the robotic capture of a non-collaborative satellite has not yet been established.

This paper proposes a idea of the satellite servicing system named the "TAKO-Flyer." This system captures a target with a multiple articulated gripper named "TAKO-Gripper", which then makes the target collaborative.

1. Purpose and Background

As the number of satellites in orbit increases, more attention is being paid to potentially hazardous collisions with drifting satellites that are in trouble or at the end of their life. In order to keep the orbital environment safe and clean, free from collisions, the technology for satellite retrieval by robotic rendezvous-docking and capture-berthing is required. Up to now, successful cases of capturing or handling a so-called *collaborative* satellite have been reported [References 1, 2]. However a satellite in trouble is tumbling and does not have specific fixtures or optical cue markers for vision based control. Such targets are called *non-collaborative*.

2. Concept of "TAKO-Flyer" and "TAKO-Gripper"

The "TAKO-Flyer" is a robotic system that makes the target collaborative in the following way: (1) approach a tumbling target and capture it by the TAKO-Gripper (see Fig. 1), (2) stabilize the tumbling motion, (3) provide grapple fixtures and optical markers. By changing the target in this manner, a tumbling satellite may be retrieved by the latest robotics technology established by the ETS-VII (the Japanese Engineering Test Satellite) [Reference 2].

M. Rycroft and N. Crosby (eds.), Smaller Satellites: Bigger Business?, 397–399.

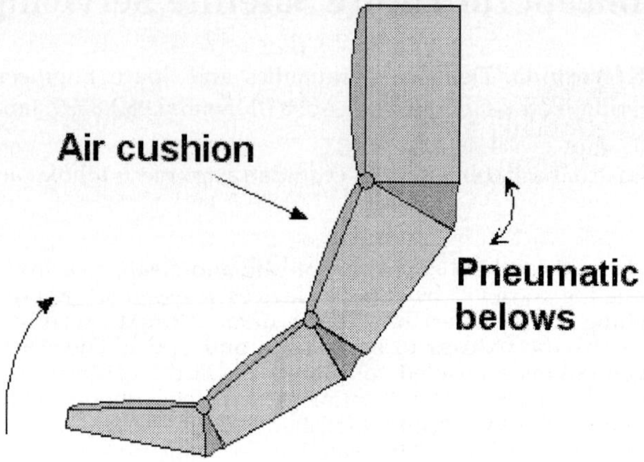

Figure 1. Concept of "TAKO-Flyer"

The word "TAKO" in Japanese means octopus. The TAKO-Gripper has several articulated fingers that work as octopus tentacles. The finger is driven by a smaller number of actuators than an articulated joint needs [Reference 3]. It can grasp an object with an arbitrary shape and in tumbling motion. In this proposal, the TAKO-Gripper is driven by gas pressure. Instead of using motors and complex wire-pulley mechanisms, the finger system makes a bending motion by pneumatic bellows. The air cushions inside the fingers hold the target gently, yet firmly.

The TAKO-Flyer will be carried by a mother-ship that has rendezvous capability, and separated from it in the vicinity of the target. The TAKO-Flyer itself has a small propulsion system to make a final approach toward the target, and then to close the TAKO-Gripper, which is as long as the target's circumference. Once holding the target, the TAKO-Flyer achieves the desired attitude stabilization by gas jet thrusters and magnetic torquers. It then provides a cooperative functionality such as grapple fixtures and optical markers for robotic capture by the mother ship. A diagram of the TAKO-Flyer is shown in Fig. 2, and configuration of whole servicing system is shown in Fig. 3. Two possible mission scenarios have been studied. The first is a mission to service a LEO satellite near the International Space Station (ISS) or Space Shuttle. The second is a mission to de-orbit malfunctioning satellites.

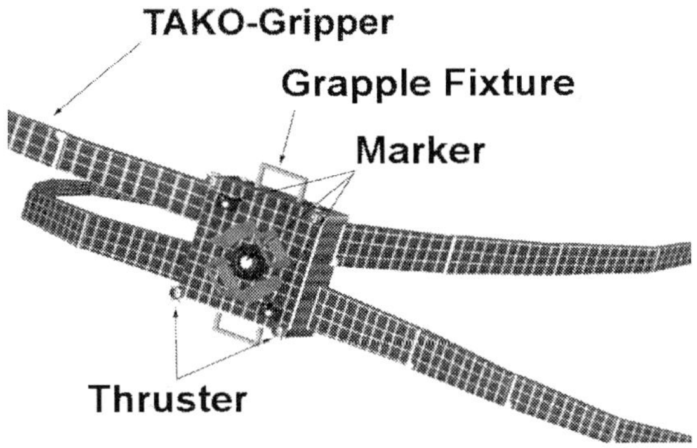

Figure 2. Overview of the TAKO-Flyer

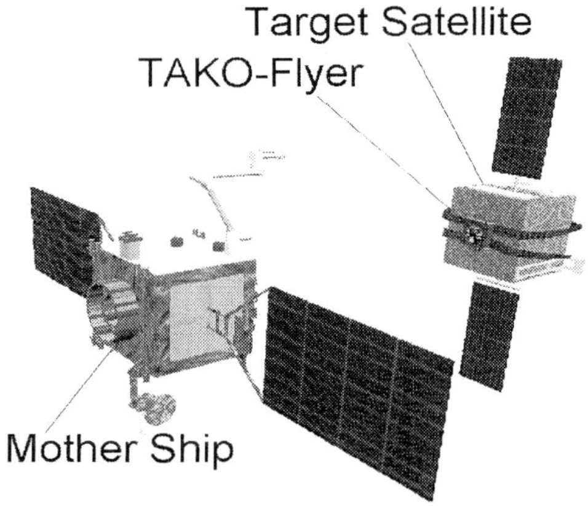

Figure 3. Configuration of servicing system

References

1. NASA: STS-72, <http://science.ksc.nasa.gov/shuttle/missions/sts-72/mission-sts-72.html>, April 28, 2001
2. Oda, M. et.al.: ETS-VII, Space Robot In-Orbit Experiment Satellite, Proc. 1996 IEEE International Conference on Robotics and Automation, pp. 739-744, 1996
3. Yoshida, K., Abiko, S. and Ota, J.F.: Development of an Adaptive Multi-finger Gripper System for Underwater ROV, Proc. 2001, IEEE/RSJ International Conference on Intelligent Robots and Systems, submitted

Status of Micro- / Nano-Satellites in NASDA

A. Noda, T. Sekiguchi , National Space Development Agency of Japan, Tsukuba Space Center, Sengen 2-1-1, Tsukuba, Ibaraki, 305-8505, Japan

e-mail: noda.atsushi@nasda.go.jp; sekiguchi.takeshi@nasda.go.jp

Abstract

The first satellite Engineering Test Satellite I (ETS-I) of National Space Development Agency of Japan (NASDA) was launched in 1975. ETS-I was a small satellite of only 85 kg. But, since then, the satellite mass has been increasing year by year, and the performance of the launch vehicle has been increasing too. Now, the mainstream of satellite research and development in NASDA is for the over 1 tonne-class satellite.

But, 50 kg-class micro-satellites and under 10 kg-class nano-satellites are also researched. Larger satellites may well have high performance, but they are not always better, because:

- No one can comprehend all the systems and subsystems of a large satellite, because it is too complex
- The large satellite will be expensive
- The development time will be too long.

Thus, in the last six years, NASDA has developed a micro-satellite (μ-Lab Sat 1) and a nano-satellite (within μ-Lab Sat 2) which are described briefly here.

1. μ-Lab Sat 1

"μ-Lab Sat 1", the first micro-satellite [Reference 1] in NASDA, is planed for launch in 2001 with the Advanced Earth Observation Satellite-II (ADEOS-II), by a H-IIA Launch Vehicle. This satellite is a piggy-back satellite in the 50 kg-class. Its objectives are:

- To achieve a low cost bus in a short development time
- To offer a chance of realizing a challenging mission
- To train young engineers in NASDA.

2. μ -Lab Sat-2

" μ-Lab Sat-2" is the next micro-satellite in NASDA; it is in the conceptual study phase. It has a 50 kg-class mothership satellite and some 1-10 kg-class daughtership satellites. The mother satellite is three-axis stabilized, and the daughter satellites will be cluster flying. It is necessary to develop a very small and low cost bus for the daughter satellite. If the bus can be made cheap enough, even high school students could make the satellites themselves.

M. Rycroft and N. Crosby (eds.), Smaller Satellites: Bigger Business?, 401–402.

μ -Lab Sat-2 has the following features:

- Simple, and optimized for micro-/nano-satellites
- Challenging advanced technology
- Low cost and short development time.

μ-Lab Sat-2 includes:

- Mothership and daughterships
- LEO and GTO
- 3-axis attitude stabilization (full time)
- Local area network in the micro-satellite
- Cluster flying by nano-satellites (daughter-ships).

Now that young NASDA engineers have experience from designing μ-Lab Sat-1, they chose 3-axis attitude stabilization and deployed solar panels to supply electric power to the attitude control subsystem. These panels do not track the Sun, but the attitude control subsystem operates in such a way that the panels trace the Sun to supply enough electric power.

μ-Lab Sat-2 is not yet an official NASDA project. In 2001, NASDA will make the conceptual design, and will manufacture it next year; μ -Lab Sat-2 will be launched two years after the start of the project.

Reference

1. "μ-Lab Sat 1": <http://giken.tksc.nasda.go.jp/abs/sentan_j/piggy/piggy.html>

Micro Spectrometer and MEMS for Space

R. Riesenberg, G. Nitzsche, A. Wuttig, B. Harnisch*, Institute for Physical High Technology, P.O.Box 100 239, 07702 Jena, Germany, * ESA/ESTEC, P.O.Box 299, 2200 AG Noordwijk ZH, Netherlands

e-mail: riesenberg@ipht-jena.de

Abstract

A novel instrument technology for a micro spectrometer is presented. A HADAMARD-transform-spectrometer (HTS), based on an optical grating and an array arrangement, is realized, by an extremely miniaturized set up by implementing MEMS. The spectrometer for the UV-VIS spectral region, with a spectral resolution (FWHM) of $\Delta\lambda$ = 1.2 nm, has a volume of about 300 cm^3.

The digital spectral resolution given by the detector pixels is increased by up to a factor of 5 using additional subpixel generation, and the throughput is increased too. The MEMS are state-of-the-art mechanical slit positioning systems with micro slit chips. The set up is qualified for space, and was realized for measurement of the ozone concentration in the upper stratosphere of the Earth.

1. Micro Spectrometer Design

The optical design of the micro spectrometer is shown in Fig. 1. The state-of-the-art of micro spectrometers for all optical spectral regions is the detector array architecture based on free space optics for high throughput with a blazed holographic imaging grating.

The limits to the spectral resolution in combination with the spectral range and the low stray-light level are given by the aberrations of the grating. That means that an entrance slit is needed, which has a width in the order of 10 μm, and also that the (virtual) size of the detector pixel is smaller than 10 μm. A slit array [Reference 1] and a double array architecture is a suitable solution [References 2, 3].

2. MEMS, Mechanical Slit Positioning System and Set-up

For a micro spectrometer MEMS, new micro-mirror-arrays [Reference 4] and shutter arrays [Reference 5] are under development. Alternative slit and fibre positioning systems are developed for advanced miniaturized spectrometers and proposed for imaging spectrometers [Reference 6].

The HADAMARD-transform spectrometer (HTS) for the visible spectral range 240 to 675 nm and a spectral resolution of $\Delta\lambda$ = 1.2 nm has a slit-grating distance of 40 mm.

M. Rycroft and N. Crosby (eds.), Smaller Satellites: Bigger Business?, 403–406.

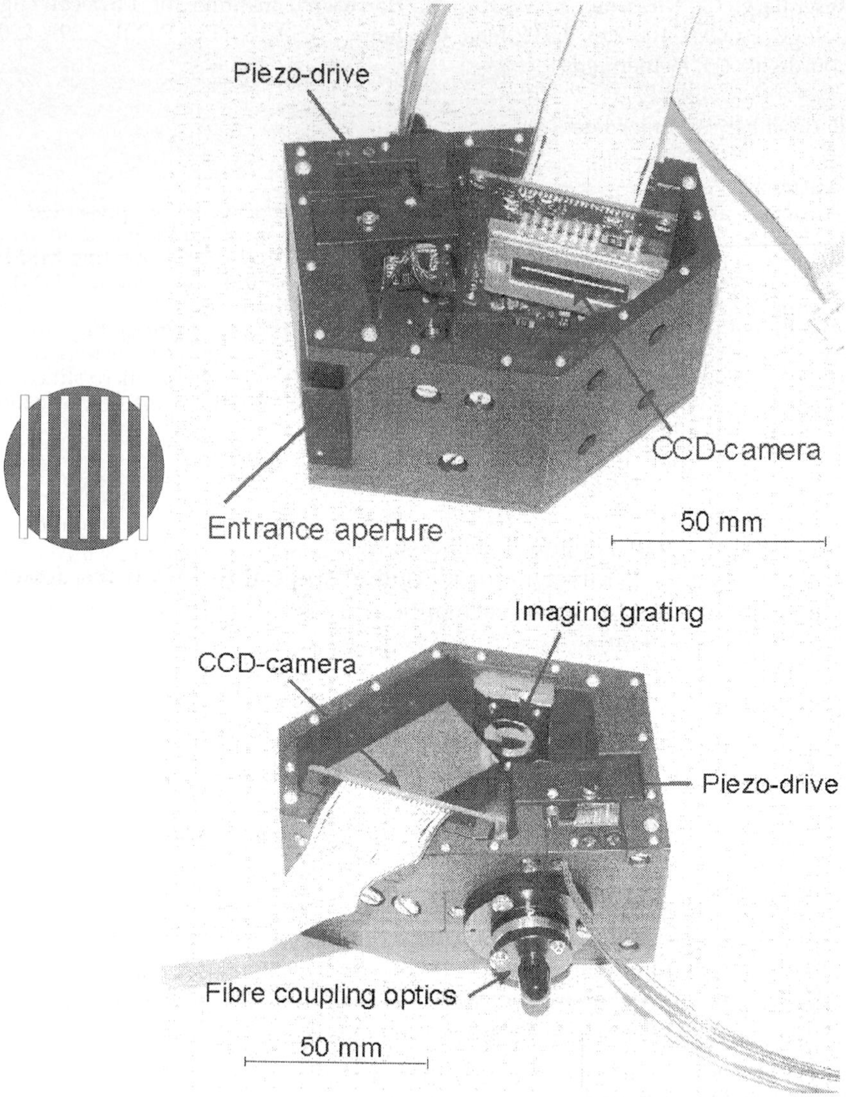

Figure 1. Set-up of the advanced micro spectrometer (HADAMARD transform spectrometer) with a micro slit positioning system with entrance aperture of 7 slits (above left)

The spectral resolution is increased up to the factor of 5 by subpixel imaging and analysis, Fig. 2, and the throughput is increased by the factor 1.5. The monochromatic stray light is 2 x 10^{-5}.

It has been qualified for space operation via tests carried out at vibrations, shocks, and temperatures cycles.

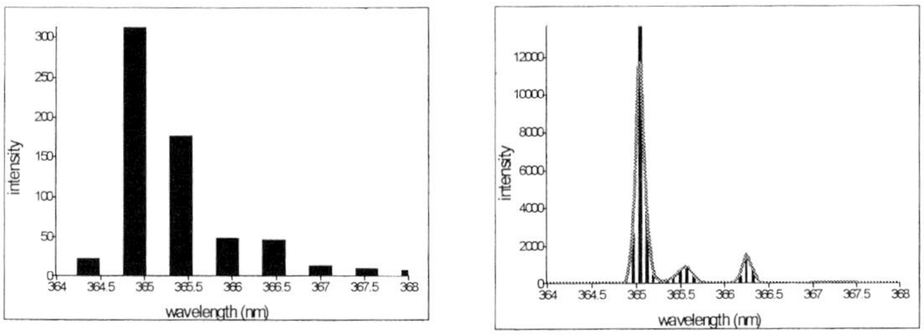

Figure 2. Image of the spectral triple line at 365 nm of a Hg-lamp, result of the commercial set-up with a single slit (left), and results by sub-pixel imaging with a linear MEMS with 7 micro slits of a width of 14 μm and super-resolution analysis (right)

3. Summary

A relatively simple fully functioning, space qualified HADAMARD micro-spectrometer (HTS) with a minimum overall size (300 cm^3) is presented. It is based on MEMS.

The new architecture optimizes the set of all spectrometer parameters realizing the world´s smallest spectrometer with excellent optical performance. According to today's knowledge it is questionable whether future detectors with smaller pixel pitch will lead to smaller architectures maintaining the same performance. With decreasing pixel pitch the crosstalk increases and the sensitivity decreases.

The MEMS are the state-of-the-art of programmable micro slit chips or mechanical slit positioning systems. Compared with well known micro mirrors, the micro slits generate no additional stray-light.

We are also going to employ novel two dimensional micro-technologies for new imaging instrument architectures based on these MEMS. The architectures are important for instruments whose detector array technology is limited by the number of pixels (infrared focal plane arrays), or an additional multi-object mode is implemented, e.g. for the Next Generation Space Telescope.

The set up is realized in the ESTEC technology program for measurement of the ozone concentration in the upper stratosphere of the Earth.

Acknowledgements

The work is supported by the DLR (50 TT 9700) and by the ESTEC (A01-3662/98/NL/WK). We also thank T. Peschel and C. Damm for the mechanical design and set-up, T. Seifert for preparing micro slit chips, and U. Dillner for helpful discussions.

References

1. Riesenberg, R., Seifert, T., *Design of spatial Light Modulator Microdevices – Micro Slit Arrays*, Proceeding of SPIE, Vol. 3680, Part One, pp. 406-414, 1999
2. Riesenberg, R. , Voigt, W., Schöneich, J., *Compact Spectrometers made by Micro System Technology*, Sensor 97, Proceedings, *Vol. 2*, pp. 145-150, 1997
3. Riesenberg, R., Lonschinski, J, *HADAMARD-Minispectrometer made by a Micro Device*, 3rd Round Table on Micro/NanoTechnologies for Space, ESTEC, Noordwijk, The Netherlands, Proceedings pp. 291 - 297, 2000
4. Garcia, E. J., Greenhouse, M. A. , Mott, B. and Kuhn, J. L., *Surface micromachined micromirror arrays for next generation space telescope*, American Astronomical Society Meeting, *Vol. 194*, pp. 9107, May 1999
5. R. Fettig, R., Moseley, S. M, Kutyrev, A. S., Orloff, J., Kuhn, J. L., and Shude, L., *Some aspects on the mechanical analysis of micro-shutters*, Materials and Device Characterization in Micromachining II, *Vol. 3875*, 1999
6. Riesenberg, R. and Dillner, U., *HADAMARD Imaging Spectrometers*, Proceedings of SPIE, *Vol. 3753*, pp. 203-213, 1999

Air Pollution Monitoring Using a Constellation of Small Satellites

W. Soh, Western Digital (M) Sdn. Bhd. Lot 3 Jalan SS 8/6, Sungei Way Free Industrial Zone, 47300 Petaling Jaya, Malaysia

e-mail: Warren.Soh@wdc.com

Abstract

A constellation of four small satellites from Surrey Satellite Technologies is selected to support a proposed Earth Observation complementary system used to monitor and track air pollution, with cost being a primary concern. A Pulsed Lidar and a Hyperspectral Imaging instrument will be used to obtain a vertical profile and to identify the gaseous pollutants. Operating at an altitude of 400km with a footprint of 6000km^2 the system will provide intermittent global coverage. Increasing onboard memory with a "store and dump" system is considered for cost-effective ground operations. The proposed scenario involves universities in several nations for the Pre-Phase A study, with a non-profit organization to facilitate and operate the entire project with the US National Oceanic and Atmospheric Administration and the International Geosphere Biosphere Program. Public Private Partnerships are suggested to meet the cost of up to US $ 30 million for development and production. Operations are estimated to be US $ 24 Million. Hence a more representative global coverage, safer observation methods and accessibility to various regions can be provided.

1. Introduction

Earth Observation (EO) systems can usefully monitor atmospheric pollutants. This paper was conducted as a study proposal for the Space Generation Advisory Council, one of the many projects that can be implemented for human sustainability using space based applications. The mission statement of this proposal is to provide a complementary system to current analytical and Earth Observation air pollution monitoring systems [Reference 1]. The designed system is effective in identifying and monitoring air pollutants from urban areas and tracking trans-boundary dispersion between adjacent nations. Finally, the system proposed is considered to be cost effective throughout its life cycle.

2. System Concept

A Micro Pulsed Lidar [Reference 2] is used to obtain a detailed vertical profile of pollutant particulates such as carbon and smog formations. For identifying the spectrum of gaseous pollutants, a low cost Hyperspectral Imaging instrument [Reference 3] will be used. The instrument operates in the 0.3 to 1 μm range, detecting the characteristic signatures [Reference 4] that are not attenuated by the Earth's atmosphere. And, with 60 bands, the spectral

M. Rycroft and N. Crosby (eds.), Smaller Satellites: Bigger Business?, 407–410.

resolution is adequate to detect small peaks across the spectrum. The spatial resolution is evaluated at 24 m.

The 'Constella' platform of Surrey Satellite Technologies Ltd. (SSTL) is ideal for this mission as it can accommodate a 60 kg payload mass in a total mass of 140 kg [Reference 5]. A constellation of four satellites in two bipolar circular orbits will be used. Such a configuration is optimum in providing a revisit of the covered area in 46 minutes, with an emphasis on the polar region for studies of ozone depletion and melting of the polar caps. The configuration also provides a reasonable coverage of the equatorial region. The spacecraft will operate at an altitude of 400 km and have a footprint of about 6000 km^2, adequate to cover the region of Athens, for example. The payload data rate is estimated to be 84 Mbps; that would require a large storage capacity for the satellites to operate in a "store and dump" manner.

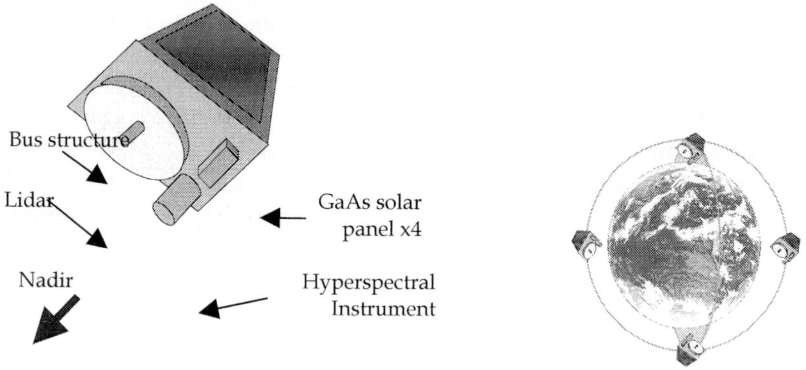

Figure 1. Constellation of four bipolar orbit satellites.

3. Business Aspects and Cost Estimation

The business strategy proposed in this study suggests that several universities in different nations carry out the Pre-Phase A study and that a non-profit organization is set up within these universities to facilitate and operate the entire project. Two basic cost estimation models were used as shown in Table 1. This was to provide a projected range as well as check the validity of each cost model. The differences between the models are based on the parameters of each model, where NASA's Parametric Cost Model [Reference 6] is based on a new one-off project working with the parameters of a previously cost evaluated system. The other cost model is based on off-the-shelf technology usage and standardized manufacturing. NASA's Spacecraft Operations Cost Model (SOCM) gave the estimated project direct [Reference 7] total for a targeted 7 years of operations as US$ 24 Million.

Cost Models	Development cost (US $ Million)	Production cost (US $ Million)
NASA Parametric	52.1	29.7
Economics of Low Cost Space System	8 .0	4.0 (Estimated)

Table 1. Range of development and production cost small satellite mission

Figure 2 reflects the costs over the entire life cycle of the system. The total of US $ 105 million includes maintaining and upgrading ground systems; mission control, tracking, telemetry, command functions, mission planning, data reduction and analysis. It does not include transportation and launch costs. Launching the constellation (based on price per kg) with the Ariane 5 launcher and two Pegasus XL launches would give a launcher price range of from US $ 7 million to US $ 23.7 million. The Pre-Phase A study can be carried out by Universities in Year 1 with the appropriate research funds. The project could then be presented to a government agency, for example ESA, who, if they approved it, would fund the Phase A, B and C of the project. The related private sector would then be invited to operate the system, providing the service to end users. This system is termed a Public Private Partnership.

4. Conclusion

The use of a constellation of four bipolar orbiting satellites with a Micro Pulse Lidar and a Hyperspectral Instrument, targeting especially the troposphere, is considered to be viable. Apart from the need for research instruments, the satellite bus would be the micro 'Constella' platform of Surrey Satellite Technology Ltd. The range of costs estimated for developing and producing four satellites is US $ 8 to US $ 82 Million. The operating cost would amount to US $ 24 Million for seven years. Hence, there is a need to develop a business plan to secure interest from government agencies to fund the development and production costs while maintaining the involvement of universities from different nations in order, eventually, to have a cleaner atmosphere.

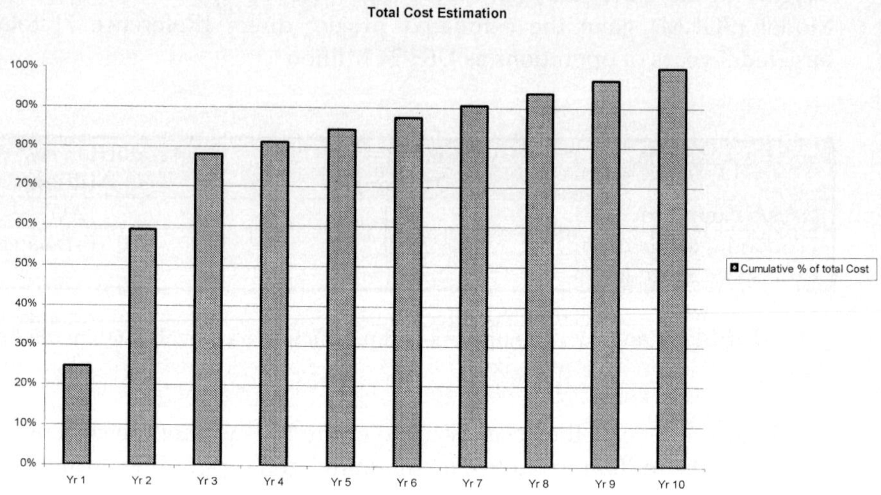

Figure 2. Cumulative development, production and operations cost

Acknowledgements

The independent study described in this paper was carried out for 'Space For Sustainable Development and Meeting Basic Human Needs' United Nations Space Generation Advisory Council Working Group. The author would like to thank I. Polat, N. Walter and J. Burke for their views and advice. Special thanks to Prof. M. Rycroft for editing the paper.

References

1. Sifakis, N., Gkoufa and A., Soulakellis, N.: *The state of the art in Earth Observation methods and systems for air pollution tracking and monitoring*, Final Report, National Observatory of Athens, University of the Aegean, Athens, 1998
2. Hwang, I., Kim, J., Ranganayakama, B., Nam, M., Lee, H., and Spinhirne, J.: *Micro Pulse Lidar for Atmospheric Monitoring*, Research Report, NASA Goddard Space Flight Center, <file:////Library pc/temp/microLidars.htm>, November 10, 2000
3. Speciale, N.: *Earth Observing-1*, NASA Goddard Space Flight Center, <http://eo1.gsfc.nasa.gov/miscPages/home.html>, November 23, 2000
4. NASA Langley Research Center, *Remote Measurement of Pollution*, NASA SP-285, Norfork, Virginia, 1971
5. Surrey Satellite Technology Ltd.: *SSTL 'Constella' Small satellite platform,* Data Sheet, <http://www.sstl.co.uk/>, November 9, 2000
6. NASA-JSC, *Cost Estimating Group: "Cost Estimating"*, <http://www.jsc.nasa.gov/bu2/ index.html>, November 16, 2000
7. Wertz, J. and Larson, W.: *Space Mission Analysis and Design*, Space Technology Library, 3rd edition, Microcosm Press, Torrance, California, USA and Kluwer Academic Publishers, Dordrecht, The Netherlands, 1999

Micro/Nano-Technologies for High Performance Spacecraft

L. Stenmark, F. Bruhn The Ångström Space Technology Centre, The Ångström Laboratory, Uppsala University, Box 534, SE-751 21, Uppsala, Sweden

e-mail: lars.stenmark@angstrom.uu.se; fredrik.bruhn@angstrom.uu.se

Abstract

The paper discusses some development projects to develop advanced micro-systems for very demanding space applications, such as interplanetary missions. However, there is a strong belief that when the new technology has proved its potential, it will be useful for many space missions or even demanding ground applications.

1. Micro/Nano-Technologies in the Near Future

Key areas for very small spacecraft for spatially distributed measurements or formation flying are several technologies which are suitable for miniaturization, from traditional cold gas propulsion systems to advanced electric propulsion, or advanced bipropellant systems with micro turbines. The ÅSTC is presently working on three projects, a Micro Propulsion Cold Gas system, a Micromechanical Xenon Feed System and Micro Pyro Matrixes for simple attitude control or station keeping.

Figure 1. A nozzle for the Hybrid Cold Gas System under development.

Two things will happen when the physical dimensions of a satellite are dramatically reduced with sustained, or nearly sustained, requirements on the performance. It can be difficult to maintain thermal control in the small satellite, as the amount of received power varies severely. This in turn calls for new

M. Rycroft and N. Crosby (eds.), Smaller Satellites: Bigger Business?, 411–412.

efficient thermal control tools such as Variable Emittance Panels, new high performance thermal coatings and thermal switches, both active and passive.

Three-Dimensional Multi Chip Modules (3D-MCM) represent an efficient way to achieve the packing density for the electronics that is required in a very small satellite. Some ÅSTC projects aim towards development of the technology for large all-silicon 3D-MCM technology, including multi functional modules, such as MCM that contain antenna electronics, integrated antenna elements and a surface covered with solar cells.

The objective is to create a state-of-the-art-technology for high-density, high-performance, low-mass electronic modules without Coefficient of Thermal Expansion (CTE) problems and with integrated means of thermal control. Such modules could be used in future space missions, where minimum size and mass are important. Internal radiation shielding around sensitive chips and internal connections with several contact points are incorporated.

When applying the new Micro/Nano technologies on nanosatellites one may be able to increase the amount of payload significantly, allowing small and cheap satellites to carry out the same scientific research as large satellites. The key elements mentioned above are all subjects to be verified and flight proven on the ÅSTC first nanosatellite, due to be launched within two years. ÅSTC is committed to put the most technologically advanced nanosatellite ever built into space by 2003, with the immediate goal to test new technologies and push scientific experiments to their very limits.

A major justification for the nanosatellite program is its educational aspects. ÅSTC is connected to a graduate research school for Advanced Micro Engineering and, as part of the university education policy, is always collaborating with other universities or industries. Many new subsystems or experiments need miniaturization for future advanced missions.

References

1. The Ångström Space Technology Hompage, <www.astc.material.uu.se>

JetGun Sat "TOPPU"

K. Ui, R. Hodoshima, N. Maeda, H. Okada, N. Miyashita, M. Iai, O. Mori, S. Matunaga, Dept. of Mechanical and Aerospace Engineering, Tokyo Institute of Technology, 2-12-1 Ookayama, Meguro-ku, Tokyo, Japan

e-mail: martini@horse.mes.titech.ac.jp, Saburo.Matunaga@horse.mes.titech.ac.jp

Abstract

In this paper, we report a conceptual design of micro-satellite for the verification of formation flying technologies. This satellite is a type of mothership-daughtership configuration called TOPPU which means a gust wind. The two satellites are connected by a tether; the mothership satellite has a docking mechanism installed, and a reel mechanism as well as gas jet gun to perform various orbital experiments. The system design of TOPPU is discussed, as is the progress of satellite system development.

1. Introduction

In recent years, research on formation flying has become a more attractive and important theme, for multi-point observations, for satellite inspection as well as for recovery of an uncontrolled satellite.

In this paper, we proposed the conceptual design of a micro-satellite, called TOPPU, for the verification of the formation flying technologies. We consider the design of the satellite bus, five experimental subsystems, and ground experiments.

2. Overview of TOPPU System and Experiments

Fig. 1 shows a conceptual overview of TOPPU, which two satellites joined by a tether, the release and docking of a daughtership using a docking mechanism and a gas jet gun. The two satellites perform the coordinated control of their relative position and attitude, with image processing using a CCD camera. TOPPU will carry out five experiments:

- Releasing and docking the daughtership
- Approaching the desired relative position stably using a gas jet gun
- Maneuvering the relative attitude using a gas jet gun
- Recharging batteries on each satellite
- Inspecting the mothership with a CCD Camera on the daughtership.

The docking sequence has two phases, the approach-grasping phase and the guiding (or releasing) phase [Reference. 1]. The reel mechanism deploys and retracts the tether under tension/length control. The gas jet gun shoots cold gas

M. Rycroft and N. Crosby (eds.), Smaller Satellites: Bigger Business?, 413–415.

at the solar paddle on the daughtership to control the position/attitude of daughtership.

Table. 1 shows the specification of TOPPU.

Attitude control subsystem	3 axis stabilization (Bias momentum method) Sensor: Sun Sensor x2, Inertial Navigation System, Magnetic Sensor Actuator: Momentum Wheel, Magnetic Torquer, Thruster (Gas jet gun)
Communication subsystem	Between Mothership and Ground Station: S Band, 256 kbps, Whip antenna Inter-satellite communication: L Band, 256kbps, Micro Strip antenna
Power Subsystem	Requirement: 55.7W > Generated power: 56.0W
Structure Subsystem	Shape: Right octagonal prism, Demention: 0.5m x 0.5m x 0.5m Weight: 49.3kg Characteristic frequency: 97.1kHz
Orbit parameter	Altitude: 350 km, Inclination: 35deg Period: 1.37 hour
Life time	3 Months

Table 1. The specification of TOPPU

We plan to conduct the following experiments, on the ground before flight.

- Functional test (Docking and reel mechanism, recharging system)
- Ground experiment using floating testbed in our laboratory [Reference. 2.] (see Fig. 2)
- Free-fall Microgravity experiment in Hokkaido, Japan.

We have already developed the docking and reel mechanism, and conducted a functional test. We shall verify the dynamics analysis and control methods by two-dimensional and Microgravity experiments.

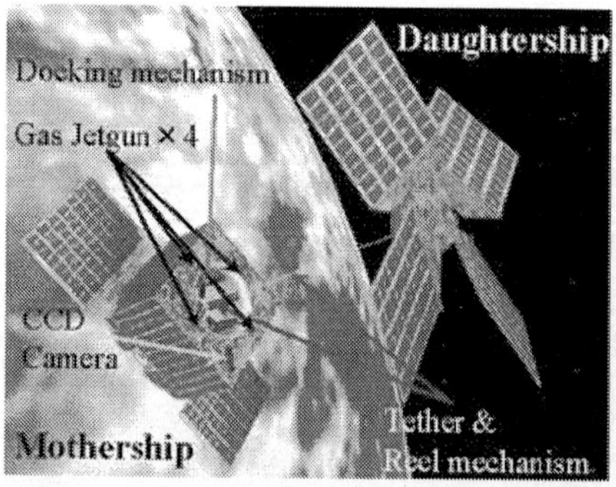

Figure 1. Overview of TOPPU

Figure 2. Ground Experiment

Reference

1. K. Ui, R. Hodoshima, M. Iai and S. Matunaga, "Research and development docking mechanism for small mothership-daughtership satellite", Proceedings of the 44th Space Science and Technology Conference, Fukuoka, October 2000.
2. S. Matunaga, O. Mori, S. Tsurumi and H. Okada, "Robot Cluster System and its Ground Experiment System", 51st Congress of the International Astronautical Federation, Rio de Janeiro, Brazil, Oct.2-Oct.6, 2000, IAF-00-A.3.06
3. LSS (Laboratory for Space Systems)'s web site; http://horse.mes.titech.ac.jp

Titech CanSat Project 2000: Report of Sub-orbital Flight and Balloon Experiment

K. Ui, K. Nakaya, K. Konoue, H. Sawada, S. Tsurumi, M. Mori, R. Hodoshima, N. Maeda, H. Okada, N. Miyashita, M. Iai, O. Mori, S. Matunaga, Dept. of Mechanical and Aerospace Engineering, Tokyo Institute of Technology, 2-12-1 Ookayama, Meguro-ku, Tokyo, Japan

e-mail: martini@horse.mes.titech.ac.jp, Saburo.Matunaga@horse.mes.titech.ac.jp

Abstract

We have participated in the ARLISS (A Rocket Launch International Student Satellite) project since 1999 to design, manufacture and operate small satellites on students' initiative. These satellites called CanSat are the size of soft drink can (350ml). We launch CanSats from the Black Rock desert in Nevada to 12,000ft using an amateur rocket, and drop them with a parachute. In this paper, we describe the subsystem design, mission and some results of the experiments on each CanSat, and report the results of an additional experiment using a balloon.

1. Introduction

The CanSat project proposed by Prof. Twiggs (Stanford University) in 1998 is a collaborative effort between Japanese and US universities to design and build simple small satellites. The primary aim of this project is educational. In less than five months we designed and built five CanSats were all launched on an amateur rocket in July 2000 (ARLISS). The launch sequence is shown in Fig. 1.

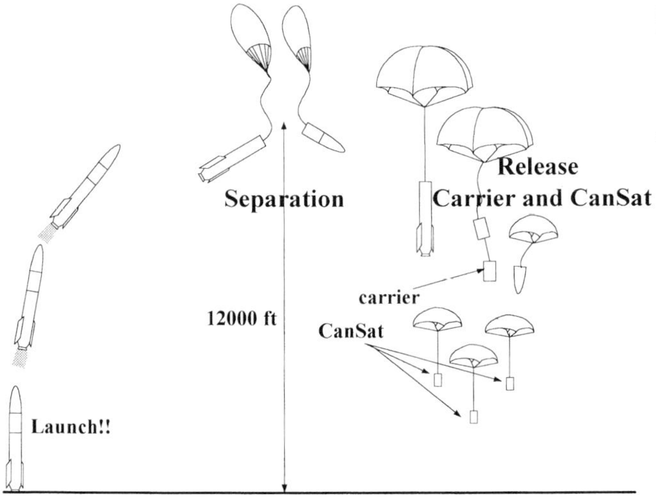

Figure 1. Experimental sequence

Figure 2. CanSat #002

2. Titech CanSat Mission

The name, mission and objectives of each CanSat are:

- CanSat #001, named C_4 (Diamond)-Sat, had a CCD Camera pointing in any chosen direction by commands from our ground station.
- CanSat #002, named AQUARIUS (see Fig. 2), stored four sensors' data in internal memory and transmitted them to our ground station.
- CanSat #003, named GNS (GPS Navigation teSt) Sat, determined its position by GPS and communication with other CanSats.
- CanSat #004, named Sat-Sat2, received the CCD Camera image via a high power amplifier.
- CanSat #005, named SSTV-Sat, took a still picture using a lo power SSTV (Slow Scan Television) system.

3. Experimental Results

We confirmed all functions of CanSat #001 with this balloon experiment. CanSat #002 stored all flight data and transmitted them via ARLISS and the balloon. Fig. 3 shows the altitude from the pressure sensor corresponding that from the GPS data from CanSat #003. #004 and #005 transmitted a very clear image continuously. In summary, we have developed five CanSats and successfully tested image remote sensing, communications and GPS techniques on this ARLISS and the balloon experiments. Please refer this web-page, http://horse.mes.titech.ac.jp/srtlssp/arliss/arliss2000/top.html.

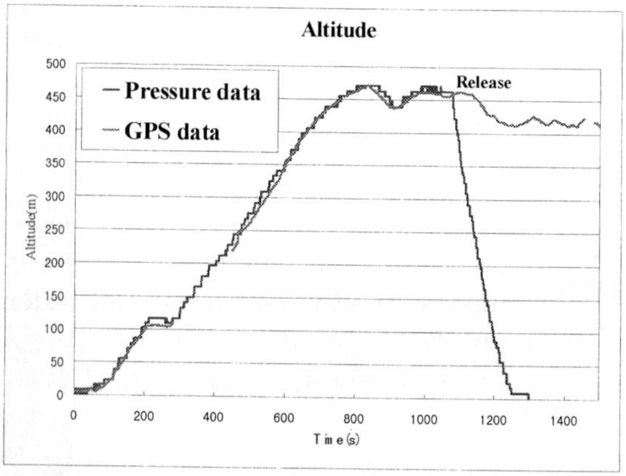

Figure 3. Altitude data from CanSat #002 and #003

Atmospheric Climate Experiment ACE a Constellation of Microsats for Atmospheric Sounding

S. M. Veldman, K. Lundahl, Swedish Space Corporation, PO Box 4207, 171 04 Solna, Sweden

e-mail: sytze.veldman@ssc.se, kaj.lundahl@ssc.se

Abstract
ACE will contribute significantly to science themes 2 (Physical Climate) and 4 (Anthropogenic Impact) for the European Space Agency Earth Explorer Programme through a unique atmospheric profiling technique. ACE will observe atmospheric parameters using radio occultations from an array of 6 micro-satellites which track the L-band signal of GPS and GLONASS satellites to map the detailed refractivity and thermal structure of the global atmosphere from surface to space. Demonstration of this technique will be based on extensions or completions of instrument and system developments already in progress within ESA. A dedicated ground segment caters for near real-time precise orbit determination, processing, and distribution of the data to the meteorological users.

1. Science Background

The instrument uses the occultation principle where the active instrument is mounted on a LEO satellite, and the monitored signal comes from a distant source, i.e. a GPS satellite. The receiver will lock on signals transmitted by selected GPS satellites and monitor the phases of these signals. As the GPS satellite sets below (or rises above) the horizon, the influence of various layers in the atmosphere can be determined in the form of vertical profiles of density, pressure, humidity and temperature. Observations for five years are probably sufficient to make well founded contributions to understanding the driving forces behind climate change.

2. The Payload Instrument

The ACE main sensor is GRAS, a GPS/GLONASS occultation receiver, which is a miniaturized version of the METOP GRAS instrument. It comprises two subsystems. The Antenna Subsystem with two occultations antennas and one navigational, zenith antenna, and the Electronics Subsystem, comprising RF Conditioning Units and the GRAS Electronic Unit with RF/IF and sampler, frequency generation, channel processing, digital signal processor and power supply.

M. Rycroft and N. Crosby (eds.), Smaller Satellites: Bigger Business?, 419–421.

3. Mission Architecture

The ACE space segment is a 6-satellite constellation, carrying the GRAS instrument, with 3 satellites each in two high-inclination, 700 to 800 km orbits. An even temporal coverage is obtained by 90° spacing of the ascending node. Redundancy of the mission is ensured on mission level by having six single string spacecraft, and a double launch. Therefore the satellites are designed to be launched in a stack of three. The launcher shortlist includes START-1 and Pegasus XL for two launches, and Rockot for a one-launch deployment. Two launches would be the simplest solution to inject into the two required orbits, and would add to the reliability of the mission.

4. The Platform

Each satellite has a mass of 80 kg, and available power of 65 W. The general layout resembles a trapezoid with five body-mounted solar panels, with a skirt of two extra solar panels and the occultation antennas. The data handling system and onboard software are based on the Smart-1 satellite from Swedish Space Corporation, and includes a CAN-bus for onboard data communication. The downlink rate is 800 kbit/s, and attitude accuracy is 1°, with a rate better than 0.025 °/sec.

5. Ground Segment

The ground segment is composed of two main facilities, the Satellite Data Acquisition and Control network (SDAC), and the Data Processing and Mission Control Centre (DPMCC). The SDAC consists of the main tracking and control station on Svalbard and the Network Management Centre (NMC) at Kiruna. The NMC supports alternative tracking and control stations especially during LEOP. The DPMCC implements three main functional groups. The first group is mission control, which is based on SCOS 2000, a generic product developed through ESA. The second group is Application Data Acquisition and Processing which includes satellite data validation, reformatting and entry, and also retrieval and similar processing of data from ground based sources such as fiducial stations. Processing is done using a precise orbit determination tool based on the Bernese tool from University of Bern. The third group is Data Distribution and Archiving, which makes different levels of data available to meteorologists, e.g., for assimilation in weather prediction models.

6. Programmatics

A proposal for ACE was submitted to ESA by 1 December 1998 by a consortium composed by Terma/DK, DMI/DK, SSC/SE and SES/SE. A feasibility study was carried out for ESA during 1999 and a phase A study is ongoing during 2001. A new proposal will be made to ESA for the second round of Earth Explorer Opportunity missions.

Acknowledgements

We thank the teams working on ACE at the Danish Meteorological Institute, Terma Elektronik AS, Saab Ericsson Space, and Fokker Space for their valuable contributions.

Leonid Meteor Observer in LEO: a University Microsatellite to Observe the 2002 Leonid from Space

K. Yoshida, H. Nakanishi, Department. of Aeronautics and Space Engineering, Graduate School of Engineering, Tohoku University, Aoba 01, Miyagi 980-8579, Japan

H. Yano, Planetary Science Division, Institute of Space and Astronautical Science, Kanagawa 229-8510, Japan

N. Ebizuka, Advanced Computing Center, The Institute of Physical and Chemical Research, Saitama 351-0198, Japan

H. Ohno, Communications System Division, Communications Research Laboratory, Tokyo 184-8795, Japan

e-mail: yoshida@astro.mech.tohoku.ac.jp

Abstract

This paper presents university-based design and development of a micro-satellite for the observation of a meteor shower from low Earth orbit. The satellite will be launched as a piggy-back payload of a commercial rocket launcher, before the 2002 Leonid meteor maximum in which tens of thousands of meteors are expected. The goal of the mission is to make observations of the meteor outburst from beyond the atmosphere, counting the meteors over a large part of the night sky and obtaining visible and ultraviolet spectra of the meteors.

1. Background of Leonid Meteor Activity

Meteor showers occur when a stream of interplanetary dust from a parent comet enters and ablates in the atmosphere on a specific day. The Leonid is one such meteor shower which attracts special attention by dramatic outbursts in history; these happen almost every 33 years, synchronizing with the encounter with the parent comet, named 55P/Tempel-Tuttle. The latest model by McNaught and Asher [Reference 1] successfully predicted the 1999 outburst with several thousands of meteors per hour. It predicts that further outbursts will occur at around 18:19 UT on 18 November 2001 and 10:36 UT on 19 November 2002.

While typical sporadic meteors are seen at around 100 km altitude, many of the Leonid maximum components are at much higher altitudes more than 150 km. This is because the Leonid meteor particles have higher velocities (about 72 km/s) than the particles of other meteor groups as they encounter the Earth, and that they contain volatile materials with low sublimation temperatures. Analyses on the recent Leonid airborne observation suggests that there are many organic features [Reference 2], which shall be particularly

M. Rycroft and N. Crosby (eds.), Smaller Satellites: Bigger Business?, 423–425.

evident in the near UV spectra. Such organic materials may have a close relation with the seeds of life.

2. Mission Outline

Observations made from low Earth orbit, looking down on Earth's atmosphere at night, will certainly be scientifically advantageous when investigating the meteor glow at near to far UV wavelengths.

Making the most of the advantages offered by small satellites - their rapid development, it is proposed that the Leonid Meteors Observer be observed. A space mission, known as the Leonid Meteor Observer is currently under the development, to maximize the scientific return in 2002. The satellite now in the design phase, uses off-the-shelf technology and components as much as possible. The size of the satellite is about a 50 cm cube, and the mass is 50 kg, small enough for a potential piggy-back launch.

The payload includes (1) a nadir-pointing and a limb-viewing pointing imager to statistical count meteors, and (2) a limb-pointing spectrometer for the spectroscopic analyses in the UV. The field of view of the imagers and the spectrometer is some 30 degrees. The satellite has three-axis stabilization. A circular Sun synchronous low Earth orbit at 700 km altitude is currently foreseen.

Figure 1. Photo at 1999 Leonid maximum by Yano and NHK

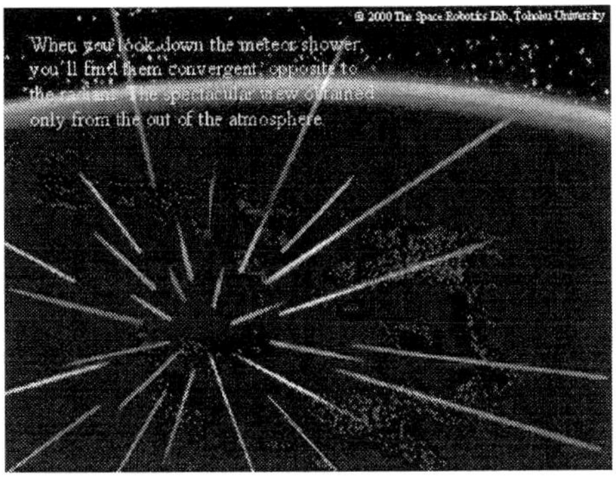

Figure 2. Artistic image of meteor storm from orbit

References

1. R.H. McNaught and D.J. Asher; "Leonid Dust Trails and Meteor Storms", WGN, J. of IMO, 27:2, 1999.
2. H. Yano, S. Abe, R. Nakamura, Y. Fujii, J. Watanabe and O. Iiyama; "Airborne Observation of the Leonid Meteor Shower with High-Definition TV Cameras in 1998-1999", Proc. IAU Coll. 181 & COSPAR Coll. 11: Dust in the Solar System and Other Planetary Systems, 2000.

Appendix

The International Space University's Small Satellite Interdisciplinary Survey (ISIS)

N. Bovet, J. Hair, G. Kennedy, P. Milani, M. Pavek, R. Schingler, International Space University, Strasbourg Central Campus, Parc d'Innovation, Boulevard Gonthier d'Andernach, 67400 Illkirch-Graffenstaden, France

e-mail: bovet@mss.isunet.edu, hair@mss.isunet.edu, kennedy@mss.isunet.edu, milani@mss.isunet.edu, pavek@mss.isunet.edu, schingler@mss.isunet.edu

Abstract

The International Space University's Small Satellite Interdisciplinary Survey (ISIS) is a general overview of small satellites from an interdisciplinary perspective. The purpose of ISIS is to address the role for small satellites in today's world, and consider whether or not they will provide benefit to the commercial, political, scientific or technological sectors of the aerospace community. First, the business aspects of small satellites are reviewed including business models, users, manufacturers, costs and potential market growth. This is followed by a discussion of the role of policy in the area of small satellites both for individual states and between states. The more technical aspects of small satellites are then discussed, starting with their applications in five main categories: telecommunications, Earth science, space science, technology demonstration and engineering training. Finally, the engineering needed to realize these applications is explored.

The conclusion reached is that small satellites are a growing area of the aerospace sector. Their low cost and rapid development times endow them with great potential in relation to larger satellites. Although there have been failures in recent attempts to exploit small satellites, there still remain untapped fields and applications in which small satellites can be used. As technological advances make smaller satellites more feasible and if cheap launch options are nurtured, small satellites have a niche in the markets of the future, especially as more countries become space-faring nations.

1. Introduction

The International Space University's Small Satellite Interdisciplinary Survey (ISIS) started as the first part of the Team Project completed by the sixth Master of Space Studies class in December 2000. It is a work completed by the 39 people who make up the class; it benefits from input from students of all cultural and educational backgrounds. The current version has been adapted to suit the Symposium forum.

ISIS is a survey designed to give the reader a general overview of small satellites in an interdisciplinary fashion. There is no doubt that there is a trend to shy away from multi-billion dollar spacecraft and also a movement in the aerospace community toward the greater use of small satellites. The purpose of ISIS is to address the role for small satellites in today's world, and to examine whether they will provide benefit to the commercial, political, scientific or technological sectors of the aerospace community.

M. Rycroft and N. Crosby (eds.), Smaller Satellites: Bigger Business?, 429–494.

2. Small Satellite Overview

In 1957, the first man-made object launched into Earth orbit, Sputnik-1, was undoubtedly a small satellite. Limited launcher capability was the main constraint on satellite size and mass at that time. In the years that followed, the size, complexity and cost of satellites grew, as did the capabilities of the launch vehicles.

With this growth in mind, we can ask ourselves: Is there still a role for small satellites today? What scientific, technological or commercial returns may be expected from them? How will they affect the global space market? Such fundamental questions are addressed in this comprehensive survey, beginning with a summary of the benefits and limitations which small satellites and spacecraft offer. A representative sample of current classifications is presented, followed by a brief programmatic review of small spacecraft missions. More in-depth reviews on the different areas of the small satellite field are given in the later sections.

2.1 *Justifications and Limitations*

Why Use Small Satellites? Today, small spacecraft are creating exciting opportunities for business and research while at the same time responding to constrained budgets. The new generation of small spacecraft has successfully proven not only to be able to support viable science programs, but also to play a key role in these programs. Small missions are further characterized by a large amount of flexibility and the ability to respond to new scientific opportunities. Shorter development periods reduce overall cost and provide additional degrees of budgetary freedom for already tight budgets.

Above all, small missions tend to revitalize the scientific community by allowing researchers, who otherwise might have waited a lifetime to analyze data, to conduct high-priority, high-quality science in a relatively short time. These programs even offer students opportunities to experience the complex nature of scientific research in all its various stages from conception and realization to analysis and publication [Reference 1].

This last characteristic ties closely into the field of international cooperation on small missions, which seems easier to accomplish among private companies or on a university level. Also, small satellites turn out to be well adapted to developing nations which have only limited space expenditures. In this context, they might be used, for example, for local projects such as disaster monitoring as well as for tools for political purposes.

With the further advance of technology, largely driven by miniaturization, there are strong indications that small spacecraft could reach unprecedented levels of performance and reliability in the near future, while at the same time maintaining mission cost at acceptably low levels [Reference 2].

Why Not Use Small Satellites? The most obvious technical challenges faced by small satellites are their limited capacity (e.g., payload, resources), their short lifetime, and the fact that they may be harder to position and control. Thus, small satellites simply cannot be adapted to some applications for which larger satellites are necessary.

Also, the size and the increasing population of small satellites in orbit are issues which lead to some questions. Might a tiny satellite be considered space debris? What happens when it gets out of control or when tracking becomes too expensive? Another issue is the lower reliability which often characterizes small satellites since they often accept the additional risks associated with new technologies before the technologies have been proven. Important gaps between space agencies in micro-technology can make cooperation on small satellite projects difficult [Reference 3]. And last, but not least, comes the question of profitability of the small satellite business. Will private companies other than the builders of small satellites and the launch companies derive profitable business from the small satellite market today or in the future?

Table 2-1 lists the main benefits and limitations of small satellites developed above. These issues are explored in depth in later sections.

Benefits	Limitations
Low cost	Limited capacity
Short time to scientific return	Limited control
Versatile	Short lifetime
Rapid development time	Questionable reliability
Revitalized scientific community	Questionable profitability

Table 2-1. Main benefits and limitations of small satellites

However, having considered both the benefits and limitations of small satellites, one should be wary of a "small versus big" mentality since both small and large satellites clearly have their own roles to play, and these roles should become ever more complementary in the future.

2.2 *Existing Small Satellite Classifications*

For this survey, the term "small satellite" is defined as a spacecraft having a wet mass of less than 500 kg, unless otherwise noted. All types of spacecraft, not just Earth-orbiting satellites, are considered. However, since various other definitions do exist, we have included a brief discussion of some them.

Mass Classification. The most obvious classification criterion is the mass of the satellite. Largely, the "wet mass" of the satellite, which includes the onboard propellant, is used rather than the dry mass since the classification division is generally made based on launcher capabilities. For example, the Small Satellite Home Page, maintained by the Surrey Space Centre [Reference 4], categorizes satellites based on their wet mass (see Table 2-2).

Class	Wet Mass (kg)
Large	> 1000
Medium	500 – 1000
Mini	100 – 500
Micro	10 – 100
Nano	1.0 – 10
Pico	0.1 – 1.0
Femto	< 0.1

Table 2-2. Surrey Space Centre wet mass classifications

Most small satellite classifications are not so specific and choose dry mass rather than wet mass. *The Cosmos on a Shoestring* [Reference 2], after reviewing other definitions, chooses to define a small satellite as one with a dry mass of less than approximately 500 kg. Simply choosing a mass threshold as the definition fails to distinguish between satellites for different applications, and ignores some of the elements of the current "faster, better, cheaper" mantra.

Program Cost Classification. In 1994, the ESA Science Directorate concluded that the best criterion for defining small missions was cost [Reference 5]. The threshold for small satellite definition was considered to be about half the cost of a medium mission budget, i.e., at the time, approximately US $ 224 million. ESA's position on small satellites has evolved somewhat, however, and current ESA missions using small satellites have even lower total costs (e.g., SMART-1: US $ 50 million [Reference 6], Mars Express: US $ 150 million [Reference 7]).

In *Reducing Space Mission Cost* [Reference 8] the overlap between small spacecraft and low-cost missions is noted; while a small mass does not necessarily imply low-cost, the authors chose to associate the two. The empirical definition of a low-cost mission used in this work (see Table 2-3) is one with a factor of three or more in program cost reduction compared with most missions in a similar class.

Class	Typical Cost (US $ million)	Low Cost (US $ million)
Low-Earth orbit	140 to 2,000+	< 50
Geosynchronous	250 to 2,500+	< 75
Interplanetary	1,500 to 3,000+	< 500

Table 2-3. Classes of low-cost missions [Reference 8]

NASA's Discovery Program [Reference 9], which targets planetary science missions, places a cost limit of roughly US $ 300 million (FY'99) on a complete mission investigation, including launch. Mission development cost, within this cap, is itself limited to US $ 190 million (FY'99).

Program Development Time Classification. In a less quantified way, the general concept of a small satellite is a mission with a very focused and usually limited goal, which generally implies a relatively short development time. A position paper from the American Institute of Aeronautics and Astronautics (AIAA) [Reference 10] firmly states that a focused mission objective is a necessity for a small satellite in order to be developed, launched, and operated in a short time frame, thus meeting the objective of decreased time to scientific return.

NASA's "Discovery Program Plan" required all Discovery missions to be ready for launch no more than 35 months following the start of Project Implementation. The German Space Agency [Reference 11] went further and suggested that, from conceptualization to launch, the mission development time for small satellites should be no more than two years.

2.3 *Historical Trends*

In the decades following the early years of space exploration, the tendency was to increase the capability of satellites; this led to an increase of their mass and power that was enabled by the increasing capacity of launchers. Recently, a renewed interest has developed in employing microsatellites (See Table 2-2) in the scientific, commercial and military market, as technology has permitted sophisticated payloads to be developed for smaller volumes [Reference 12].

In the early 1990s, the complete or partial failure of several high profile, US $ multi-billion science spacecraft put NASA's spacecraft design practices under public scrutiny. Coupled with the "faster, better, cheaper" mantra of the new Administrator, Dan Goldin, the reversal had truly begun. A paradigm shift occurred when NASA started to judge spacecraft by their "specific performance", or performance per unit cost. Combined with NASA's shrinking budget caused by federal government downsizing, the smaller satellite ability to accomplish a mission of slightly reduced scope at greatly reduced costs became very attractive to NASA, and thus produced the downturn in satellite mass and power [Reference 13].

Fig. 2-1 presents the number of small satellites launched during the past two decades. Early peaks are dominated by small disposable tactical military communications satellites launched by the former Soviet Union. Peaks in 1990 and 1991 are largely due to a sudden increase in interest on the part of the US military, with budget cuts causing the subsequent dip. The significant rise in 1998 is due to the launch of 18 satellites in the ORBCOMM constellation heralding the start of commercial service provision through microsatellites, a trend that continued with the launch of Globalstar satellites in 1999 and 2000.

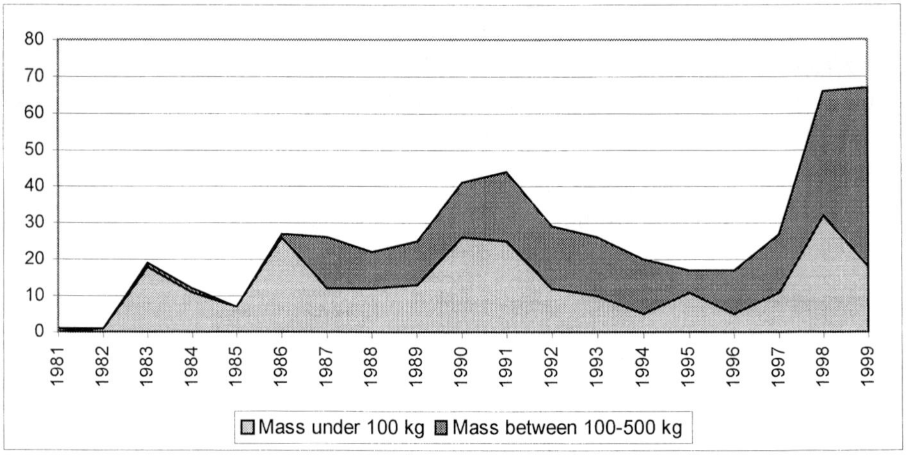

Figure 2-1. Number of small satellites launched during the last two decades [Reference 12]

In terms of mass distribution, there has been an increase in the number of minisatellites (100-500 kg), and the trend is towards ever lighter spacecraft, such as microsatellites (10-100 kg) and even nanosatellites (< 10 kg), the first modern versions of which were launched at the end of the 1990s.

Concerning small satellite users, a clear trend can be seen moving from military to civil small satellite missions as the primary market. As well, commercial interest in small satellites has increased. Finally, the government share of small satellites has slightly increased, especially during the last five years.

Changes in the execution of small satellite programs will be necessary for the continued success of these programs in the future. These changes include reduced government supervision of contractors, reduced paperwork, and fewer reviews. These changes have inherent risks, however, as shorter development times may encourage companies to take shortcuts and choose higher risk options without thorough analysis. As a result, small satellite developers will be required to adopt prudent risk management practices.

3. The Business of Small Satellites

In 2001, humankind celebrated 44 years of space activity. For the first twenty years, governments poured money into utilizing space as a resource and a tool, mainly for military purposes. As a by-product, these services can now be applied to new fields and thus become commercially viable. In today's global economy, space is an area of increasing commercial activity, focusing on economics rather than on government dominance.

This section addresses the shift of space activities from government control to industry involvement, and focuses on a number of key questions about small satellite markets: Who are the procurers and users of small satellites? Who are the manufacturers of those satellites? Who launches small satellites? What is the market potential for small satellites? Whilst it focuses on the business-to-business aspect of the small satellite market, it should be noted that most small satellites are still built under contract to government entities.

3.1 The New Space Age

Barrett [Reference 14] divides the opening of new environments and industries into three phases (see Fig. 3-1). The transition between each phase is not often obvious, as the progression occurs continuously rather than in discrete steps. Each phase has different drivers and rationale (see Table 3-1), which establish the mindset and rules for conducting business.

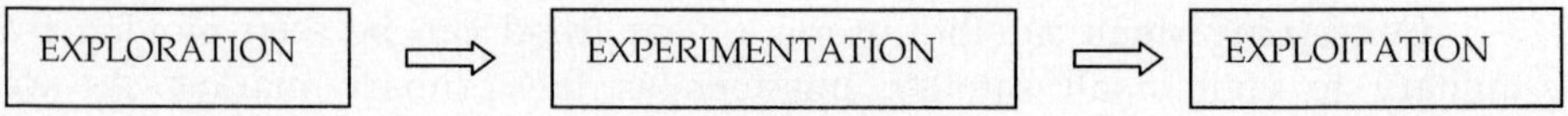

Figure 3-1. Three phases of a developing industry

At each of the three UNISPACE conferences, held in 1968, 1982, and 1999, the discussions fit the characterizations of each of these phases. The UNISPACE example shows the progression of the emerging commercial space industry.

Exploration	Experimentation	Exploitation
Characterized by:		
- Journey through unknown - High cost, high risk - Driven by government agendas - Almost exclusively government funding	- Initiated by successes in exploration phase - High cost, lower risk - Focussed on understanding new environment and operating in new domain - Driven by government prestige, military, and economic agendas - Almost exclusively government funding	- Technology/science in new environment well understood - Demonstrated value of operating in new medium - Unique services often available - Higher cost, higher return - Industry uses government infrastructure or builds own for commercial exploitation
Business Paradigm:		
- What government wants, at government pace, by government rules	- What government wants, at government pace, by government rules - Industry aware of opportunities, but expects government to lead	- Industry takes the lead and invests with or without government - Market forces determine investment

Table 3-1. Characterization and business paradigm of an emerging industry [Reference 14]

3.2 Business Trends

During the 1990s, telecommunications constellations were seen as a major area of growth in the small satellite industry. Mass produced satellites would drive down costs, while at the same time driving the development of the launcher industry. Several companies tapped the capital markets to implement ambitious plans, and many satellites were in fact launched. The bankruptcy filing of Iridium put a chill on the telecommunications constellations market, and capital became difficult to find. Consequently ORBCOMM, Globalstar and ICO have all experienced severe difficulties, with ORBCOMM and ICO filing for bankruptcy, and Globalstar requiring financial assistance from one of its owners. For the moment, it seems, voice telecommunications will not be a driving force in the evolution of the small satellite market, making many of the forecasts in the last few years for Little LEO growth overly optimistic. This can be seen by analyzing the markets, which responded favorably in 1998 with the initial optimism that greeted the launch of the constellations. However that optimism soon evaporated, leaving Orbital and Loral with share prices valued at less than half of what they were five years previously. The larger firms, such as Lockheed Martin and TRW, are down about 10% over the same period, reflecting the general difficulties in the satellite sector [Reference 15].

Nevertheless, associated services, such as two-way paging, fax and messaging services, will continue to grow. The most profitable area appears to be data packet store-and-forward services. A number of the smaller companies are exploring this market, which covers such broad applications as tracking shipments, trucks and railcars, utility meter reading, medical e-mails and collecting climatological data. As more applications are elaborated, this area will likely experience demand driven growth, both from the private and the government sectors.

Another area being explored by private enterprise is the provision of imaging and remote sensing data. ORBIMAGE is one example of small satellite efforts in this area. As with the telecommunications sector, however, more demand needs to be found to expand the client base and drive profitability. This is restricted by the fact that many governments also provide remotely sensed data and images, often at prices that are below those in the private sector. Export licenses for high resolution imagery can still be hard to obtain although the acceptable threshold is constantly being refined and one meter imagery may soon be easily licensed. Barring policy changes, this sector is likely to remain constrained in the near future.

3.3 *Procurers, Users and Operators*

Before briefly listing the different users of small satellites and the main applications which they cover, it is important to stress the distinction between the procurers, the actual users, and the operators of a satellite. Although very often these three actors are the same for a given project, there are several scenarios where this is not the case. An example could be a situation where NASA procures a satellite, which would be operated by a private company but used by a university. Having noted this distinction, the procurers/ users/ operators of small satellites fall into three main categories: government, universities and private companies.

Government users can be separated into civil and military users. Military users mainly focus on technology demonstration, signal intelligence, communications, and Earth observations. Civil users can be further considered as the main established space agencies (NASA, RAKA, ESA, NASDA, CSA...) on the one hand, and developing countries on the other.

The main applications of small satellites used by space agencies include remote sensing, space science, technology demonstration, inspection satellites, and telecommunications. In the space science field, the US and Europe are the main players, but Russia and Japan are also involved.

A few developing countries, such as Argentina and Brazil, use small satellites for communications and remote sensing. Meteorological data relay, ocean and Earth resource monitoring are important in Brazil (in both cases driven by INPE), Mexico and South Africa. In the field of technology demonstration, South Korea and Argentina are notably active.

3.4 *Small Satellite Business Models*

The business initiatives of some of the companies using small satellites to offer commercial products/services in the private sector are considered. The companies presented, both in the commercial remote sensing and the telecommunications markets, are large companies with a number of international partners. The space segment is only part of the business model, however, since the management and distribution of the information downlinked from the satellites must be sharply focused on identified potential customers. Thus, in addition to the benefits of a company's small satellite technology, each company's business model must also be based on good marketing capabilities, and efficient operations and management in order to succeed. This area is probably one of the weakest links in today's small satellite business.

Commercial Remote Sensing Companies

ImageSat International [References 16-19]

- *Estimated Price (of Product)*: US $ 1,500 for a 12.5 x 12.5 km scene
- *Niche*: global commercial high resolution (1.8 m or better) imagery of any point on Earth, available for purchase on the Internet within hours of acquisition, and subsequent product delivery for customers anywhere
- *Technical*: constellation of eight small satellites planned, with EROS-A1 launched on 5 December 2000 as the first of the series; two satellites will be launched per year, with planned launches continuing indefinitely according to market demand
- *Partners:* joint venture between Israel Aircraft Industries Ltd., Electro Optics Industries Ltd., Core Software Technology, AAD (Acquisition, Archiving, and Distribution) ground center partners, and ImageNet (online "department store" of geospatial information)
- *Initiative/Strategy*: cost-effective technology married with international partnering strategy (partners include manufacture of satellite constellation, ground distributor, online commerce system, and consumer) with the potential to have several images per day of the same scene from different viewing angles
- *Potential Target Markets*: Defense departments worldwide, urban planners, individual consumers of high-resolution imagery
- *Note*: Low development cost as EROS-A1 is based on Israel's Offeq-3 surveillance satellite.

Orbimage [Reference 20]

- *Estimated Price (of Product):* US $ 500 for 1500 x 2800 km scene
- *Niche:* daily color imagery available either directly to customer via purchase of downlink license or from Orbimage's online service
- *Technical:* OrbView-2 launched in 1997, 1 km resolution, downlinked in real time
- *Partners:* affiliate of Orbital Sciences Corporation
- *Initiative/Strategy:* integrated global system of imaging satellites, ground stations and Internet-based sales channels to collect, process and distribute imagery data at low cost
- *Potential Target Markets*: mapping/surveying, the natural resource sector, value-added consumer applications, and national security agencies
- *Note*: anchor customer NASA uses OrbView-2 data to study global warming trends.

Factors affecting the commercialization of remote sensing

- The Internet, which allows low cost, mass market access to satellite image databases, and develops multiple distribution channels
- The use of space-proven, military spin-off, small satellite technology reducing the initial investment in development costs
- Commercialization of high-resolution imagery still an unproven market, although ImageSat International estimates potential revenue on the order of US $ 2 Billion
- Image resolution still constrained by government policy.

Commercial Telecommunications Companies

Teledesic [Reference 21]

- *Estimated Cost of System*: US $ 9 billion
- *Niche*: global, broadband Internet-in-the-Sky™ network providing computer networking, broadband Internet access, interactive multimedia and high-quality voice services
- *Technical*: constellation of 288 small satellites, designed to support millions of simultaneous users using standard dish and modem
- *Partners*: primary investors include Craig McCaw, Bill Gates, Motorola, Saudi Prince Alwaleed Bin Talal, the Abu Dhabi Investment Company, and Boeing
- *Initiative/Strategy*: alliances with service providers worldwide, as well as marketing services directly to consumers
- *Potential Target Markets*: Internet services, global area networking and supply-chain networking enable organizations to connect and communicate seamlessly with internal divisions, subsidiaries, suppliers and partner users, plus under-served, developing/remote markets
- *Note*: service targeted to begin in 2005, with costs comparable to terrestrial options.

Globalstar [References 22-24]

- *Estimated Cost of System*: US $ 3-5 billion
- *Niche*: global-access telephony services ("Globalstar extends your reach")
- *Technical*: constellation of 48 small satellites
- *Partners*: consortium of international telecommunications companies led by Loral Space and Communications Ltd. and Qualcomm Inc.

- *Initiative/Strategy*: complementing terrestrial networks, leveraging the marketing, operating and technical capabilities of strategic partners, with system architecture designed to minimize cost and technological risk
- *Potential Target Markets*: highly mobile travelers/cellular users, and under-served, developing/remote markets
- *Note*: basic service already available worldwide.

Factors affecting the commercialization of telecommunications constellations

- Enormous up-front costs to create large constellations of small satellites which demand a strong solid financial base
- Immense systems costs, on the order of US $ several billions, with revenue needing to be generated to show positive operating cash flow and to service debts in an unproven market [Reference 25]
- Similar technological concepts (global coverage with LEO small satellites), but each initiative develops different capabilities to target different markets (Internet/cellular).

3.5 *Manufacturers of Small Satellites*

In the business realm of small satellites, one of the key questions is: who is making small satellites? Of the approximately 200 small satellites launched in the 1990s, over half have been made by private industry. Overall, Surrey Satellite Technology Ltd. and Orbital Sciences Corporation are responsible for half of the global production with a rate of 2 to 5 satellites per year. The rest have a production rate of about one satellite every 2 to 4 years.

In North America, the US manufacturers can be broken down into three main areas: specialty, medium-sized (mid-sized) firms, and large firms. The specialty firms make up the most interesting group. These tend to be very small companies which are developing or testing targeted technologies, although they also provide satellites for specific applications, most often under contract to government agencies. The Faisat-1 satellite, though, is an example of an emerging trend in the private sector towards data packet store-and-forward services. Some North American manufacturers are listed in Table 3-2.

For specialty and mid-sized firms, involvement in the small satellite sector can be seen to be either the primary focus, as in the case of Orbital, or as an adjunct to their involvement with larger satellites, as in the case of Aerospace Corporation and Ball Aerospace. Again, the main thrust of these satellites has been technology testing, with a secondary emphasis on Earth science missions, mostly under government contract. All are dependent to some extent on government contracts (currently 34% in the case of Orbital). Of particular note has been Orbital's involvement with the communications sector. However, Orbital has experienced its own share of difficulties with ORBCOMM, going from a smaller than expected client base and increased capital needs by the subsidiary to a bankruptcy filing in September 2000. Nevertheless, ORBCOMM is one of the market definers for the store-and-forward application.

The larger firms involved with small satellites are well-recognized names. They focus on those areas in which they have established competencies and developed infrastructure, and they tend to build the largest of the small satellites. Additionally, they have extensive experience and relations with the main contractors for satellite services, be they government agencies/departments or companies. The satellite business of most of the companies has been predominantly derived from government contract work, with telecommunications, and the associated store-and-forward market, being the notable exceptions.

Companies	Number of Employees	Number of Satellites	Average Mass (kg)	1999 Gross Revenues (US $ million)
SPECIALTY FIRMS				
AeroAstro LLC	20	2	121	2.5
Interferometrics	39	1	12	n/a
Spectrum Astro	270	5	215	30
Swales Aerospace	700	1	529	90
MID-SIZED FIRMS				
Aerospace Corporation	3,100	1	0.5	346
Ball Aerospace	2,352	3	202	383
Orbital Sciences Corporation (Space and Ground Infrastructure)	4,800	65	72	751
LARGE FIRMS				
Lockheed Martin (Space Systems)	9,663	4	330	5,825
Space Systems/Loral	3,100	48	222	1,000
TRW Space and Electronics	9,000	11	352	1,900

Table 3-2. North American small satellite manufacturers

The other regional focus on small satellite development is in Europe (see Table 3-3), where corporate structures are tied more closely to governments and universities. This fact is due in part to the historically socialist bias of government ownership of strategic industries, which has only begun to change in the 1990s. Most of the companies working on small satellites are focused in France, Germany and Italy, with Surrey Satellite Technology Ltd. in the UK and the Swedish Space Corporation in Sweden being two notable exceptions.

Although the European companies share with their US counterparts the bias towards government contract derived business, what is notable is the larger number of much smaller companies involved, with only a few larger companies in the small satellite market. European companies also seem to be involved to a much greater extent in working with universities. The main private sector activities, as with the US, are in telecommunications and, more notably, in data packet store-and-forward applications.

For the rest of the world, the companies involved in small satellite development tend to be closely linked with their respective governments. One example is Israeli Aircraft Industries; INVAP plays the same role in Argentina, working closely with CONAE. By the same token, Russian companies, such as Lavochkin Research and Prikladnoi Mekhaniki, are former state-owned companies still closely involved with government and science activities.

Companies	Number of Employees	Number of Satellites	Average Mass (kg)	1999 Gross Revenue (US $ million)
Alcatel Aerospace	17,400	3	158	3,220
Alenia Aerospazio	2,800	3	469	662
Swedish Space Corporation	353	4	135	41
Kayser-Threde GmbH	207	2	31	30
OHB-System GmbH	150	4	191	11
Surrey Satellite Technology, Ltd.	100	18	58	8 (Year 1998)
Carlo Gavazzi Space SpA	70	1	170	4
Gruppo Meggiorin	n/a	2	35	n/a

Table 3-3. European small satellite manufacturers

By and large, the small satellites were built either by the government agencies themselves, or by universities as training exercises. In this regard, the business benefit derives at the component supplier level, rather than at the manufacturer level. There have been a number of private/university partnerships, with Surrey Satellite Technology Ltd. once again being the most notable example. This reflects in part the need, especially in smaller firms, to develop the skill base necessary to produce small satellites. These skills can be derived at the larger, more experienced firms, from existing competencies and infrastructure.

3.6 *Manufacturers of Small Satellite Launchers*

At present the users of typical small satellites have at their disposal approximately 30 launch vehicles and various permutations, provided by private companies and governmental organizations on an international level. With the growth of the satellite industry, the space transportation industry has been growing rapidly since 1990. Many factors have contributed to this growth, such as growth of the GEO commercial market, and implementation of the LEO commercial constellations. Other factors include growth in the LEO civil governmental market and emergence of a commercial MEO/HEO market. Even though the GEO commercial communications satellite market seems to be heading towards heavier satellites weighing 2,700 - 3,700 kg at launch, smaller satellites continue to be ordered due to price advantages [Reference 26].

Launch vehicles range from the famous and heavily used Proton and Ariane family of launch vehicles to the lesser-known Tsiklon and Mu launchers. Typically these launchers have carried small spacecraft as either auxiliary or piggyback payloads. For example, Ariane 4 has to date orbited 25 spacecraft weighing less than 500 kg, with a total of 9 launches using its ASAP capability [Reference 27]. In some cases, however, as with the Tsiklon launch vehicle, these spacecraft were designated as the primary payload due to the use of a multiple spacecraft dispenser, which in this case has the capability to launch up to seven identical spacecraft at once.

Both governmental and private launch providers are actively present in both the commercial and non-commercial satellite launch markets. However, with the growth of commercial satellites, private launch providers seem to be replacing the role of governmental organizations by providing more flexibility and cheaper prices.

In future, the primary driving factor for the users of small satellites will be the cost of launching spacecraft into the desired orbit and the reliability of inexpensive launchers. In response to this trend, the launch vehicle manufacturing industry has begun to develop launchers to serve this sector of the market. Examples include the development by Orbital Sciences Corporation of the Pegasus launcher, as well as the Taurus and recently inaugurated Minataur launchers [Reference 28]. Existing launchers/rockets are being modified for use with small satellites. An example of this is Israel's development of the Leolink 1 and 2 launchers from a military missile. Other examples include development of the VLS by the AEB, the Brazilian Space Agency, the H- series of launchers at NASDA, the National Space Development

Agency of Japan, and potentially SpaceLift, a commercial venture in Australia proposing to use Russian START missile technology.

There is a general trend that the small spacecraft user relies on the use of launchers designed for much larger satellites. With development of small, cheaper and reusable launchers, the present day medium and heavy lift launchers are also undergoing developmental changes which will allow them to launch small spacecraft at affordable prices, the best example being the Titan IV and the Atlas family of launch vehicles.

3.7 Small Satellite Market Growth

In addition to geostationary fixed satellite services, GEO and LEO systems providing worldwide personal, portable and mobile telephony and data communications through small satellites will profoundly impact the satellite market. Little LEO services will tap into a range of markets. Messaging is expected to be a high-demand service, and will include e-mail, limited Internet access, two-way paging and fax. Remote data communications will also be an important area, especially for emergency services.

Recognizing the value of the applications, major international space companies (both satellite and launcher manufacturers) have acknowledged the dramatic potential of the small satellite market. Between 1996 and 2000, investments of US $ 350 million moved to Little LEOs [Reference 29].

The tenth Department of Trade and Industry (DTI) report from the UK Committee on Trade and Industry states that "it is very difficult to see a case for market failure in small satellites – this part of the space business is one of the few areas of the space industry where a successful open market does exist" [Reference 30]. In 1994, a Department of Financial Institutions (DFI) report [reference 31] predicted that commercial small satellite programs would outpace US government efforts over the 10-year period ending in 2004.

The activities of private-sector firms lead to a number of lessons being learned concerning value creation and delivery. While maintaining a culture of customer focus, continuous improvement, and investment in the future, firms must first choose the value (small satellite sub-market segmentation, selection and positioning), then concentrate on selling the product (design, development, pricing and distribution), and finally communicate the value (channels, promotion and advertising). The engineering supply-side mentality of simply "make and sell" without sufficient concern being paid to market demand, competition, and time to market, accounts for many of the problems facing satellite firms today.

Only a few firms enjoy absolute cost advantages from proprietary learning curves involving low-cost product design. The scale benefits of long and standardized production runs are not reaped to any extent. Substantial initial capital investment and working capital requirements serve as massive barriers to market entry. Even in the face of attractive applications, the combination of this initial investment, high cost of capital, and pressure to produce profits all contribute to a rather unattractive risk profile.

3.8 Small Satellite Cost

It is relatively easy to claim that a concept is low-cost; however, small satellite cost analysis yields the conclusion that it is more difficult to prove such a notion. Firstly, access to cost data is difficult as such information is closely guarded by most organizations, and rarely distinguishes between satellite, launch, operation or overall cost. Furthermore, it is often complex to estimate the cost, and especially the savings, of implementing automation concepts and other technological advancements. In addition, comparison of small satellite systems and missions is intricately dependent on specific mission requirements and applications.

Small satellites do, however, offer opportunities for low cost missions, although very low costs are experienced only with simple spacecraft performing limited missions. To the extent possible, the costs of 27 missions and 14 small satellite systems were investigated, and a sample of the results obtained is presented in Table 3-4. The costs of small satellite systems given include all impacts associated with systems engineering, payload, bus, satellite integration, testing, and program management costs.

Small spacecraft can be relatively expensive when they are required to meet demanding science objectives. The cost of developing the advanced technology required for a mission adds significantly to the true cost of the small satellite system used. Nonrecurring costs can add 150 percent to the total if a complex, new, mission-unique bus must be developed, while the cost is substantially less if a previously developed small satellite can be adapted [Reference 32].

Satellite	Launch Year	Wet Mass (kg)	System Cost (US $ Million)	Application
ALEXIS	1993	115	4.0	Space science
TERRIERS	1999	125	6.1	Space science
FAST	1996	191	45.0	Earth science
TOMS-EP	1996	295	39.4	Earth science
TMSAT	1998	55	2.0	Earth science
SNOE	1998	114	4.4	Earth science
Clementine	1999	50	18.0	Earth science
KOMPSAT	1999	465	32.0	Earth science
SurfSat	1995	55	3.0	Technology
KITSAT-3	1999	110	6.0	Technology
HETE-1	1996	128	15.0	Astronomy
WIRE	1999	250	45.0	Astronomy
Ofeq-3	1996	189	50.0	Surveillance
SunSat	1999	60	2.5	Communications

Table 3-4. Small satellite system costs [References 33, 34]

The true cost of a small satellite system depends on several factors [Reference 35]:

- Complexity of the payloads
- Nature of the mission
- Development state of the mission elements
- Capability of the bus
- Technological heritage
- Details of the program management and bus production.

Traditionally, cost modelers have used cost estimating relationships based on mass to predict satellite development costs. However, cost estimating - and cost reduction - require accurate forecasting of technology development costs.

Advanced scientific satellites have complex requirements in order to meet demanding scientific objectives, such as pointing accuracy, power, and processor speed, thus retaining a relatively high cost in spite of their mass-based size classification. Table 3-5 presents a comparison of some costs between space science and Earth science missions launched between 1992 and 1999.

Satellite	Wet Mass (kg)	Mission Cost (US $ million)	Launch Year	Application
SAMPEX	158	77.0	1992	Space science
ALEXIS	113	17.0	1993	
HETE-1	128	32.0	1996	
Lewis	385	82.0	1997	
TRACE	250	49.0	1998	
ACRIMSAT	115	32.0	1999	
TERRIERS	125	12.0	1999	
FREJA	214	19.0	1992	
Astrid	26	1.0	1995	
FAST	191	62.0	1996	
TOMS-EP	295	111.0	1996	Earth science
FORTE	210	15.0	1997	
SeaStar	372	44.0	1997	
SNOE	132	12.0	1998	
KOMPSAT	510	32.0	1999	

Table 3-5. Small satellite space science and Earth science mission costs [Reference 34]

One of the major contributors to small satellite mission cost is the launch cost. With both government and private launch providers in the market, there is a wide range of launch prices available to the users. The estimated cost per launch varies from US $ 10 million for Start, a Russian launcher, to US $ 300 million for Space Shuttle as shown in Table 3.6.

The launch cost is dependent on various factors such as the launcher type (size and technology), its reliability and the risk involved with the launcher. This risk is important since a low launch price can be more than offset by a partial – or a total – launch failure.

Launcher	Kosmos	Minataur	Pegasus	H-II	Taurus
Estimated cost/launch (US $ million)	12	12.5	12-15	10-15	18-20
Average payload mass (kg)	6.5-450	0.25-120	13.5-430	180	1-500
Total number of launches - Small satellites only	11	9	40	4	8
	Various Russian Launchers	**Delta**	**Ariane 4**	**Shavit 1 and 2**	**Space Shuttle**
Estimated cost/launch (US $ million)	10-95	45-90	65-125	165-170	300
Average payload mass (kg)	12-317	5.5-530	11-260	2543	38-518
Total number of launches - Small satellites only	19	15	26	1	12

Table 3-6. Estimated launch costs [Reference 27]

Thanks to globalization and the easing of Cold War tensions, the merging of business and technology from the East and West gives users more flexibility in launcher and price options nowadays [References 26, 36, 37]. More changes seem to be required to reduce launch costs further, such as the use of off-the-shelf technology in the development of future launchers, as well as investing in re-usable launchers.

Current strategies for achieving lower mission costs [Reference 38] include:

- Streamlining procurement practices, using suitably modified commercial products (hardware and software), avoiding expensive technology/processes
- Having low overhead rates, with as small an organization as possible, limiting the budget from the outset
- Improving design and development methods, having compatibility with more than one launch vehicle
- Reducing operations costs, and involving operations personnel in the team as early as possible.

3.9 *Insurance / Risk Aspects of Small Satellites*

Small satellite technologies are still developing and their relative probabilities of failure make them more risky than traditional big satellites. The insurance market for small satellites should be developed since small satellites are often not insured at the current time. The space insurance market in general has faced some tough times recently due to several closely placed launcher and satellite failures.

3.10 Conclusions

The trend towards smaller satellites is undeniable. Governments have been reducing their military and space budgets, while on the user side there have been demands for faster development of projects. Coupled with this has been an increased supply of launchers, and a number of business plans to exploit Low Earth Orbit. The telecommunications sector, led by Iridium, but joined by ORBCOMM, Globalstar and ICO, thundered into LEO on grand dreams of a wireless world. Unfortunately, the markets did not develop as fast as the constellations did, and they have all experienced severe financial difficulties. This has put a damper on the ability of other firms to raise capital for their own projects, and slowed the development of the commercial small satellite market to a crawl. In spite of this, there are areas which show promise for commercial development, especially data packet store-and-forward applications, which have been extensively explored by both governments and private enterprises. Another potential area is remote imaging, although this sector is hindered by the large amounts of existing government data often sold at below-market prices. This could be countered in part by the addition of value-added data interpretation to raw data sales.

For small satellite launchers, the anticipated growth in the telecommunications sector has led to a glut of small satellite launch vehicles. Some are derived from ICBM technology, using decommissioned military missiles for peaceful purposes, though this represents a limited supply. The most likely scenario is a consolidation in the industry over the next five years, as marginal companies succumb to capital pressures, and larger companies seek to gain market share. This sector also remains vulnerable to the development of reusable launchers, although this threat is likely to be minimal over the next several years, especially following the recent cancellation of projects such as the X-38.

Given the impressive pool of potential applications, cost seems to be the largest barrier inhibiting the development of the small satellite market [Reference 39, 40, 41]. As a result of these prohibitive costs and despite the wide range of applications, it is concluded that the small satellite market has rather limited short-term potential. Assuming decreased cost to orbit and further miniaturized component cost reduction, strong promise can be predicted, in the medium-term.

4. Policy and Law

4.1 Political Considerations of Small Satellites

Until the early 1990s, the trend of the space industry was to use bigger satellites to provide such services as better communications with the ground or images of deep space. Only a few countries were able to utilize space as a resource, as bigger satellites came at a very expensive price. However, small satellites have always been a key component in military space programs. The military use of small satellites revolved around testing new technologies in order to minimize the possibility of failure on much larger missions. This was the primary role for small satellites at the time, with scientific payloads as secondary objectives. As time progressed, however, the scientific community recognized the benefits of small satellites, and scientific payloads became the primary objective, pushing the testing of new technology to a secondary benefit [Reference 2].

In 1986, the Space and Earth Science Advisory Committee (SESAC) stated that "science is not best served by exclusive emphasis on major missions." They also suggested that "NASA should reexamine its approach toward implementation of flight projects with the intent of reducing overall mission cost" [Reference 42]. In 1988, NASA's Office for Outer Space Affairs (OOSA) agreed with the scientific community, and they recognized that small satellites "stimulate the research community, particularly at universities, with exciting new opportunities, which will attract new scientists and engineers to space science" [Reference 43]. Significant budget cuts within agencies such as NASA were soon to follow and then accelerate the trend toward the increasing use of small satellites.

In the mid-1980s, two new players entered the space arena - private companies and universities. Private companies soon caught on to the concept outlined by the scientific community, and commercial small satellite companies, such as Orbital and AeroAstro, were formed. These companies targeted the emerging commercial audience by offering quick, cost-effective access to space. Creating a new industry outlet is beneficial since new companies create new jobs and contribute to economic growth. In this case the emerging aerospace industry also provided competition, driving an improvement in quality.

4.2 Why Governments are Interested in Small Satellites

Governments around the world have been interested politically in the implementation of small satellites. Budget considerations, military interests, and the educational value of small satellites hold particular interest for governments.

Budget Considerations. The 1990's were a decade in which space sector funding remained relatively stagnant. In the larger space agencies, cheaper and shorter term small satellite missions are chosen to fill the yearly end of budget surplus to maintain funding allotments. In smaller space agencies, government institutions, private companies and educational institutions, small satellites are used because they are relatively cheap. The AIAA has noted that "small satellites may offer benefits to the nation's space endeavors that far exceed their costs" [Reference 44].

Military Interest in Small Satellites. While the militaries around the world have long been using small satellites, only recently have they become an explicit area of focus. Small satellites have been used primarily as technology demonstration tools, especially the DARPA satellites in the United States, due to budget restraints. Some reports cite small satellites as more reliable and flexible than their larger relatives, and advocate the construction of a fleet of storable, single purpose satellites [Reference 45]. Furthermore, the US Air Force has launched many small satellite packages in co-operation with NASA, with access to such Space Shuttle programs as Get-Away Specials and Hitchhiker packages. Both the US and Russian governments have been attempting to generate an international small satellite market, with varying degrees of success.

Israel's Offeq series of satellites reflects its investment in minisatellites. "[Their] strategy is to build satellites small enough and to develop micro-electronics and other technologies that will provide a sensible, low-cost solution to [their] requirements in this field" [Reference 46]. This demonstrates the dual-use technologies available for development with small satellites, as well as being an example of a small country's ability to develop a small satellite and small payload launcher based on missile technology.

Educational Value of Small Satellites. Small satellites cater to a wide base of applications, have short development times and low risks, and are suited to entry-level programs in the space sector. Universities around the world are creating their own small satellite programs to develop specialist technology and a knowledge base. Students are usually creative with very limited budgets, and they work closely with industry. Emerging graduates enter the work force with knowledge of the industry, thereby creating a better space industry. Industry can benefit from this process, not just by recruiting these new talented graduates, but also by sponsoring programs and testing out new technologies themselves. Government agencies are also pleased with the new educational programs as the funding requirements for such programs are relatively low and they can share in the direction of new technology and conceptual research. Such programs have had a significant impact in developing countries, as engineers from China, South Korea, and many other nations continue to flock to Surrey University, John Hopkins University, Utah State University, and others to learn about small satellite technologies, and to bring their new knowledge back to their home country. The United Nations is helping to encourage these exchanges in the light of environmental monitoring and management, as is the case with Portugal's PorSat [Reference 47].

4.3 National Perspective

NASA's "faster, better, cheaper" program was implemented to build "smaller more capable spacecraft to improve the performance and lower the cost of future space missions" [Reference 48]. The program has had its successes, such as the Mars Pathfinder and Lunar Observer. However, with two major failures (Mars Observer and Mars Explorer), NASA has had to re-evaluate this approach.

The ESA Science Program encompasses small relatively low-cost missions, in addition to the ambitious Cornerstone and medium-sized missions, recently dubbed "flexi-missions". These have been given the generic name of SMART, or Small Missions for Advanced Research and Technology, with the objective to test new technologies for spacecraft and instruments that will eventually be used on bigger projects [Reference 49]. For example, SMART-1 has the primary objective to flight test Solar Electric Primary Propulsion; ESA's Bepi Colombo mission to explore the planet Mercury could be the first to benefit from SMART-1's demonstration of electric propulsion.

In the August 1996 Australia Budget Statement, the Federation Satellite 1 (FedSat) program started and changed the prevailing administrative arrangements for space. FedSat is to be a 50 kg satellite to be launched in 2002 by the Japanese Space Agency NASDA as a piggyback payload. FedSat is a perfect example of how a nation with the technological capability and emerging space industry can use a small satellite to enter the international arena. Utilizing ties with private companies in the UK and Australia, and with universities in the US and Australia to build the bus and payload, the FedSat program will also help the government, the emerging space-related companies, and the Australian public become aware of and realize Australia's capabilities.

Japanese experience of small satellites is obtained via the Institute of Space and Aeronautical Science (ISAS). The intent is to foster close cooperation between NASDA and ISAS, in addition to enhancing technology and developing relationships with other organizations. The JAS 1a and JAS 1b small satellite ISAS missions are being developed in cooperation with NASDA and are aimed at enhancing technology. This is yet another political use of small satellites to foster closer cooperation between organizations within the same country [Reference 50].

4.4 *International Co-operation*

Small satellites open the way to cooperation at regional and international levels. Small satellites allow developing countries a certain level of technological independence, which enables them to develop more specific programs for the future. The Charter of the United Nations offers each country the right to have the opportunity to participate in space activities, and small satellites can certainly help realize this principle [Reference 51]. In 1994, the United Nations General Assembly endorsed a recommendation that the Scientific and Technical Subcommittee should emphasize "the utilization of micro or small satellites for the expansion of low cost activities, taking into account the special needs of developing countries" [Reference 52]. The Sub-Committee on Small Satellites for Developing Nations [Reference 53], created in 1993, has been involved in the International Astronautical Congresses, with a session for papers presenting small satellite activities in developing nations. Some of the papers presented include examples of international cooperation such as the Argentina-Brazil satellite for information on water, agriculture and environment.

NASA's shift to a "faster, better, cheaper" (FBC) policy, and the resulting emphasis on small satellites, has left questions about international cooperation. Organizational changes within NASA and ESA introduce new challenges as well as improvements for cooperation. One change is the US declassification of Strategic Defense Initiative technologies and their use in small civilian satellites, which has generated a technological gap between the US and Europe as far as miniaturization is concerned. The FBC policy, and the resulting shorter time for responding to Announcements of Opportunity (AOs), may render US cooperation with ESA and European countries more difficult in terms of small satellites [Reference 54].

Benefits of cooperation between space-faring nations can be seen within the French space agency, CNES, as they propose to design one or more micro-satellite constellations with international partners within a cooperative framework. This cooperation would be as flexible as possible and open to any organization ready to contribute to system implementation, maintenance and management, and to data exploitation. Each organization or group of organizations would provide an element of the system (satellite, launcher, ground or mission control segment, etc.). Project participants would thus have access to a wider range of services in return for a relatively small individual contribution. Furthermore, benefits are felt well beyond the national agencies developing the space technologies involved, as they allow users easier access to satellite data, thus encouraging the use of space applications around the globe.

Some small satellite projects could naturally be opened to space-faring and non-space-faring partners through cooperation agreements. A good example of this cooperation is found in the FASAT-BRAVO [Reference 55] microsatellite, an atmospheric ozone monitoring satellite developed jointly by the Chilean Air Force and Surrey University in the UK. This type of cooperation could be particularly interesting for missions focusing on global issues such as environmental protection or natural risk mitigation. A joint venture between China's Tsinghua University and the University of Surrey is the Tsinghua system, a constellation of seven minisatellites with 50-meter resolution remote sensing payloads. In addition, China participates in the Asia-Pacific Small Multi-Mission Satellite Project as part of the Asia-Pacific Multilateral Cooperation in Space Technology and Applications Program, which includes Iran, Pakistan, Thailand, Mongolia, South Korea, and Bangladesh [Reference 56]. The Thai microsatellite (TMSat), which has been jointly developed and built by Mahanakorn University, the UCOM Group in Thailand and the University of Surrey, is a good example in which an industry player, an educational institution, and a developing space country have worked together to the benefit of all concerned.

4.5 *Legal Aspects of Small Satellites*

To date, there is no document unique to small satellites in international law or any national law. This means that regardless of the satellite's payload, cost, weight, or manufacturer, all satellites are considered under the same legal regime. Recently, the hazard of man-made debris, of all shapes and sizes, in space has attracted international attention. On the one hand, the decreased volume of small satellites lowers collision probability. Simultaneously, however, the increasing number of small satellites contributes to the problem, especially when universities could soon be launching pico-satellites with sub-kilogram weights and lifetimes on the order of days.

The issue of space debris has been on the agenda of the UN COPUOS Scientific and Technical Sub-Committee since 1994, but progress seems to be blocked by the lobby of powerful emerging private sector entities with interests in space [Reference 57]. The legal sub-committee has not yet dealt with the issue, only vaguely concluding that applicable legal instruments are non-existent, and that the legal problems which debris pose are substantial. Increasingly, corporations which operate LEO constellations will respond only to market-based incentives. As they struggle in search of profits, increased shareholder value and respectability, obstructive standards are ignored. Consequences are rarely considered as long-term destructive behavior becomes standard industry practice. In a legal regime that drives itself on customary law, the issue of space debris will end up generating another piece of international law if industry, government, and educational players do not change their practices.

4.6 *Conclusion*

Small satellite policies diverge from the larger and encompassing space and commercial policies due to their lower costs and easily available technologies. In the United States and the European Union, small satellites often serve political needs, such as budget considerations, military objectives, and educational goals and are not necessarily seen as technology drivers. However, reduced costs and flexible mission planning have been enabling space-developing countries and previously shunned or excluded countries to circumvent the heavily controlled launcher and manned space systems markets. The result is more widespread and individual contributions to the space sector, leading towards more equality in this field. Small satellites could also provide some directly beneficial solutions to problems of developing countries [Reference 58]. In addition, the small satellite sector is fostering international co-operation for and between developed and developing countries.

Despite the promise of small satellites, the problems of space debris must be taken into account. The onus rests on all parties involved to make the international scientific community and the legal community move to mitigate the unacceptable collision probabilities that are forecast for the future.

5. Applications

The applications of small satellites can be broken down into five areas – telecommunications, Earth science, space science, technology demonstration, and engineering training. Small satellites can be used for all of the primary areas of general satellite applications, except for one notable exclusion, navigation.

5.1 *Telecommunications*

Since the dawn of communications using electromagnetic waves, i.e. radio, telephone and television, users have struggled with the problem of transmitting those waves over long distances. Satellites have provided a solution to this problem since their inception, and small satellites themselves have played a major role.

There are many applications for small satellites in telecommunications, ranging from simple point-to-point relays to mobile communications. Current and future tasks for these small satellites, in order of applicability to small satellites, include:

- Store and Forward data relay
- Real time data relay
- Telephone service to remote areas
- Broadband Internet
- Electronic mail
- Personal messaging and paging
- Remote monitoring and asset tracking
- Vehicle and animal tracking
- Transmission of television, telephone, telegraph, and radio messages
- Recovery of stolen property
- Maritime communications
- Video conferencing
- Tele-education
- Tele-medicine
- Broadcast radio/TV
- High definition television broadcast

Some examples of civilian small satellites which perform some of the tasks listed above are given in Table 5-1. Small satellites are usually only used in Low Earth Orbit (LEO) because the signal carrying capacity requirements of Geosyncronous Earth Orbit (GEO), a popular communications orbit, requires a high spacecraft mass.

LEO applications, where each satellite can service only a small portion of the Earth, employ two methods of operation: store and forward data relay and LEO constellations. Store and forward data relay satellites pass over the entire globe once or twice a day, store information, and then forward the stored information to a ground station. LEO constellations utilize many satellites to provide continuous global coverage for broadband, e-mail, messaging, tracking, and telephone services. The satellites pass transmissions on to each other until one satellite is in view of the intended recipient.

The Way Forward. LEO constellations have emerged to overcome the main drawbacks of large GEO satellites, such as latency time of communication, high launch risk, and the time required to develop spacecraft. They provide continuous global coverage with relatively low power and low mass satellites

[Reference 68]. The signal loss to a LEO satellite is also reduced compared to GEO, which reduces antennae size and transmission power requirements. Using many identical satellites allows cost savings through the introduction of batch production. Many proposals exist for using small satellite constellations to meet telecommunications needs.

Name/ Country	Year of launch	Services	Mass(kg)/ No. of Satellites	Ref.
UoSAT (Various)/UK	1981-Present	Amateur radio, Store and Forward data relay	50/1	59,60
Healthsat 2/ US	1993	Store and Forward data relay for Internet	48/1	61
SCD 1,2/ Brazil	1993-1998	Real time relay of environmental data	115/1	61
FAISAT/US	1995-2002	Personal messaging, e-mail, asset tracking, remote monitoring, and file transfer	150/32	62
TUBSAT/ Germany	1998	Store and Forward data relay to track vehicles, migrating animals, marine buoys	8/1	62
SAFIR/ Germany	1998	Store and Forward data relay	60/1	62
ORBCOMM/ US	1998-1999	Relay of alphanumeric data packets for two-way paging and e-mail	43/26	63-65
Globalstar/US	1998-1999	Telephone service to remote areas	450/24	27, 66
Teledesic/US	1998-2003	Broadband, Internet	70/288	67

Table 5-1. Examples of small telecommunications LEO satellites

Technology advancements in transponders, antenna systems, and optical communication technology using lasers have allowed the communications payloads to become smaller, making the entire spacecraft smaller in turn [Reference 69].

In conclusion, small satellites are valuable to telecommunications in the following applications:

- Low data rate, messaging systems
- High data rate, broadband and Internet services
- LEO constellations for continuous global coverage
- Store and forward data relay
- Telecommunications in developing countries.

5.2 Earth Science

Earth science includes the observation of the Earth and the near-Earth environment. Small satellites cover most areas of Earth science, or will in future small satellite missions.

Some examples of applications [Reference 32] include observations of:

Atmosphere and Meteorology
- Cloud properties
- Precipitation
- Radiative fluxes
- Atmospheric chemistry/temperature/humidity
- Aerosol properties
- Atmospheric dynamics and wind

Land
- Vegetation
- Land cover
- Land use change
- Surface temperature
- Resource exploration
- Fire and flood monitoring
- Volcanic effects
- Snow and ice cover, polar monitoring

Ocean
- Surface temperature
- Phytoplankton and dissolved organic matter
- Surface topography
- Wind fields

Near-Earth environment
- Magnetosphere
- Ionosphere
- Radiation belts
- Solar irradiance

Solid Earth
- Gravimetry
- Geodesy
- Magnetism

Some selected examples of Earth science small satellite missions are presented in Table 5-2.

Small Earth observation satellites are generally in circular, highly inclined, sun-synchronous, LEO orbits, although specific applications may require unique parameters. Most small imaging satellites have sophisticated onboard processing capacity, which maximizes the information per byte of downlinked remotely sensed data [Reference 85]. When necessary, data can be stored and forwarded to a ground station.

Name/ Country	Year of Launch	Maximum Resolution (m)	Purpose	Mass (kg)	Ref.
Near-Earth, Atmospheric Science					
IMAGE/ USA	2000	n/a	Magnetosphere study	494	70
BADR-B/ Pakistan	2000	250	Cloud monitor; Earth observation and communications payload	70	71
Ørsted/ Denmark	1998	n/a	Radiation environment measurements ; geomagnetic field	61	72
Ocean Science					
GANDER/ UK	>2002	~kms	Ocean observation; surface winds and wave heights	90	73
OrbView-2/ US	1997	~kms	Ocean color (fish location)	245	74
SMOS/ France	>2002	60km	Soil moisture and ocean salinity	475	75
Land Surface Remote Sensing					
UoSAT 12/ UK	1999	10	EO, panchromatic, multispectral	315	13
FASat-bravo/ Chile	1998	120	EO, land and atmospheric sensing	50	76
TMSat/ Thailand	1998	100	EO, data store and forward	150	77
KITSAT-3/ S. Korea	1999	15	Testing of multispectral sensors	110	78
TUBSAT-C/ Morocco, Germany	1999	6	High resolution Earth remote sensing	45	79,80
EROS-A, B Constellation/ International	2000-2004	<1	Commercial multispectral high resolution images	250-350	81
Orbview-4/ USA	2000	1	Multispectral, high resolution	270	82
Tiungsat 1/ Malaysia	2000	70	Pollution, forestry and agriculture, meteorology, disaster study	50	83
Tsinghua/ UK, China	2000	36	Disaster management	70	76
Solid Earth					
GRACE/ US, Germany	2001	n/a	Earth's gravity field	432	84

Table 5-2. Examples of Earth science small satellites

The Way Forward. In addition to the use of lightweight materials and ingenious packaging designs, the miniaturization of small satellites continues to be made possible by technological advances [Reference 32, 86] in the following areas:

- Electronic cameras
- Telescope optics
- Focal plane detectors
- Electronics.

Fundamental limitations restrict the degree of miniaturization possible for some instruments, especially for SAR and for optical detectors, which require large apertures for high resolution [Reference 87].

Also, task-specific sensors may reduce the complexity of the system, and focus the definition of the mission requirement. These dedicated sensors may be placed on satellites flying in formation and gathering data in a temporally and/or geographically coincident manner. This technique lowers risk, and adds mission flexibility. Consecutive strings of small satellites allow the comparison and verification of measurements from instruments on different satellites [Reference 88].

Small satellites are useful to Earth science for the following applications:

- Observation of the Earth and the near-Earth environment
- Dedicated missions
- Multiple missions in consecutive strings for data comparison and verification
- Increasing self-sufficiency in developing countries.

5.3 *Space Science*

Space science is one of the few areas where small satellites have always played a major role. Current and future scientific applications are mainly deep space probes, microgravity sciences, and astronomy and astrophysics. The main types of space science missions [Reference 89] are listed below and Table 5-3 provides a few examples of these missions.

Deep Space Probes
- Fly-by probes
- Orbiters
- Landers
- Impactors
- Penetrators

Astronomy and Astrophysics
- Telescopes
- Interferometers

Microgravity Sciences
- Fly alone spacecraft

Currently there are very few proposals for small satellites for microgravity applications. One of the projects being undertaken is the 125 kg, Dutch Sloshsat-FLEVO (Facility for Liquid Experimentation and Verification in Orbit) [Reference 101], which was to be launched in the year 2000, but has been delayed. The satellite will be used to investigate the behavior and management of fluids in space by monitoring water inside a tank.

However, microgravity experimentation is an application for small satellites that should not be overlooked. To aid in the development of future small satellites for such applications, Fig. 5-1 shows a requirement tree that will help developers decide how to design their system.

The Way Forward. Small space science missions are launched by many different nations and have, to date, all been procured by academic or government institutions. The US has launched most of the recent small interplanetary probes but in the upcoming years Japan and Europe will be more active in this field. More countries will become involved in the coming years, and the sector may even see commercialization. The major driving factor towards smaller satellites in this area is the often referenced reduction of funding for space programs in general. Examples of increased activity are ESA's plans to release two small landers from bigger spacecraft and to start using small probes to test technologies such as solar electric propulsion. Russia has been active in this sector as well, but the masses of their probes are high, given the nature of their technology, and therefore do not fit the scope of this survey.

Improvements in micro-technology, propulsion systems, and power systems will allow space science satellites to become more efficient. The main enabling technologies for these missions include high specific impulse propellants, thermal shields, efficient radioisotope thermal generators (RTGs), long life electronics, and long-range communications [Reference 102]. Constellations of small space science satellites for purposes such as interferometry or coincident measurements will become a common practice.

Name	Country/ Launch Date	Destination/ Purpose	Mass (kg)	Ref.
Deep Space Probes				
Hiten	Japan/1990	Moon orbit	185	90
Mars Pathfinder	US/1996	Mars landing	264	
Nozomi	Japan/1998	Mars orbiter	258	
Deep Space1	US/1998	Asteroid fly-by and Technology Development	495	
Stardust	US/1999	Comet fly-by	339	
Smart1	Europe/2002	Moon orbiter and Technology Development	300	
Beagle2 (from Mars Express)	Europe/2003	Mars lander	60	
Huygens (from Cassini)	Europe/2004	Titan lander	318	
Europa Orbiter	US/2004	Europa orbiter	<500	
Astronomy and Astrophysics Missions				
Ulysses	Multiple/1990	Heliosphere structure	370	91
SARA	France/1991	Radiometry of Jupiter	39	92
Yohkoh	Japan/1991	Corona dynamics	390	93
Alexis	USA/1993	UV/X-ray survey	115	94
TRACE	USA/1998	Photosphere, corona dynamics	250	95
WIRE	USA/1999	Galaxy evolution	250	96
HETE-II	Multiple/2000	Gamma ray burst location	130	97
ODIN	Sweden/2001	Star composition, nebulae	250	98
DIVA	Germany/2003	Optical interferometer	380	99
STEREO	USA/2004	Stereoscopic solar imaging	250	100

Table 5-3. Examples of recent and future small space probes

Unless cheaper and more technologically advanced systems are developed, there might be no long-term future at all for small satellites for microgravity applications. Yet there is no doubt that the use of small satellites in this area remains an untapped resource.

Thus, in the future small satellites will be important for:

- Interplanetary probes
- Astronomy and astrophysics missions
- New mission architectures such as constellations
- Large apertures, especially for interferometry
- Microgravity science.

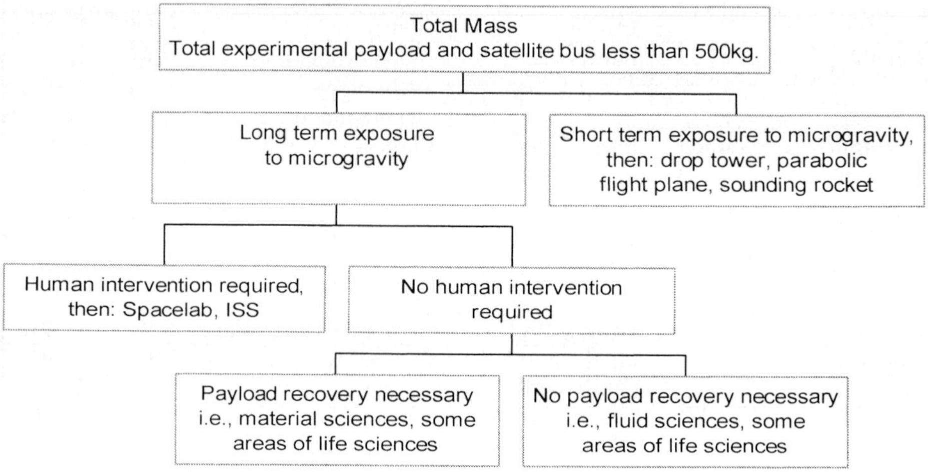

Figure 5-1. Flow chart illustrating requirements for using small satellites for microgravity applications

5.4 *Technology Demonstration*

Small satellites have demonstrated the ability of various technologies to function in the space environment since the beginning of the space age. This purpose suits small satellites because they are a fast and cheap method for testing technology before using it in larger, more expensive and/or more sophisticated spacecraft. Small satellites are capable of testing many new technologies at once, with innovations being demonstrated in payload and bus subsystems. Almost every agency or company that has ever built a spacecraft has used technology demonstration satellites.

Small satellites demonstrate numerous technologies, from space qualifying existing hardware to experimenting with completely new ideas. A few example missions can be seen in Table 5-4. Many technologies being developed by the Soviet Union have been delayed due to the collapse of Russia's economy, which explains their absence in the table.

The Way Forward. Technology demonstration missions will continue to use small satellites for the foreseeable future. In addition to validating improvements for existing technologies, small satellites will be used to demonstrate new technologies such as interferometry [Reference 110], tether propulsion [Reference 111], solar sails [Reference 112], inflatable structures [Reference 113], and constellation flying for various applications. We can expect the number of technology demonstration satellites to increase in direct relation to the increasing number of space-faring countries and the accelerating rate of technological developments.

Name	Country/ Launch Date	Technology Demonstrated	Mass (kg)	Ref.
Debut	Japan/ 1990	Extend and contract boom and aerodynamic brake	50	59
Tubsat-B	Germany/ 1994	Test of spacecraft attitude control system	40	59
Spartan 207/IAE	US/1996	Inflatable structures (antenna, truss, solar array)	60	103
REX-2	US/1996	Electronic radiation hardening, GPS, ADCS	85	104
Deep Space-1	US/1998	Ion propulsion	495	105
Mightysat 1	US/1998	Test of composite structure, advanced solar cells, advanced electronics, and shock device	320	106
TSX-5	US/2000	MWIR optics, laser communications, materials testing, near space environment detection	250	107
OPAL	US/2000	Picosatellite launcher, accelerometer, magnetometer	23	108
Three Corner Sat	US/2002	Constellation flying and re-configurable resources	18	109

Table 5-4. Examples of small technology demonstration satellites

5.5 *Engineering Training*

Affordable, yet sophisticated technology increasingly makes it possible for universities to be involved in small satellite development, fabrication and operations as a part of their engineering training. The satellites are projects in which the engineering students can apply what they have learned and gain hands on experience. This allows engineering training programs to accept a number of characteristics typically shunned in commercial spacecraft development. Such characteristics include high risk, low level of component redundancy, low precision control, non-optimal designs, and inefficient spacecraft operations. The mission objectives in general are very limited, and there are no contractual or long-life service obligations to be met.

In educational programs, the design, construction and operation of the satellite are the main objectives. In most cases though, the satellite also serves as a platform for low-power, -volume and -mass experiments developed by the universities. Primary tasks for small engineering training satellites include:

- Satellite development training
- Satellite fabrication/production training
- Satellite operation training
- Low-cost, simple payload integration.

The satellites permit rapid access to space and unique opportunities for low-cost payload iteration, making them especially suitable for technology demonstration as well as other mission goals.

An example of the success of an active university approach is the current UoSAT series, started in 1979 by the University of Surrey [Reference 114]. The series demonstrates that small satellites built within short time spans can be used to perform sophisticated missions. The Surrey Satellite Technology Limited (SSTL) company was formed by the University of Surrey in 1985, using personnel and facilities from the UoSAT Unit.

Table 5-5 gives an overview of some examples of the satellites developed for engineering training purposes; it is a representative sample of activities taking place in many countries.

The Way Forward. The main use of small satellites in universities is for:

- Training engineers
- Testing technologies developed by universities.

Name	University Country	Launch Year	Primary objectives	Secondary objectives	Ref.
UoSAT	Surrey, UK	1981	Servicing state-of-the-art payloads.	Various	114
SURFSAT-1	CalTech, USA	1995	Undergraduate student design, manufacture and testing	New NASA ground tracking facilities tests	115
SQUIRT Series	Stanford, USA	1998	Yearly student engineered satellites for state-of-the-art payloads	IR sensor, digital camera and voice communications	116
SUNSAT	Stellenbosch, South Africa	1999	Built by postgraduate engineering students	NASA experiments, radio amateur communications, HR imager	117
CanSat series	Arizona State, USA	2000	Student involvement in all aspects of the project	Environment detection	118
BLUEsat satellite	New South Wales, Australia	2002	Student involvement in all aspects of the project	Experimental GPS, imager	119
3^SAT	Arizona State, USA	2002	Student involvement in all aspects of the project	Stereo imaging, formation flying, communications	120

Table 5-5. Some examples of engineering training satellite programs

5.6 *Military Uses*

Since the beginning of the space age, military space programs have led the industry. Military small satellites are discussed here due to their pioneering role in developing space technologies and because commercial contracts are awarded by military customers. Due to the secrecy of military programs, however, few examples are discussed.

Military small spacecraft are engaged in technology demonstration, signal intelligence, communications, and observations. Most of the satellites are French, Russian, and American satellites that perform store and forward data relay and photo-surveillance. Some examples are the American GLOMR satellites, the Russian Strela program, and the French Zenon series.

Though many other applications use small satellite constellations, the role of constellations within military programs is still unclear. The French Délégation

Génerale pour l'Armement is developing Essaim, a military small satellite constellation system for telecommunications. However, Michel Courtois, Technical Director of Alcatel Space Industries, contends that constellations of small satellites would be the efficient way to perform military remote sensing [Reference 118].

5.7 Other Applications

Two applications that do not fit into the other categories are inspection and service satellites, and public participation space missions.

In 1999 the US Air Force, together with JPL, studied the use of microsatellites to inspect and service target satellites [Reference 121]. Inspector satellites orbit the target satellite while storing data and subsequently transmitting them to Earth. Service satellites are more complicated and require docking with the target, at which point they could then take control of the target or unjam structures that have not deployed. The study spawned a new application for satellites in the US; however, it was on Mir that a first generation inspector satellite, built and developed by Astrium specifically for a space station, was successfully tested for a month in 1997 [Reference 122]. Then, in June 2000, SSTL launched its 6.5 kg SNAP 1 satellite to test its capability to rendezvous with another spacecraft and inspect it with an array of video cameras [Reference 123]. There have been no regular servicing missions as of yet.

An example of a public participation mission is the Celestis program, a fully commercial program aimed at sending cremated human remains into space on small satellites for a considerable fee. Celestis offers three different services [Reference 124]:

- Earthview service – launches cremated remains into LEO
- Lunar Service – places the cremated remains in lunar orbit or on the surface
- Voyager service – launches the cremated remains into deep space.

The Way Forward. Astrium is now, in cooperation with RSC Energia, in the process of developing second generation inspectors designed especially for the ISS [Reference 122]. These satellites will also operate as servicing satellites.

Small satellites are practical for use as inspector and service satellites because of their maneuverability and close proximity safety, and for public participation space missions because of their small, single use payloads, affordable launch, and rapid and easy manufacture. Public participation space missions are a relatively new space application, and predicting a future trend is difficult; they may likely expand, which would be good for the industry as a whole by interesting the general public in space.

5.8 *Conclusion*

The survey of applications for small satellites shows that there are many applications for small satellites, some of which are in highly commercial areas. The areas of space science, Earth science, and telecommunications show the most advancement and the best prospects for the future for small satellites. In these categories, there is an overall push towards constellations and the most active area is the space science sector, with new, exciting missions being planned for both planetary exploration and astronomy using interferometry.

6. Engineering

The ever-increasing number and diversity of applications of small spacecraft have increased the desire to improve their efficiency and capabilities. Thus, specific spacecraft subsystems have been targeted for improvement (or, in some cases, re-invention), including certain areas of systems engineering. Specific changes are most obvious in the area of solid-state devices in on-board computers and data handling systems, and in attitude determination and guidance and control systems. These developments have led to increased autonomy and more integrated design solutions, with many redundant and interacting systems.

Large improvements and innovations in the area of propulsion subsystems, lightweight structures and advanced thermal control methods have also occurred.

State-of-the-Art vs. Off-the-Shelf technology. There are two categories of small satellites. One uses miniaturized state-of-the-art technologies for specific, dedicated missions. These satellites are not cheap since the development of miniaturized components is a costly operation. Small spacecraft which use off-the-shelf technology keep costs to a minimum.

6.1 *Systems Engineering*

There are three key differences between small satellite systems engineering and larger satellite projects:

- shortened life-cycle

- concurrent engineering
- mass production.

Concurrent Engineering. Concurrent engineering is a systematic approach for simultaneously carrying out processes of design, product, operation, etc. In this approach all elements of life cycle cost, such as quality, time and requirements, are considered at the beginning of the design process. This approach is performed in an integrated and iterative fashion.

Mass Production. For small satellites in constellations and formation-flying, assembly line manufacture and testing will be used. Validation of mass produced small satellite constellations can be done by validating random samples.

6.2 Launching Systems for Small Spacecraft

In addition to the large range of launch vehicles available, several different methods exist for both the integration of the spacecraft to these launch vehicles and for launch/separation. There are essentially three launch configurations available to the small spacecraft operator:

- Piggyback (auxiliary) payload
- Multiple spacecraft dispenser
- Secondary payload.

A fourth, typically financially unrealistic, option is that of being the primary, or sole, launch vehicle customer.

Piggyback (auxiliary) launch. In this most commonly used method, the spacecraft is launched alongside the primary payload in the excess volume that, in most launchers, would contain a dummy payload to create the appropriate mass distribution within the launcher. This means that such a launch is generally very cost effective. However, the major drawback of this type of launch is that the primary payload often determines the orbital injection point of the satellite sitting in the piggyback position. The Ariane 4 launch vehicle in the form of the ASAP (Ariane Structure for Auxiliary Payloads), which is capable of carrying four 50 kg spacecraft, is a good example of such a launch system [Reference 125]. A similar system is being developed for the Ariane 5, to be capable of carrying eight 80 kg spacecraft [Reference 126].

There is also an opportunity to piggyback on lighter launchers. Probably the best example of this is the Eurockot family of launch vehicles. A variation of this method also exists for the Space Shuttle. With the GAS (Get Away Special) system and an appropriate payload ejection system installed, it is possible to launch several small spacecraft from the Space Shuttle's cargo bay [Reference 127].

Multiple Spacecraft Dispenser/Mission Specific Dispenser. Via a mission specific dispenser, several spacecraft can be integrated and launched on a single launch vehicle. An example of such a system is to be employed by the Eurockot and it will be used to launch two identical spacecraft [Reference 128].

Such a system also exists for the Russian Proton and Tsiklon launchers, which have been used to launch seven Iridium and six Kosmos/Gonet satellites. There is an added risk involved in placing a large number of spacecraft on one launch vehicle, i.e., placing all your eggs in one basket; notably this risk was realized in 1999 when 12 Globalstar spacecraft were to be launched onboard a Proton launcher. The Proton subsequently failed with the total loss of all 12 spacecraft.

Secondary payload. A third method involves two payloads, which are typically stacked one on top of the other within the launcher fairing, as is available for the Ariane 4, Ariane 5, Pegasus XL and Proton launch vehicles. This stacking can take several forms; good examples exist in the Ariane and Delta families of launchers, which use dispensers for dual or triple launches.

6.3 Attitude Determination and Control / Guidance, Navigation and Control

Even though the functions of the ADC/GNC [Reference 129] systems are the same as for larger satellites, there is a definite difference in terms of the devices incorporated within the small satellite in order to enable the efficient and correct operation of these systems. The primary area of research presently being performed for such systems is concerned with miniaturization and autonomy.

Attitude Determination and Control. Small spacecraft tend to use the same systems for attitude determination and control as their larger cousins, at least in theory. Table 6-1 lists the three basic methods of ADC and their properties, whilst Table 6-2 lists examples of the hardware required for each system.

Control	Gravity gradient and magnetic	Momentum biased	3-axis reaction wheel
Sensors	Sun, magnetic	Sun, magnetic	Sun, star, magnetic
Cost	Low	Medium	High
Complexity	Low	Medium	High
Lifetime	High	Medium	High
Mass	Low	Medium	High
Volume	Low	Medium	High
Accuracy	Pitch, roll 1°, yaw 5°	Pitch, roll, yaw ≤5°	Pitch, roll, yaw 0.1°
Drift rates	≈0.05° /s	≈0.005° /s	≈0.001° /s

Table 6-1. Attitude Determination and Control methods

Stabilization and Control Type	Torquing Device/Method	Attitude Sensing Determination
Spinner	Reaction jets Magnetorquers Nutation damper	Sun, Earth sensor Magnetometers
Magnetic Stabilized	Magnetorquers	Sun, Earth sensors
3-Axis	Reaction wheel Gas jets Magnetorquers	Sun, Earth, star sensors, gyros, accelerometers
Momentum Biased	Biased momentum wheel Gas jets, wheel, Magnetorquers	Sun, Earth, sensors, gyros
Dual Spinners	Gas jets Nutation damper	Sun, Earth, star sensors

Table 6-2. ADCS methods and associated devices

Guidance, Navigation and Control. There are typically three GNC systems available to the user of small spacecraft - ground radar, ranging and GPS (Global Positioning System). The main properties of each of these are listed in Table 6-3.

	RADAR	Ranging	GPS
Cost	High	High	Medium
Accuracy	1 km error	< 100m error	High
Availability	Depends on NORAD	Independent	Depends on GPS System
Suitability	Ground station tracking	Orbit determination before and after phasing	Autonomous orbit determination and control

Table 6-3. Guidance, Navigation and Control systems

Trends in Attitude Determination and Control and GNC. The capabilities of current ADC and GNC systems for small spacecraft match or even exceed those of their larger cousins [Reference 130]. Two new approaches have emerged:

- Miniaturization by the use of microsystem technology (MST)
- GPS compatible attitude sensors.

These developments are most prevalent in the area of sensors, such as solid-state accelerometers, solid-state Earth horizon scanners, star sensors and single chip magnetometers. An example is the system onboard the Danish satellite Ørsted; it incorporates an advanced stellar compass containing a CCD star sensor whose accuracy is approximately 20 arcseconds.

GNC functions have generally been performed from the ground. However, with the use of an onboard GPS receiver, some autonomy can be transferred to the spacecraft. As a result of this, the obvious thing to do is combine the ADC and GNC of a small satellite into one system. An example of a spacecraft that incorporates this unified system is the Romulus and Remus spacecraft; however, it is only an experimental mission to assess the accuracy and viability of such a system.

Table 6-4 shows the various ADC/GNC methods presently used on small satellites. The table shows a trend from the very simple spin and gravity gradient stabilization used in the past (TIROS-2 and RADCAL) towards full three-axis control in the last decade (Meteor-1 and Thai-puhtt).

Satellite Launch/(kg)	Country	Type of Attitude Control	Orbit Determination	Application
TIROS-2 1960/127	NASA and the US DOD	Magnetic control	-	Meteorological and military applications
Nimbus-1 1963/376	US Navy and weather services	Computer controlled to within 1 degree employing gas jets and reaction wheels	-	Meteorological and infrared images
ATS-2 1967/370	-	Gravity gradient stabilized	-	Direct satellite TV broadcasts between Australia, Japan and the US
Skynet 1 1969/422	British military	Spin stabilized	-	Secure voice, telegraph and fax links
AEROS 1 1972/127	German ministry and NASA	2x 20N hydrazine thrusters	-	Upper atmosphere and F region of ionosphere
RADCAL 1993/87	Defense Systems Inc	Gravity gradient boom, MHRD and magnetorquers and magnetometer	GPS receivers	Space-based radar cross-sectional area calibration for more than 70 radars operating
Meteor-1 1995/363	NASA	3-axis controlled and inertially pointed	GPS receiver	Microgravity research
Thai-puhtt 2002/70	Thailand	3-axis magnetorquers, momentum wheel (pitch axis) dual redundant	GPS receiver	Earth observation

Table 6-4. Typical small spacecraft ADC and GNC

6.4 *Propulsion Systems*

Many concepts exist which small satellites could use either as primary or secondary propulsion systems. For many commercial, scientific and military missions, on-board propulsion systems make up between 10 and 50% of the total spacecraft mass [Reference 130]. The propellant itself can be up to 90% of the propulsion system weight. High performance advanced chemical, electrical and *in-situ* propulsion systems with superior specific impulse (I_{sp}) will be beneficial, as less propellant means decreased overall spacecraft mass.

Electric Propulsion. Electrical energy from solar cells or nuclear-electrical reactors can heat propellant to produce much higher exhaust velocities (V_e) than those produced by chemical reactions. Spacecraft employing electric propulsion also have reduced propellant mass; however, the acceleration is small and has to be applied for an extended period. Such systems are therefore ideal for small deep space probes, though there must also be an electrical power supply to heat the propellant. Table 6-5 considers the properties and characteristics of some electric propulsion devices [Reference 131].

Thrusters	Power Range (W)	Specific Impulse (s)
Electrothermal – uses electrical energy to heat propellant		
Resistojets	10^2	300 – 400
Hydrazine	10^3	500 – 600
Hydrogen	10^4	900 - 1,200
Ammonia	$10^3 - 10^4$	600 – 800
Electrostatic – uses potential difference to accelerate propellant ions		
Gridded Ion Engines	$10 - 10^5$	2,000 – 10,000
Stationary Plasma Thrusters (SPT)	$10^2 - 10^4$	1,000 - 2,500
Thruster with Anode Layer (TAL)	$10^2 - 10^4$	1,000 - 4,000
Magnetoplasmadynamic (MPD) – uses electromagnetic forces to accelerate plasma		
Pulsed	10^3	1,000 - 4,000
Steady-State	$10^5 - 10^6$	3,000 - 7,000
Pulsed Plasma Thruster	$10 - 10^2$	1,000 - 1,500
Pulsed Inductive Thruster	10^4	3,000 - 5,000
Electron Cyclotron Thruster	$10^3 - 10^4$	2,000 - 4,000

Table 6-5. Electric thruster comparisons [Reference 131]

In the future:

- Both conventional chemical systems and progressive electric systems have a part to play in near- to mid-term small satellite propulsion applications
- As electric propulsion matures, these systems will replace conventional chemical concepts.

6.5 Power Systems

Although many power sources for satellites exist, the ones most suitable for small satellites are:

- Solar Arrays - GaAs, Si, and emerging technologies such as Indium-Arsenide, Indium Phosphide, Cleft Gallium Arsenide (C/GaAs) and Copper Indium Diselenide (CIS)
- Batteries - NiCd, NiH_2, NiMH, and more recently Li-ion and Li-C
- Radioisotope Thermal Generators (RTGs) - suitable for interplanetary missions only.

Solar Arrays. Solar cell technology has advanced rapidly with improvements in efficiency, radiation tolerance, cost and weight. There are three main classes of solar cells under production today - thin film, single junction and multiple junctions. Although multi-junction cells are most efficient, they are also the most expensive, and the most difficult to handle and install. Lowest cost thin film cells have relatively low densities. However, they may be the most attractive spacecraft option if a lightweight mounting structure is used [Reference 132].

Table 6-6 contains a summary of the state-of-the-art solar cell efficiencies.

Cell Type	Cell Efficiency
Gallium Arsenide (GaAs)	25%
Silicon (single crystal)	23%
Silicon (polycrystalline)	18%
Indium Phosphide (InP)	18%
Copper Indium Diselenide (CIS)	17%
Cadmium Telluride (CdTe)	16%
Silicon (amorphous)	12%

Table 6-6. Solar cell efficiencies [Reference 133]

Inflatable structures may be utilized for the deployment and structural support of solar cell blankets. Optical concentrators could reduce the number of photovoltaic cells required [Reference 134]. NASA has initiated the SCARLET project whereby high efficiency solar arrays with linear solar concentrators are being developed for use in the New Millennium program [Reference 135]. This technology provides high performance (50 to 70 W/kg under the current design) with minimal weight penalty when additional shielding against space radiation is required.

Batteries. Primary battery systems have demonstrated excellent performance and no further advancements in performance are anticipated in the near future. However, there are significant developments in secondary batteries, especially lithium-based designs and nickel-hydrogen batteries.

Rechargeable lithium batteries seem most promising for superior energy density and specific energy; however, nickel based systems are still considered the baseline battery systems for space missions today.

RTGs. Radioisotope power sources [Reference 136] are generally considered when the mission is:

- Too far from the sun to make solar power feasible
- In a space radiation environment too harsh to allow sustained use of solar cells
- In a region where the use of solar power is insufficient, impractical or impossible
- In long night environments.

Numerous potential space science missions utilizing RTGs are under study. Some examples of future interplanetary missions using RTGs include the Europa Ocean Explorer and Lander, and the Pluto Express.

Looking ahead. In the future, it seems that:

- GaAs solar arrays will be used
- Although thin-film solar cells have lower conversion efficiencies and have not been proven in space, they have the potential for high specific power
- Inflatable solar arrays are an enabling technology for large solar arrays
- Optical concentrators reduce the number of photovoltaic cells required
- Rechargeable lithium batteries seem most promising for superior energy density and specific energy; however, nickel-based systems are still considered the baseline battery systems
- RTGs will only be used as a source of power for interplanetary missions.

6.6 Thermal Control Systems

Small satellites contain fewer parts than their larger counterparts, but they require efficient thermal control systems due to high component densities. Their power dissipation requirements will also be quite high. Table 6-7 lists typical operating temperatures for different satellite subsystems [References 137-140].

Subsystem	Maximum (°C)	Minimum (°C)
Communications	65	-5
Beacon	60	-0
Transmitter	50	10
Antenna	70	-5
TT&C	50	-5
Solar array mechanical assembly	90	-80
Power system	60	0
Attitude control system	60	0
Reaction control system	50	5
Batteries	20	0

Table 6-7. Working temperature ranges for different subsystems

Proper choice of coatings and insulation can perform 85% of the spacecraft's thermal requirement. Passive control systems are generally lighter, require less electrical power, and are less costly than active subsystems; due to the fact that they lack moving parts, they are also more reliable. These factors make passive systems ideal for lighter, more efficient and cheaper small satellite designs. Table 6-8 shows the different types of thermal control systems that could be utilized for small satellites.

A typical small satellite thermal control system would conduct heat from a heat generating subsystem to passive radiators (simple ones that rely on second surface mirrors or optical-surface reflectors, or complex ones which have louvers) via a heat-conduction path, perhaps with the assistance of heat pipes. Surface materials and insulation would also be required.

Some evolving techniques for thermal control are fluid loops (where heat is absorbed by the liquid from a cold plate or a heat exchanger and dissipated via a radiator) and micro heat-pipes. Other future thermal control systems are the so-called variable emissivity surfaces. These vary the effective emissivity of surfaces in response to low voltage signals. They have long life and accord important weight and power savings.

Description	Advantages/Disadvantages
Coatings	
Paints, chemically or physically produced conversion coatings. Characterized by thermo-optical properties (absorptivity, emissivity, reflectivity and transparency)	Degradation caused by operating environment and contamination induced by ground handling or space operations. Absorptivity is the most affected characteristic
MLI (Multi Layer Insulation)	
Mylar or Kapton coated on one or both sides with a layer of metallic material separated by sheets of Dacron. Used when surface coating is insufficient to prevent large heat loses or gains	Needs electrical energy
Heat Pipes	
Active thermal control system, consisting of sealed aluminum tubes containing a capillary structure on the inside and filled with an appropriate fluid such as ammonia Different types according to the amount of heat transported: Constant Conductance Heat Pipe(CCHP), Variable Conductance Heat Pipe (VCHP), Loop Heat Pipe (LHP)	Expensive but capable of conducting large amounts of heat
Electric Heaters	
The easiest means of providing heat to spacecraft equipment. Electric heaters can be continuously powered or can be switched on and off according to the temperature of the element being monitored. Used to maintain battery temperatures during eclipses and to keep the reaction control system temperature within specified limits	Simple, controlled by thermostats (active thermal control system)

Table 6-8. Different thermal control systems

In conclusion, it is noted that:

- Due to the component and subsystem size reduction in small satellites, the heat rejection requirements will also be reduced
- For low power levels, thermal management can be carried out via passive devices such as radiators, heaters and insulation
- For high power levels, capillary fluid loop systems could be used, though mechanical systems (e.g., pumps) create vibrations harmful to small satellites
- Higher component densities in small satellites require more efficient thermal conduction systems.

6.7 *Communication Systems*

The factors that most affect small satellite communication systems [References 141, 142] are efficiency, mass, volume, reliability, power consumption and cost.

Transponders. There are two different kinds of transponder amplifiers – solid-state and traveling wave tube (TWTA). Solid-state amplifiers are preferred for power outputs up to 5 to 10 Watts, as found in many small satellites. Low voltage solid-state amplifiers are more reliable than the higher voltage traveling wave tubes [Reference 143].

Solid-state amplifiers must evolve towards higher powers and frequencies as their size and mass is reduced. For instance, a state-of-the-art deep space X-band transponder developed by AeroAstro for NASA's Trailblazer Mission [Reference 144] will use one-fourth the voltage and half the power of proven technology; its mass is 12 times less and it is nine times smaller than present systems. A transponder with a 750 kbps capability has a mass of only 0.3 kg [Reference 145].

Future small satellite systems will require high-speed switching networks to allow communications between satellites flying in formation. The NASA Lewis Center and Sierra Monolithics developed a state-of-the-art 10 Gbps/port switch. This promising silicon-germanium (SiGe) technology enables the design of high-speed circuits at very low power consumption. The benefits of this state-of-the-art technology are that satellite and terrestrial systems will become fully interoperable, bandwidth efficient systems will improve capacity 100 times over existing systems for commercial applications, and the complexity of Earth stations will be reduced [Reference 146].

Antennas. Current small satellites use low frequencies for communications, causing them predominantly to use ordinary wire antennas, either as strata wires or helix antennas. Less commonly used antennas include aperture and microstrip antennas. While most small satellites use custom designed antennas, some off-the-shelf antennas are available from Surrey Satellite Technology Ltd.

New antenna materials are under development, using high temperature superconductors, carbon fiber, and advanced substrates [Reference 147]. One such design is the integrated antenna-solar cell concept SOLANT, a combined solar cell and antenna [Reference 148].

6.8 On-Board Computer and Data Handling

Basically, small satellite on-board computer and data handling systems (OBCDH) do not significantly vary from those on larger satellites [Reference 149]. However, small satellite limitations give the opportunity to exploit new technologies and create mass-produced satellites. Due to their smaller structures and overall weights, small satellites commonly use solid-state memory, allowing for higher storage densities, and easier and faster digital memory access than most tape, optical and hard-disk technologies.

Small satellites may also have higher speed processors, including 486s and Pentiums. Due to the shorter lifetimes and the location in LEO orbits (beneath the Van Allen belts), these more fallible technologies are protected from harmful radiation. In addition, small satellites may begin to use neural networks and parallel computing systems.

While software systems have historically been one of the more expensive aspects of satellites (they had to be independently written for each mission), the trend toward cheaper and mass produced small satellites has brought about a change in the software industry. Software can now be prepared for a satellite bus, with add-ons made specifically for individual missions. This has the potential to reduce costs significantly.

6.9 Small Satellites Structures

Small satellite structures [Reference 150, 151] are commonly constructed from aluminum and aluminum alloys, but there is an increasing use of composite materials. In deep space probes, more specialized matrix materials, such as boron-aluminum metal matrix and Timetal 21s are used. In future small satellite structures, it is anticipated that secondary structural elements will become integrated into the primary structure, with instrumentation casing, thermal systems, etc., being designed as load bearing members. Higher precision systems, including interferometric satellites will need to use active dampening systems and smart structures to maintain their tolerances. Volume will continue to be a concern as well, especially for solar cell surface area. Inflatable structures might be one solution to this problem.

Inexpensive small satellites may require new, simpler, yet reliable and deployable designs. One of the present thrusts of development efforts involves the use of inflatables, which may be cheaper and more dependable than articulated structures. Smart structures will become essential as scientific and other payloads become more sensitive, and as pointing requirements and dimensional precision constraints become more severe.

6.10 Microelectromechanical Systems

Micro-Electro-Mechanical Systems (MEMS) [Reference 152] is the integration of mechanical elements, sensors, actuators, and electronics on a common silicon substrate through the utilization of micro fabrication technology. While the electronics are fabricated using integrated circuit process sequences, the micro mechanical components are fabricated using compatible micro machining processes that selectively etch away parts of the silicon wafer or add new structural layers to form the mechanical and electromechanical devices.

MEMS promises to revolutionize nearly every product category by bringing together silicon-based microelectronics with micro machining technology, thereby making a complete systems-on-a-chip (SOAC). The system would include sensors, a microprocessor, data acquisition, a radio transceiver and batteries all on one component about the size of a postage stamp.

One example of this revolutionary quality is the fact that one of the goals described by NASA's Glenn Research Center (GRC) [Reference 153] and its partners in the field of MEMS is to develop microwave MEMS devices and integrate them with miniature microwave (RF) transmission lines and components to build the low loss RF distribution networks required for system-on-a-chip and phase array antennae. These novel, low loss, miniature RF components will be fabricated using multi-layer processing. In addition, GRC will build and demonstrate a MEMS-based, phased array antenna with 30 dB of gain and 35% efficiency at 26.5 GHz.

6.11 Conclusions

The design drivers, from an engineering viewpoint, of small spacecraft are as follows:

- Many satellites of same type/application are produced and tested simultaneously; assembly line operation is used in development
- Not all satellites will be tested completely; rather, sampling testing will be carried out to guarantee reliability
- New launch vehicles are being developed for "faster, better, cheaper" spacecraft
- Sharing a launch vehicle is common practice, at least among the developed space powers
- For developing countries the use of new small lift launchers or possibly converted InterContinental Ballistic Missiles (ICBMs) is foreseen

- There will be complete spacecraft autonomy in regard to Attitude Determination and Control as well as Guidance, Navigation and Control
- Reliability levels match those of present large spacecraft
- Small satellites of more than 100 kg warrant some form of propulsion system for station keeping, using proven technology with chemical propellants
- Rechargeable lithium batteries seem most promising for superior energy density and specific energy; however, nickel based systems are still considered the baseline battery systems for space missions
- Future small satellite systems will require high-speed switching networks to allow communications between satellites in formation flying applications
- Deep space probes will employ new technology such as ion engines
- Due to the component and subsystem size reduction in small satellites, the heat rejection requirements will also be reduced
- Inexpensive small satellites may require new, simpler, yet reliable, deployable designs
- Emphasis is focused on the use of MEMS and solid-state technologies.

7. Conclusions

In the second paragraph of Section 2 of this survey three questions were posed. In answer to them, the survey has shown that:

- There is a definite role for small satellites today. Reduced funding and improvements in technology have made small satellite use necessary and realistic
- There are a number of scientific, technical and commercial returns that can come out of using small satellites. Market demand still needs to be researched and aggressively pursued, but a nascent market is already in place
- Small satellites will affect the global space market by allowing new players to enter the field as well as creating opportunities for and demands on international cooperation.

While the traditional players are taking advantage of the qualities of small satellites to achieve their goals, private industry's ability to develop and exploit a market is essential to growth. The industry is still mired in the mentality that space is a hard to reach frontier while, at the same time, the public has started to take the presence of satellites and their services for granted. For the space industry to achieve the development it seeks, it will need to adopt more standard industry practices, including a more user focused regime. The general public needs to be convinced that they need satellites in their lives to provide certain services. The only area of space that has successfully come close to public awareness to date is telecommunications; however, value added remote sensing seems to be a market that has potential for expansion, and we may see the emergence of commercial navigation satellite services.

Small satellite communications constellations went a good distance towards stimulating the market, but recent failures have caused the market to stumble. Although these early business plans have failed, they should not be dismissed but rather learned from. Iridium is one of the most public failures, and yet the company has been rescued by an international consortium composed of the international gateway partners. A contract from the US DoD has given them some cash flow and Boeing is operating the ground systems and mission control operations. They have reactivated their commercial operations with cheaper handsets and lower per minute rates, a move they can afford now that the company is no longer weighted down by debt. Finally they will be launching two replacement satellites late in 2001.

The increased user base created by the capability to produce viable small satellites shows that interest in these satellites exists. Although things are moving slowly at the moment as programs are pushed through government processes or past public scrutiny, there are a plethora of satellites waiting to be built. Space is an industry in transition and small satellites may be the way through the current difficulties facing the development of a sustainable commercial space industry.

One of the major areas that needs focus is promotion of the space industry; small satellites offer an excellent opportunity for focussed or individualized programs. Such programs are most likely to be successful in the private sector, but non-commercial uses and successes could be used to promote the industry as a whole.

In space law, space debris is a problem area. There is a need to create a binding and enforceable international law on space debris in the interest of our planet and the solar system. Further, there is the need to protect commercial interests which manifest themselves in an operational satellite, especially with the increasing number of satellites in orbit thanks to the use of small satellites.

Technology, on the other hand, is constantly improving performance and capability and has a habit of responding to needs, so the technical area is more of a catalyst than a hindrance for the small satellite industry. That said, research still requires funding, but this funding will become more easily accessible if the programs are deemed to be worthwhile.

Currently the MSS6 class is working on the follow up to the present survey. In the course of reaching a decision on a specific topic, the class discussed ideas varying from furthering the idea of a standardized bus, echoing work done at Surrey, to employing small satellites to de-orbit space debris, to developing a plan to aid a developing country in founding a space program or in dealing with its internal needs, i.e., disaster management or telecommunications. At the time of publication, the focus was on a interplanetary science mission involving a constellation of small satellites developed by various international entities. Such a project could have multiple benefits in that it could provide further scientific information about the chosen planet, promote international cooperation, and promote interest in space by giving individual users a stake in the program.

Acknowledgements

The sixth Master of Space Studies class would like to thank faculty and staff for their help and support during the writing of Part One of the Team Project and would like to particularly thank Michael Rycroft for his assistance in modifying our work to fit the current format.

References

1. National Research Council: *Assessment of Mission Size Trade-Offs for NASA's Earth and Space Science Missions*, <http://books.nap.edu/books/0309069769/html/index.html>. National Academy Press, Washington, D.C., November 2000
2. Sarsfield, L.: *The Cosmos on a Shoestring: Small Spacecraft for Space and Earth Science* (1998), <http://www.rand.org/publications/MR/MR864/>. RAND, Critical Technologies Institute, Santa Monica, CA. November 2000
3. European Space Science Committee, Committee on International Space Programs: *US-Europe Collaboration in Space Science* (1998), <http://books.nap.edu/books/0309069769/html>. National Academy Press, Washington, DC, November 2000
4. da Silva Curiel, R.A.: *SSHP - Satellite Classification, Small Satellites Home Page*, <http://www.ee.surrey.ac.uk/SSC/SSHP/sshp_classify.html>. Surrey Space Centre, November 2000
5. Dale, D.C. and Whitcomb, G.P.: *Small Satellite Missions in the Context of the ESA Scientific Programme*, ESA Bulletin, No. 80, p. 1, 1994
6. Racca, G.D., Whitcomb, G.P., and Foing, B.H.: *The SMART-1 Mission*, ESA Bulletin, No. 95, p. 35, 1998
7. Schmidt, R., McCoy, D., Whitcomb, G.P., and Chicarro, A.: *The Mars Express Mission Concept – A New Management Approach*, ESA Bulletin, No. 95, p. 30, 1998
8. Larson, W., Wiley J., and Wertz, J.R., ed.: *Reducing Space Mission Cost*, Microcosm Press, Torrance, CA, 1996

9. NASA Office of Space Science: *Discovery Program Plan* (October 1999), <http://discovery.larc.nasa.gov/discovery/Discovery_Program_Plan.pdf>. NASA, November 2000
10. AIAA: *Small Satellites: Neglected Opportunity* (August 1994), <http://www.aiaa.org/policy/papers/small-satellites.html>. American Institute of Aeronautics and Astronautics Public Policy Position Paper, AIAA, November 2000
11. Rabenau, E., Klein, W., and Fischer, H.J.: *Operations Concepts for Small Satellites – Results of a Workshop*, paper from proceedings of 10th Annual AIAA/USU Conference on Small Satellites, September 1996
12. da Silva Curiel, R.A.: *SSHP - Intro, Small Satellites Home Page*, <http://www.ee.surrey.ac.uk/SSC/SSHP/micro/micro_intro.html >. Surrey Space Centre, November 2000
13. Jilla, C.D. and Miller, Dr. D.W.: *Satellite Design: Past, Present and Future Small Satellites Home Page*, <http://www.ee.surrey.ac.uk/SSC/IJSSE/issue1/cjilla/cjilla.html>. Surrey Space Centre, November 2000
14. Barrett, W.: *Doing Space Business in Australia – An ASICC Perspective*, presented at the 8th International Aerospace Conference, Adelaide, South Australia, September 13, 1999
15. Bloomberg.com: *Five year share performance*, <http://www.bloomberg.com/>. December 2000
16. Jonsson, H-O.: *EROS*, Remote Sensing Satellites (September 15, 2000) <http://satellites.satellus.se/eros.asp>. Swedish Space Corporation Group, December 2000
17. ImageSat International: *ImageSat International*, <http://www.imagesatinternational.com/>. ImageSat, December 2000
18. ImageNet: *West Indian Space, Ltd.*, <http://www.westindianspace.com/imagenet.html>. ImageSat, December 2000
19. Barzilai, A.: *Civilian Satellite as Military Backup*, <http://www3.haaretz.co.il/eng/htmls/kat10_3.htm>. Haaretz Internet Edition, December 2000
20. ORBIMAGE: *ORBIMAGE*, <http://www.orbimage.com/>. Orbital Imaging Corporation, December 2000
21. Teledesic: *About Teledesic: Fast Facts*, <http://www.teledesic.com/about/about.htm>. Teledesic, LLC, December 2000
22. Globalstar: *About Globalstar*, <http://www.globalstar.com/pages/about.html>. Globalstar, December 2000
23. Schay, A.: *Fool on the Hill, an Investment Opinion: Globalstar's Crash Course* (September 10, 1998), <http://www.fool.com/EveningNews/1998/EveningNews980910.htm>. Fool.com, December 2000
24. GMPCS: *Globalstar Frequently Asked Questions,*, <http://www.gmpcs-us.com/satellite_telephones/globalfaq.html>. GMPCS Personal Communications, December 2000
25. Mann, B.: *"Bandwidth Glut?" Shut Up!*, <www.fool.com/news/2000/ene001201.htm>. Fool.com, December 2000

26. Euroconsult: *Launch Services Market Survey, World wide Prospects: 1996-2006*, Euroconsult, Paris, 1996
27. Isakowitz, S.J, Hopkins Jr., J.P, Hopkins, J.P, *Space Launch Systems, 3rd ed.*, AIAA, 1999
28. Orbital: *Launch Vehicles*, <http://www.orbital.com/LaunchVehicles/>. Orbital, December 2000
29. Dykewicz, P.: *The Global Satellite Market, Emerging Trends and Future Prospects*, <http://www.telecomweb.com/samples/report.htm>. From the proceedings of the American-European Satellite Communications Seminar, 1998
30. UK Committee on Trade and Industry: *Trade and Industry – Tenth Report*, Memorandum submitted by the British Interplanetary Society, Appendices to the Minutes of Evidence, February 2000
31. DFI International: <http://www.dfi-intl.com/>. DFI International, Washington, D.C., 1994
32. Committee on Earth Studies, Space Studies Board, National Research Council: *The role of small satellites in NASA and NOAA Earth observation programs*, National Academy Press, Washington, D.C., 2000
33. JPL: *Quicklook Index*, <http://msl.jpl.nasa.gov/browse/quicklook/indexA.html>. JPL Mission and Spacecraft Library, November 2000
34. TSE: *The Satellite Encyclopedia*, <http://www.TBS-satellite.com/tse/online/>. Tag's Broadcasting Services, December 2000
35. Wertz, J.R., and Larson, W.J: *Reducing space mission cost*, Space Technology Library, 1996
36. Koelle, D.: Cost Engineering – The New Paradigm for Space Launch Vehicle Design, *Journal of Reducing Space Mission Cost*, Vol. 1, 1998
37. Kistler Aerospace: *Kistler Aerospace Vehicles*, <http://www.kistleraerospace.com>, Kistler Aerospace, December 2000
38. Joint Committee on Technology for Space Science and Applications: *Reducing the Costs of Space Science Research Mission: Proceedings of a Workshop*, Aeronautics and Space Engineering Board and Space Studies Board, National Research Council, 1997
39. Space News: *Space News: Directory of Worldwide Space 2000*, Space News Press, 2000
40. International Space Business Council: *State of the Space Industry 2000*, International Space Business Council, 2000
41. Vadas, D.H., *The Race for Space: A General Survey of the Commercial Space Market*, Aerospace Industries of America, Inc., 1998
42. NASA, Advisory Council, Space and Earth Science Advisory Committee: *The Crisis in Space and Earth Science*, Washington, D.C. November 1986
43. NASA, Advisory Council, Space and Earth Science Advisory Committee: *A Strategy for Leadership in Space Through Excellence in Space Science and Applications: An OOSA Strategic Plan*, Washington, D.C. November 1998
44. AIAA Space Systems Technical Committee: *Small Satellites: Neglected Opportunity* (Aug. 1994), <http://www.aiaa.org/policy/papers/small-satellites.html>. American Institute of Aeronautics and Astronautics, November 2000

45. Bille, M. and Kane, R.: *The High-Low Mix: A New Concept in Military Space*, <http://www.sdl.usu.edu/conferences/smallsat/proceedings/11/tech11.pdf>. Proceedings of 11th AIAA/USU Small Satellite Conference, 1997, Utah State University, North Logan, UT, November 2000
46. Opall-Rome, B.: *Israel Invests Research Funds to Develop Light Minisatellites*, Space News, January 24, 2000
47. Horais, B.: *International Activities in Small Satellites; Comparisons with US Programs*, Proceedings of 6th AIAA/USU Small Satellite Conference, Utah State University, North Logan, UT, 1992
48. National Science and Technology Council: *National Space Policy*, The White House, September 19, 1996
49. ESA: *Europe's first lunar adventure*, <http://sci.esa.int/content/doc/10/2320_htm>. ESA, November 2000
50. Space Activities Commission: *Fundamental Policies of Japan's Space Activities. rev. 24*, January 1996, p. 13.
51. United Nations: *Small Satellite Missions: Background Paper 9*, Third United Nations Conference on the Exploration and Peaceful Uses of Outer Space, May 1998
52. United Nations: *International Cooperation in the Peaceful Uses of Outer Space*, <http://www1.umn.edu/humanrts/resolutions/50/27GA1995.html>. G.A. res. 50/27, UN Doc. A/RES/50/27 (1995), University of Minnesota Human Rights Library (filed June 2000), November 2000
53. International Academy of Astronautics (IAA): *Subcommittee on Small Satellites for Developing Nations* (September 2000), <http://www.iaanet.org/commite/9b.htm>. IAA, November 2000
54. Committee on International Space Programs, European Space Science Committee: *US-Europe Collaboration in Space*, Science National Academy Press, Washington D.C, 1998.
55. FASat: *Microsatelite Experimental FASAT-Bravo*, <http://www.fasat.cl/>. Programa Satelitál FASAT, November 2000
56. US Department of Defense (DOD): *Annual Report on the Military Power of the People's Republic of China* (June 2000), <http://www.freerepublic.com/forum/a3958439a559e.htm>. Free Republic, November 2000
57. Rossi, A.: *A Survey of the Debris Activities*, <http://apollo.cnuce.cnr.it/~rossi/publications/iaf_pechino/node5.html>. CNUCE Institute, November 2000
58. Dhanji, N.: *Prospects for Involvement in Space Activities in Developing Countries via Small Satellites*, poster paper presented at International Space University's Smaller Satellites: Bigger Business? Symposium, May 21-23, 2001
59. Wade, M.: *Spacecraft by Objective, Encyclopaedia Aeronautica* (July 2000), <http://www.friends-partners.org/~mwade/craft/spactive.htm>. Friends and Partners in Space, November 30, 2000
60. University of Surrey: *On Board Data Handling* (2000), <http://www.ee.surrey.ac.uk/SSC/SSHP/list/list_obdh.html>. University of Surrey. November 28, 2000
61. Baker, D.: *Jane's Space Directory, 16th ed.*, Sentinel House, Surrey, 2000
62. Global System: *Global Systems – Final Analysis Inc* (2000),

<http://www.finalanalysis.com/global/main.htm>. Final Analysis, November 26, 2000
63. ORBCOMM: *Our Technology* (2000), <http://www.orbcomm.com/frames/frame3.htm>. ORBCOMM, November 31, 2000
64. GMPCS: System Specs: *Mobile Satellite Systems Specifications*, <http://www.cellular.co.za/gmpcs-specs.htm>. iTouch, 3 December 2000
65. Patel, B.T.: *On-orbit Performance of the ORBCOMM Spacecraft Constellation*, Proceedings of 13th AIAA/USU Small Satellites, Utah State University, 2000.
66. Globalstar: *Products and Services*, <http://www.globalstar.com/pages/products.html>. Globalstar, November 29, 2000
67. Teledesic: *Technology*, <http://www.teledesic.com/tech/tech.htm>. Teledesic, November 30, 2000
68. Elbert, B.R.: *The Satellite Communications Applications Handbook*, Artech House, Boston, 1997
69. Elbert, B.R.: *Introduction to Satellite Communications. 2nd ed.*, Artech House, Boston, 1999
70. NASA-GSFC: *IMAGE Science Center* (August 2000), <http://image.gsfc.nasa.gov/>. NASA-GSFC, November 28, 2000
71. RAL: *BADR-B – Technology for Micro-satellites* (July 2000), <http://www.ssd.rl.ac.uk/news/cdrom/eo/badr.htm>. Rutherford Appleton Laboratory-SSTD, December 6, 2000
72. Toffner-Clausen, L.: *The Ørsted Satellite*, The Ørsted Satellite Project (December 2000) <http://web.dmi.dk/projects/oersted/mission/mission.html>. Danish Meteorological Institute, December 6, 2000
73. GANDER Investing: *GANDER web pages: Investing* (2000), <http://www.satobsys.co.uk/GANDER/webpages/ganinv.html>. GANDER Investing December 6, 2000
74. Orbital Science Corporation: *Orbital Sciences Corporation OrbView-2 Satellites* (July 2000), <http://earth.esa.int/spgorbview2>. ESA, December 6 2000
75. SMOS: *SMOS*, <http://www-sv.cict.fr/cesbio/smos/>. ESA, December 6, 2000
76. SSTL: *Missions*, <http://www.sstl.co.uk/missions/>. Surrey Satellite Technology Ltd., December 6, 2000
77. Tag's Broadcasting Services: *The Satellite Fact Sheet*, The Satellite Encyclopaedia (August 2000), <http://www.tbs-satellite.com/tse/online/sat_kompsat_1.html>. Tag's Broadcasting Services, December 6, 2000
78. ISRO: *KITSat-3*, <http://www.isro.org/kitsat3.htm>. ISRO, December 6, 2000
79. Leitenberger, B.: *Deutsche Satelliten*, <http://home.knuut.de/Bernd.Leitenberger/deutsche-satelliten.htm>. Bernd Leitenberger's Info Site, December 6, 2000

80. SSTL: *Companies and Organisations in the small satellite industry – Africa*, Small Satellite Home Page, <http://www.ee.surrey.ac.uk/SSC/SSHP/sso_africa.html>. Surrey Satellite Technology Ltd., December 6, 2000
81. ImageSat International: *ImageSat International Satellites – EROS Satellites*, <http://www.imagesatinternational.com/satellites_1.html>. ImageSat International, December 6, 2000
82. Orbimage: *High-Resolution Imagery and Hyperspectral Data*, <http://www.orbimage.com/corp/orbimage_system/ov4/>. Orbimage, December 6, 2000
83. Institute of Engineers – Malaysia: *Tiungsat-1 Malaysian first Microsatellite* ,The Monthly Bulletin of the Institution of Engineers Malaysia (May 2000), <http://www.yellowpages.com.my/tiungsat/tiungsat.htm>. yellowpages.com, December 6, 2000
84. CSR/TSGC: *Welcome to GRACE (Gravity Recovery and Climate Experiment)* (November 2000), <http://www.csr.utexas.edu/grace/mission/>. CSR/TSGC, December 6, 2000
85. Fouquet, M., and Sweeting, M.: *Remote Sensing Using the University of Surrey's Microsatellites*, Presented at SCRS'95 Conference, Xi'an, China, <www.ee.surrey.ac.uk/CSER/UOSAT/papers/xian_mf/xian95.htm>. SSTL, December 6, 2000
86. Krasilshikov, M. and Jacobson, M.: New Technology Of Satellite Navigation: Positioning Based On Observation Of Ground Landmarks, *Small Satellites for Earth Observation: Digest of the International Symposium of the International Academy of Astronautics*, edited by. H.P. Röser, R. Sandau, and A. Valenzuela, pp.43-46, Walter de Gruyter, Berlin, November 4-8, 1996
87. Tobias, A.: The (ESA) Earth Explorer Candidates: Nine Focused Missions, All Satellite Classes, *Small Satellites for Earth Observation: Digest of the International Symposium of the International Academy of Astronautics*, edited by. H.P. Röser, R. Sandau, and A. Valenzuela, pp.11-14, Walter de Gruyter, Berlin, November 4-8, 1996
88. Reigber, C. and Schwinter, P.: *CHAMP Gravity Field Recovery* (30 June 2000), <http://www.gfz-potsdam.de/pb1/CHAMP/scientific_objectives.htm>. GFZ Postdam, December 4, 2000
89. Yahoo ! Geocities : *Space probes*, <http://www.geocities.com/CapeCanaveral/Hangar/2671/spacep.htm>. Yahoo! Geocities, November 30, 2000
90. SSTL: *Future Interplanetary Small-satellites*, Small Satellite Home Page <http://www.ee.surrey.ac.uk/SSC/SSHP/interpl.html>. SSTL, November 29, 2000
91. JPL: *Ulysses Home*, (August 28, 2000), <http://ulysses.jpl.nasa.gov/index.html>. Jet Propulsion Laboratory, December 5, 2000
92. SSTL: *SARA*, Small Satellite Home Page, <http://www.ee.surrey.ac.uk/SSC/SSHP/data_sara.html>. Surrey Satellite Technology Ltd., December 5, 2000
93. ISAS: *ISAS/Missions*,

<http://www.isas.ac.jp/e/enterp/missions/astro-e/index.html>. Institute of Space and Astronautical Science, December 5, 2000

94. Los Alamos: *ALEXIS Project* (January 6, 1998), <http://nis-www.lanl.gov/~asocops/alxhome/>. Los Alamos National Research Laboratory, December 5, 2000
95. Watzin, J.: *TRACE* (November 21, 1997), <http://sunland.gsfc.nasa.gov/smex/trace/index.html>. Goddard Space Flight Center. December 5, 2000
96. Watzin, J.: *WIRE* (July 22, 1999), <http://sunland.gsfc.nasa.gov/smex/wire/>. Goddard Space Flight Center, December 5, 2000
97. MIT: *High Energy Transit Explorer (HETE-2)* (November 15, 2000), <http://space.mit.edu/HETE/>. Massachusetts Institute of Technology, December 5, 2000
98. Swedish Space Corporation: *The ODIN Satellite* (1999), <http://www.ssc.se/ssd/ssat/odin.html>. Swedish Space Corporation, December 5, 2000
99. Ulrich, B.: The small astrometric interferometry satellite DIVA, *The Message of the Angles - Astrometry from 1798 to 1998. Proceedings of the International Spring Meeting of the Astronomische Gesellschaft, Gotha. 11-15 May 1998*, edited by P. Brosche, W. Dick, O. Schwarz, R. Wielen, Deutsch: Thun, Frankfurt am Main, 1998
100. GSFC: *STP-STEREO Mission*, <http://stprobes.gsfc.nasa.gov/stereo.htm>. Goddard Space Flight Center, December 5, 2000
101. Nice, A. : *Mini Satellites Future* (December 1998), <http://www.ee.surrey.ac.uk/ssc/sshp/mini/mini_future.html>. University of Surrey, November 2000
102. IAA: *Fourth IAA Conference on Low-Cost Planetary Missions*, IAA, May 2000
103. L'Garde, Inc.: *Inflatable Antennae Experiment*, <http://www.lgarde.com/programs/iae.html>. L'Garde Inc. December 5, 2000
104. US DoD: *REX-2*, SMC/TE (August 7, 2000), <http://www.te.plk.af.mil/stp/rex2/rex2.html>. U.S. Dept. of Defense, December 5, 2000
105. NASA Glenn: *NASA Glenn Deep Space 1*, <http://www.grc.nasa.gov/WWW/PAO/ds1.htm>. NASA Glenn Research Center, Dec. 5, 2000
106. JPL: *Surfsat-1*, Mission and Spacecraft Library, <http://samadhi.jpl.nasa.gov/msl/QuickLooks/surfsatQL.html>. JPL, December 5, 2000
107. US DoD: *Tri-Service Experiments Mission (TSX-5)* (August 7, 2000), <http://www.te.plk.af.mil/stp/tsx5/tsx5.html>. US Air Force, December 2000
108. Stanford Univ.: *The OPAL Pages*, <http://aa.stanford.edu/%7Essdl/projects/squirt2/intro.html>. Stanford University, December 5, 2000
109. NASA: *Three Corner Sat*, ASU NASA Space Grant Student Satellite Program (April 28, 2000), <http://nasa.asu.edu/asusat/3CS/index.html>. NASA, December 5, 2000

110. GSFC: *MAXIM Pathfinder*, <http://maxim.gsfc.nasa.gov/pathfinder.html>. Goddard Space Flight Center, December 5, 2000
111. Michigan Technic: *STEP-AIRSEDS Information*, <http://www.airseds.com/stepairsedsinfo.html>. Michigan Technic Corporation, December 5, 2000
112. JPL: *Space Technology 6*, New Millennium Program (November 1, 2000), <http://nmp.jpl.nasa.gov/st6/>. Jet Propulsion Laboratory, December 5, 2000
113. Yamazoe, Y., Mizuno, T., Nomura, K., and Hiroshi, I.: *Inflatable Technology Demonstration Satellite 'ASAGAO'*, <http://forth.aero.cst.nihon-u.ac.jp/~yamazoe/ISTS/index.htm>. Department of Aerospace Engineering, College of Science and Technology, Nihon University, December 5, 2000
114. da Silva Curiel, A. (Ed.): *UoSAT & SSTL History* (1996), <http://www.ee.surrey.ac.uk/CSER/UOSAT/history.html>. Surrey Satellite Technology Limited, December 2000
115. Caltech: *GALEX: The Galaxy Evolution Explorer A UV Space Imaging and Spectroscopic Mission*, <http://www.srl.caltech.edu/galex/>. Caltech, November 2000
116. Stanford Univ.: *Welcome to Stanford University Space System Development Laboratory*, <http://aa.stanford.edu/~ssdl/>. Stanford Aeronautics and Astronautics, November 2000
117. Univ. of New South Wales: *BLUEsat PROJECT*, <http://bluesat.unsw.edu.au/main.html>. University of new South Wales, November 2000
118. Thomas, B.: *Alcatel Space Industries, une utilisation pragmatique*, CNES Magazine, No. 9, p. 30, 2000
119. ESL: *Sunsat Homepage*, <http://sunsat.ee.sun.ac.za/index.html>. Electronic System Laboratory, November 2000
120. NASA: *Student Satellite Program*, <http://nasa.asu.edu/asusat/index.html>. Arizona State University, November 2000
121. Moser, C., Das, Feber, Javin, Madison, Smith, Stallard: *Novel Missions for Next Generation Microsatellites*, SSC99-VII-1, Proceedings of the 13th annual AIAA/USU Small Satellites Conference, Utah, September 1999
122. Astrium: *Inspector – the monitoring system*, <http://www.matra-marconi-space.com/www/push/0000000001/activities_and_projects/en/html/2_1_8_1_5_inspector.html>. Astrium, November 2000
123. SSTL: *SNAP-1*, <http://www.sstl.co.uk/missions/mn_snap_1.html>. SSTL, November 2000
124. Celestis: *Celestis*, <http://www.celestis.com/>. Celestis Inc., November 2000
125. Arianespace: *Ariane 4 ASAP Platform for Auxiliary Payloads*, <http://www.arianespace.com/launcher_4_payloads3.html>. Arianespace, December 2000
126. Arianespace: *Ariane 5 Payloads: The Ariane Structure for Auxiliary Payloads (ASAP)*,

<http://www.arianespace.com/launcher_5_payloads3.html>. Arianespace, December 2000
127. NASA: *Shuttle Small Payloads Project*, <http://sspp.gsfc.nasa.gov/>. NASA, December 12, 2000
128. EUROCKOT: *EUROCKOT Launch Services*, <http://www.eurockot.com/>. EUROCKOT, December 12, 2000
129. Hughes Prof P.C.: *Spacecraft attitude dynamics, 1986*, proceedings of the fourth ESA International Conference of Spacecraft Guidance, Navigation and Control Systems, October 18-21, 1999
130. Diani F.: *New Technologies for Small Satellites and an Analysis of Their Application*, Scientific and Technical Report STM-256, ESA, August 1997
131. JPL: *Introduction To Electric Propulsion*, Advanced Propulsion Concepts, <http://sec353.jpl.nasa.gov/apc/Electric/00.html>. JPL, December 12, 2000
132. Hsiao, Fei-Bin (ed.): *Microsatellites as Research Tools*, Proceedings of COSPAR Colloquium on Scientific Microsatellites, held in Tainan, Taiwan, December 14-17, 1997
133. NASA, Office of Space Science: *Solar Energy Conversion*, Advanced Technology and Mission Studies, (May 7, 2000), <http://spacescience.nasa.gov/osstech/solar.htm>. NASA, December 12, 2000
134. Andreev, V.M., Rumyantsev, V.D. and Shvarts, M.Z.: *Photovoltaic Modules with Sunlight Concentrators*, proceedings of IWRFRI 1999 conference, St. Petersburg, Russia, May 24-26, 1999
135. NASA, Office of Space Science: SCARLETT Concentrator Solar Arrays, Advanced Technology and Mission Studies (April 26, 2000), <http://spacescience.nasa.gov/osstech/scarlet.htm>. NASA, December 12, 2000
136. Hagen, R., *Nuclear Powered Space Missions - Past and Future*, (August 1998), <http://www.globenet.free-online.co.uk/ianus/npsm4.htm>. Global Network Against Weapons and Nuclear Power in Space, December 12, 2000
137. Thornton, E.A.: *Thermal Structures for Aerospace Applications (AIAA Education Series)*, AIAA, Arlington, VA, 1996
138. Sellers, J.J.: *Understanding Space: An Introduction to Astronautics*, McGraw-Hill, 1993
139. ESA: *First ESA/ESTEC Workshop on Thermal Protection Systems*, ESA-WPP-053, 1996
140. Karam, R.D. and Zarchan P.: *Satellite Thermal Control for Systems Engineers*, part of proceedings for ESA-WPP-053 First ESA/ESTEC Workshop on Thermal Protection Systems, AIAA, Reston, VA, 1998
141. TBS: *LEO Telecommunication satellites*, The Satellite Encyclopedia, <http://www.TBS-satellite.com/tse/online/mis_telecom_defilement.html>. Tag's Broadcasting Service, December 12, 2000
142. Gregory, C.: *WinOrbit's Satellite List* (OSCAR series, K8CG), <http://www.sat-net.com/winorbit/help/idxsatellites.html>. Sat-Net, December 12, 2000
143. Chan V.W.S.: *Intersatellite Optical Heterodyne Communication Systems*, The Lincoln Laboratory Journal, volume 1, pp. 169-183, 1998
144. NASA Tech Briefs: *Miniature Satellites to Test Space Technology*, <http://www.nasatech.com/NEWS/AUG99/ntb.nasa823.html>. NASA, December 12, 2000

145. AeroAstro: *AeroAstro to Provide Advanced X-Band Transponders for NASA's New Millennium Nanosatellite Constellation Trailblazer Missions*, <http://www.aeroastro.com/Hot_News/hotnews082400.html>. AeroAstro, December 12, 2000
146. NASA: *Silicon-Germanium Fast Packet Switch Developed for Communications Satellites*, Research & Technology (1998), <http://www.lerc.nasa.gov/WWW/RT1998/5000/5650quintana.html>. NASA, December 12, 2000
147. Lecompte, A., and Pogarieloff, D.: *Modelisation and simulation of dielectric and metallic antennas*, Proceedings of AP2000 Millennium conference on Antennas and Propagation, Davos, Switzerland, session 4A7, paper 0611, SP444, pp. 13-16, April 9-14, 2000, ESA, Publications Division, 2000
148. Vaccaro, S., Torres, P., Mosig, J.R., Shah, A., Skrivervik, A.K., Zürcher, J-F, de Maagt, P. and Gerlach, L.: *Integration of antennas and solar cells for satellite communications*, Proceedings of AP2000 Millennium conference on Antennas and Propagation, Davos, Switzerland, session 2A6, Paper 0954, SP444, pp. 1-4, April 9-14, 2000, ESA, Publications Division, 2000
149. Fortescue P. and Stark J.: *Spacecraft Systems Engineering*, Wiley Publishers, 1994
150. ESA: European Conference on Spacecraft Structures, Materials and Mechanical Testing, ESTEC, Noordwijk, The Netherlands, SP428, ESA, Publications Division, November 29 - December 1, 2000
151. CERT: *Fourth European Symposium on Spacecraft Materials in Space Environment*, CERT, Toulouse, France, September 6-9, 1988
152. NASA: *MEMS*, MED Skills and Services, <http://www.lerc.nasa.gov/WWW/MED/skills/mems/>. NASA, December 12, 2000
153. NASA: *Low Loss, Minature Components for System on a Chip and Phased Array Antennas*, Electron Device Technology, <http://spacecom.grc.nasa.gov/technologies/electron/lowlossm.asp>. NASA, December 12, 2000